EXPERIMENTS IN
GENERAL CHEMISTRY

FIFTH EDITION

Eugene Weiner
University of Denver

Harcourt Brace Jovanovich, Publishers
and its subsidiary, Academic Press

San Diego   New York   Chicago   Austin   Washington, D.C.
London   Sydney   Tokyo   Toronto

PREFACE

Chemistry is an exciting subject, and one does not have to be a chemistry major to enjoy it. The fact that the introductory course almost always includes considerable laboratory experimentation should make it all the more attractive. Laboratory work is enjoyable and rewarding when its relevance is clear and its accomplishment challenging, but not frustrating.

Since the fourth edition, there has been a significant increase of concern with laboratory safety and the expense of running a meaningful laboratory course. This fifth edition of a long accepted and popular laboratory manual has been retitled "Experiments in General Chemistry" to emphasize that it has been completely rewritten and newly illustrated to address these concerns. With the help of teaching assistants, students, and many reviewers, every experiment has been reevaluated and modified to make it work better and read more clearly, always with concern for safety and expense. A few new experiments have been added and a few old ones have been dropped. The qualitative analysis experiments are completely new and much improved. The familiar features of Prelaboratory Exercises and prominent display of safety precautions have been expanded. Altogether, this edition is a new book that retains the character and experiments that made previous editions so popular, but has eliminated the problems that sometimes frustrated students and teachers alike.

Prelaboratory exercises are designed to drill students in the calculations and concepts needed for each experiment. They are on tear-out sheets and are intended to be turned in. Safety precautions that are appropriate for each experiment are collected and prominently displayed where they will receive the most attention, at the beginning of the experimental procedure section. Especially important concepts and procedural details are in boldfaced type, serving to highlight the text just as an astute student might highlight with a colored marker. The section on Laboratory Methods describes general procedures for most freshman laboratory operations, and each experiment begins with references to the paragraphs which review the special techniques needed for the experiment. The Laboratory Methods section includes discussions of statistical deviation and the theory and instructions for pH meters and spectrophotometers.

Experiments have been grouped by topic to fit conveniently into most lecture schedules. In particular, Experiments 1 and 2 use the first three laboratory periods to introduce basic laboratory techniques and quantitative measurements, without requiring any theoretical support from the lecture course. This helps students feel more comfortable in the laboratory before the experiments begin to illustrate lecture material. All experiments include a discussion of the theoretical and practical importance of the experimental topic. Worked examples illustrate the calculations that are needed for evaluating the experimental data.

Students learn best when they recognize, from their own perspective, the nature and importance of the subject matter. We hope this manual will serve them as the basis for chemistry laboratory experiences that establish an appreciation for the value and satisfactions of experimental science, and that help instill a quantitative and critical approach to understanding the chemical world around us.

The contributions of Frank Brescia, John Arents, Herbert Meislich, and Amos Turk are greatly appreciated and deserve special mention. The first two editions of this manual were theirs alone. They created the basis for a laboratory experience that challenged the student to learn more than how to follow a recipe. By explaining the reasons for many procedural details, they avoided giving students the impression that special arcane knowledge is needed to make an experiment really work. The third, fourth, and this fifth edition continue their basic concept that a laboratory experience is more satisfying if all the "magic" is explained, revealing the logic and beauty of the experimental procedure.

Eugene R. Weiner
Denver, Colorado

TO THE STUDENT

The purpose of laboratory studies is to carry out operations and make measurements that will give us information about the physical world in which we live. Our ideas or theories can thus be supported, refuted, modified, or given a quantitative basis.

The objective of the laboratory part of a beginning chemistry course is, first of all, to illustrate the empirical nature of science. Your laboratory course emphasizes, by example, that all of our scientific knowledge, both "facts" and theories, must be based on, and continually tested by, objective observations. You learn, also, something about the practice of scientific observation and how best to design, set up, carry out, and interpret experiments, with a view to obtaining the most information for the effort expended. This manual is structured to first provide instruction in those techniques of laboratory operations that you must learn and master in detail, before you can work independently and even creatively in the laboratory. With these techniques as a basis, you are encouraged to regard the later experiments somewhat as a first investigator might:

    (a)   What aspect of chemical behavior do I wish to investigate?
    (b)   What measurements might give some of the desired information?
    (c)   What techniques and equipment are needed for the measurements?
    (d)   How can the experimental data be interpreted, qualitatively and quantitatively?

It is essential that you understand what the experiment is about and what you are expected to do, when you start the laboratory period. *This means that you must study each experiment before coming to the laboratory,* referring, if necessary, to your textbook and lecture notes for the theory and calculations involved in the experiment.

Each experiment contains a **prelaboratory exercise** which is intended to test your understanding of the principles involved. Your instructor may require that this exercise be completed and turned in before beginning the experiment. In any case, you will find the experiment easier to understand and the calculations much easier to perform if you already have done the prelaboratory exercise. Where feasible, you are expected to make duplicate measurements, because it is unrealistic to accept, with confidence, the result of a single experiment as accurate. In some cases, all the measurements from the entire class will be analyzed statistically to yield a more precise result than a single student's work could give; an approach that parallels much "real-life" data analysis. Your laboratory grade will be determined mainly by your method and neatness of experimentation, the accuracy of your results, and your knowledge of what you are doing and why, as shown by your ability to answer questions included in the experiment or posed by your instructor.

A SPECIAL NOTE ON SAFETY

In a well run laboratory, with careful and thoughtful workers, there will be very few accidents. This state of affairs, of course, is to be sought after in every way. But, it can establish a condition of overconfidence, especially in inexperienced students. It sometimes is difficult to take troublesome precautions seriously if you never have observed the accidents they are meant to prevent. Nevertheless, this is exactly what you must do; laboratory accidents can have outcomes that are too tragic to allow them to serve as learning experiences. Utilize your teacher's experience in this matter and rigorously follow every safety procedure described; they are not suggestions, they are mandatory. Every chemical must be regarded as potentially hazardous. Even so innocuous a substance as sodium chloride might have become contaminated by a careless worker, or by merely sitting exposed in a location where it absorbed toxic vapors from the air. Each experiment contains appropriate safety precautions in a special **SAFETY** box preceding the **PROCEDURE** section.

The safety precautions for each experiment must be read and carefully considered before beginning the experiment.

Some additional safety rules may have been established for your particular laboratory. For your own safety, and that of your coworkers, please pay close attention to these regulations.

CONTENTS

Solutions and Solubility

Acids and Bases

Oxidation–Reduction Reactions

Ion–Exchange Analysis

Thermochemistry

Influence of Structure

Organic Chemistry

Reaction Kinetics

Electrochemistry

Qualitative Analysis

ASSIGNMENT 1
Safety Lesson

Your first assignment in this course is:
1. Read the Special Note on Safety on the previous page.
2. Learn the following rules of laboratory safety.
3. Sign the statement that says you have read and agree to comply with these rules.
4. Complete the safety quiz which follows the rules.

The signed statement and a satisfactorily completed safety quiz are to be turned in to your instructor before beginning any laboratory work.

SAFETY RULES

The most common laboratory accidents are burns, cuts, eye damage, and damage to skin and clothing by corrosive chemicals. These happen because people accidently drop reagent bottles, absent-mindedly touch hot objects, thoughtlessly create the conditions for an explosion, fire, or boil-over, or are carelessly bumped-into while manipulating their equipment. Occasionally, an accident is not anyone's immediate fault. For example, an accident might be the result of an unanticipated equipment failure, such as the bursting of a pressurized flask.

The common factor to *all* accidents is unpredictability. The correct strategy is to take precautions beforehand to prevent or minimize any damage.

In an automobile, you wear seat belts and drive carefully. In the laboratory, you wear eye protection and suitably protective clothing, and work carefully. This requires that you follow all the safety precautions below.

1.　　　In all general behavior, "play it safe." Your *first* concern in the laboratory is safety. Whether or not you learn anything, whether you enjoy the lab or hate it, avoid the unnecessary tragedy of physically damaging yourself, or your friend across the bench. The potential for serious injury is always present in a chemistry laboratory, just as it is on a busy highway. The same sort of continual alert care is required in both cases.

2.　　　Your eyes are the most important part of your body to protect in the laboratory. Eye protection must be worn whenever explosion, boiling or chemical splash hazards exist.

In the laboratory, interpret this to mean *all* the time.

Suitable goggles or protective glasses will be supplied by your instructor. If you wear contact lenses, be sure to wear splash-proof goggles with side protection. Otherwise, splashed chemicals may become trapped between the contact lens and your cornea, making it difficult for tears or an eye wash to quickly rinse away the irritant.

PROTECT YOUR EYES. WEAR GOGGLES OR PROTECTIVE GLASSES.
KNOW THE LOCATIONS OF THE EMERGENCY EYE WASH FACILITIES.
KNOW HOW TO USE THEM.

3.　　　Long hair must be tied back or up, where it will not get into flames or beakers, or brush against equipment and bottles. In a laboratory, free long hair has a great propensity for being where it shouldn't be.

4.　　　Wear shoes with closed upper parts. When a flask or bottle of reagent is dropped to the floor, your feet are right down there where the glass and chemicals are flying around. Do yourself a favor and keep your feet covered. Sandals protect the bottom, but not the tops, of your feet, and are not adequate protection.

5.　　　Do not bring any food into the laboratory. Exposing food to the vapors and chemicals that pervade a working laboratory can contaminate it far beyond the toxic effects of trace food additives or residue pesticides. In fact, you should keep all objects out of your mouth while in the lab. This includes fingers, pipets (you will learn how to use a rubber pipet safety bulb) and other glassware, rubber and plastic tubing, or anything else. Smoking is absolutely forbidden in the laboratory.

6.   Protect your good clothing: don't wear it. Nearly everyone will lean against a countertop contaminated with acid, splash an indelibly colored solution, or intercept a flying bit of corrosive chemical at least once during the course. Protect your bare legs against stains and scars. Do not wear shorts unless additional leg protection is worn also. Wear aprons or lab coats, and always wear old clothing that is not too loose, especially at the sleeves. Then the stains and holes you collect in your lab clothing can become badges of honor among your labmates, but badges you do not have to wear in the outside world.

7.   Know what to do in case of skin burns from heat or chemicals. If anyone's skin is exposed to a splash or spill of corrosive chemicals, rush them under the emergency shower and turn it on quickly to flush away the chemicals.
KNOW WHERE THE EMERGENCY SHOWER FACILITIES ARE.
For minor skin contact, on a finger or arm, flush the chemicals away with large amounts of cold, running water and remove any contaminated clothing. If skin is burned from a flame or hot plate, treat it only with cold water until a doctor has seen it.
Be sure to inform the instructor about all accidents that occur.

8.   Do not create hazards for others out of carelessness or laziness. Clean up chemical spills right away and wipe dribbles off of reagent bottles before the next person can touch them. Many toxic chemicals can be absorbed through the skin and, of course, many chemicals will burn or irritate skin. Your instructor will explain how to neutralize acid and base spills, and what to use for wiping up spilled chemicals.
Do not leave broken glass around. Warn your labmates if you do anything that might release toxic or flammable vapors, or cause a boil-over or explosion. Use the fume hood when told to do so. The fume hood draws off toxic and irritating fumes from your experiment and vents them to the outside, with great dilution.
You will use many chemicals that are flammable. Know which these are and never use them near an open flame. Know how to extinguish a fire if one should start. A fire in a beaker or other vessel is best extinguished simply by closing the opening with a watch glass or stopper, depriving the fire of oxygen. Other fire control techniques may be demonstrated by your instructor.
KNOW WHERE THE FIRE EXTINGUISHERS AND FIRE BLANKETS ARE KEPT.
Be prepared to use these items if it ever should be necessary.

9.   *Never* attempt to do an original, unauthorized experiment. While originality and enthusiasm are commendable, foolishness is not. There are many potential, hidden hazards in chemistry laboratory work, and once you stumble onto one, it is too late to repent. If you want to try something novel, you must have the approval of your instructor before proceeding.

10.   Pay close attention to the instructions and demonstrations that tell what to do if certain emergencies occur. Know where the safety and first-aid equipment is located and how to use it. You suddenly may find yourself to be the person best able to act first to stop a fire or give first aid. Be prepared for this, your friends are depending on you.
KNOW THE LOCATIONS OF THE FIRST-AID BOXES.
Your instructor will explain what they contain and when they should be used.

11.   Report all accidents and injuries immediately to your instructor.

In spite of the alarmist tone of the safety rules above, very few accidents ever occur and those that do happen are seldom serious. Working in the chemistry laboratory is similar to driving in rush-hour traffic: danger is always an inattentive moment away, and so there are certain rules you **never** break. In addition, you always must watch out for the occasional idiot. If you have confidence that everyone else is being careful, you can relax a bit, and enjoy your laboratory experience.

Sign and date the following statement. Cut it out and turn it in to your instructor.

I have read the safety rules for my chemistry laboratory. I understand and agree to comply with them.

Print Name: ____   Student No.: ____________

(Signature): ____   Date:_________________

LABORATORY SAFETY QUIZ

Name: ____   Student No.:___________

Section: _________________   Date: __________________

1.   Sketch the general floorplan of your laboratory in the space below. Show the positions of all exits, emergency telephones, fire extinguishers, fire blankets, eye washes, emergency showers, and fume hoods. Include the outside hall if some safety equipment is there.

2.   What are the four most common laboratory accidents?

3.   When should eye protection be worn?

4.   Why are sandals and shorts not suitable for wearing in the laboratory?

5.   Why should contact lenses not be worn in the laboratory?

6.   What are the laboratory showers for?

7.   What should you do if a chemical is splashed into your eyes?

8.   What is wrong with bringing a food snack into your laboratory?

9.   Are any special precautions needed with long hair? If so, what?

10.   What is the proper first-aid treatment for a heat burn?

11.   What is the proper first-aid treatment for a chemical burn?

13.   What should you do if some chemicals spill on the work bench or on the side of a reagent bottle?

14.   Why should you not apply pressure or suction to glass or flexible tubing with your mouth?

15.   What is the most important part of your body to protect in the laboratory? What is the best means of protection?

CHEMICAL UNITS

The table below is intended to provide a quick reference to the different chemical units used in chemical calculations. Notice that there is just one fundamental unit for quantity but there are many different units for mass and concentration. Unfortunately, this complication cannot be avoided, because it is impossible to devise just one all–purpose measure for either of the quantities of mass or concentration that will serve all the different situations arising in chemical calculations. Each of the different units in the table has its particular applications. Using the proper units in the right application will make your calculations simpler and help avoid errors. Detailed descriptions and examples of calculations will be found in your textbook.

Unit of Quantity

Unit	Symbol	Definition	Comments
mole	n or mol	1 mole is the quantity of a substance whose mass in grams is numerically equal to its atomic or molecular mass.	1 mole of any substance contains Avogadro's number, $N_A = 6.022 \times 10^{23}$ mol^{-1}, of particles of that substance. N_A is chosen to be the number of carbon atoms in exactly 12 grams of pure carbon-12.

Units of Mass

Unit	Symbol	Definition	Comments
atomic mass unit	amu	The mass of one atom expressed in amu is numerically equal to the mass of one mole of the atoms expressed in grams.	To convert from amu to grams, divide by Avogadro's number. $1\ amu = (6.022 \times 10^{23})^{-1} = 1.661 \times 10^{-24}$ g
atomic mass	(none)	The atomic mass of an element is its relative mass compared to the mass of the carbon-12 isotope, which has been chosen as the standard reference mass. The mass of one atom of carbon-12 is defined to be exactly 12 amu.	Because atomic masses are ratios, they are dimensionless.
molecular	(none)	The molecular mass of a molecule is the sum of the atomic masses of all the atoms in it.	Like atomic masses, molecular masses mass are ratios and are dimensionless.
formula mass	(none)	Formula mass is the molecular mass of the smallest repeating molecular unit in substances with no clearly defined molecular mass, such as ionic crystals, polymers, and molecular aggregates.	
equivalent	$equiv$	**Acid-base reactions:** equivalent mass of a reactant is the mass, in grams, that reacts with, or produces, one mole of H_3O^+ in a neutralization reaction. **Redox reactions:** equivalent mass of a reactant is its molecular mass divided by the absolute value of the oxidation number change of that one of its atoms which changes its oxidation number.	The equivalent mass of a reactant depends upon the reaction being considered, which always must be specified. The concept of equivalent mass is used mainly with acid-base and oxidation-reduction (redox) reactions.

Units of Concentration

Unit	Symbol	Definition	Comments
molarity	M	$\dfrac{\text{moles of solute}}{\text{liters of solution}}$	M varies with temperature, because the volume changes with temperature.
molality	m	$\dfrac{\text{moles of solute}}{\text{kg of solvent}}$	m is independent of temperature. For dilute solutions, $m \approx M$.
normality	N	$\dfrac{\text{gram equivalents}}{\text{liters of solution}}$	N depends on the particular reaction, see equivalent mass.
mole fraction	X	$\dfrac{\text{moles of component}}{\text{total moles of sample}}$	X is dimensionless and cannot exceed 1. The sum of the mole fractions of all the components in a sample must be equal to 1.
volume percent	$vol\%$	$\dfrac{\text{volume of component x 100}}{\text{total volume of sample}}$	Volumes of each component in a solution must be measured before mixing, because mixed volumes are not additive.
mass percent	$mass\%$	$\dfrac{\text{mass of component x 100}}{\text{total mass of sample}}$	It is important to specify whether a percent composition is mass % or vol %.
parts per million	ppm	$\dfrac{\text{mass of component x } 10^{6}}{\text{total mass of sample}}$	ppm is analogous to mass % (mass % is parts per hundred).
milligrams per milliliter	mg/mL	$\dfrac{\text{milligrams of component}}{\text{milliliters of solution}}$	Commonly used in biochemistry as a convenience in preparing solutions.

CHEMICAL REFERENCE DATA
RELATED TO THE LABORATORY EXPERIMENTS

TABLE 0.1. VAPOR PRESSURE OF WATER

Temp. (°C)	Pressure (torr)	Temp. (°C)	Pressure (torr)	Temp. (°C)	Pressure (torr)	Temp. (°C)	Pressure (torr)
-5 (ice)	3.0	25	23.8	55	118.0	85	433.6
-4 (ice)	3.3	26	25.2	56	123.8	86	450.9
-3 (ice)	3.6	27	26.7	57	129.8	87	468.7
-2 (ice)	3.9	28	28.3	58	136.1	88	487.1
-1 (ice)	4.2	29	30.0	59	142.6	89	506.1
0	4.6	30	31.8	60	149.4	90	525.8
1	4.9	31	33.7	61	156.4	91	546.1
2	5.3	32	35.7	62	163.8	92	567.0
3	5.7	33	37.7	63	171.4	93	588.6
4	6.1	34	39.9	64	179.3	94	610.9
5	6.5	35	42.2	65	187.5	95	633.9
6	7.0	36	44.6	66	196.1	96	657.6
7	7.5	37	47.1	67	205.0	97	682.1
8	8.0	38	49.7	68	214.2	98	707.3
9	8.6	39	52.4	69	223.7	99	733.2
10	9.2	40	55.3	70	233.7	100	760.0
11	9.8	41	58.3	71	243.9	101	787.6
12	10.5	42	61.5	72	254.6	102	815.9
13	11.2	43	64.8	73	265.7	103	845.1
14	12.0	44	68.3	74	277.2	104	875.1
15	12.8	45	71.9	75	289.1	105	906.1
16	13.6	46	75.7	76	301.4	106	937.9
17	14.5	47	79.6	77	314.1	107	970.6
18	15.5	48	83.7	78	327.3	108	1004.4
19	16.5	49	88.0	79	341.0	109	1038.9
20	17.5	50	92.5	80	355.1	110	1074.6
21	18.7	51	97.2	81	369.7	111	1111.2
22	19.8	52	102.1	82	384.9	112	1148.7
23	21.1	53	107.2	83	400.6	113	1187.4
24	22.4	54	112.5	84	416.8	114	1227.3

TABLE 0.2. PROPERTIES OF COMMON STOCK SOLUTIONS OF ACIDS AND BASES

Reagent	Formula	Molarity	Density	Percent solute
Acetic acid, glacial	CH_3COOH	17 M	1.05 g/mL	99.5 %
Acetic acid, dilute		6	1.04	34
Hydrochloric acid, concentrated	HCl	12	1.18	36
Hydrochloric acid, dilute		6	1.10	20
Nitric acid, concentrated	HNO_3	16	1.42	72
Nitric acid, dilute		6	1.19	32
Sulfuric acid, concentrated	H_2SO_4	18	1.84	96
Sulfuric acid, dilute		3	1.18	25
Ammonium hydroxide, concentrated	NH_4OH	15	0.90	58
Ammonium hydroxide, dilute (Also called "ammonia solution": NH_3)		6	0.96	23
Sodium hydroxide, dilute	NaOH	6	1.22	20

See reagent bottle label for exact concentrations.

TABLE 0.3.   GENERALIZATIONS CONCERNING SOLUBILITIES OF SALTS IN WATER

Mainly Water-Soluble

NH_4^+	All ammonium salts are soluble
NO_2^-	All nitrites are soluble.
NO_3^-	All nitrates are soluble.
$C_2H_3O_2^-$	All acetates are soluble.
ClO_3^-	All chlorates are soluble.
ClO_4^-	All perchlorates are soluble.
Cl^-	All chlorides are soluble except $AgCl$, Hg_2Cl_2, and $PbCl_2$.
Br^-	All bromides are soluble except $AgBr$, Hg_2Br_2, $HgBr_2$, and $PbBr_2$.
I^-	All iodides are soluble except AgI, Hg_2I_2, HgI_2, and PbI_2.
SO_4^{2-}	All sulfates are soluble except $BaSO_4$, $SrSO_4$, and $PbSO_4$. $CaSO_4$, Hg_2SO_4, and Ag_2SO_4 are sparingly soluble.

Mainly Water-Insoluble

S^{2-}	All sulfides are insoluble except those of the IA and IIA elements and $(NH_4)_2S$.
CO_3^-	All carbonates are insoluble except those of the IA elements and $(NH_4)_2CO_3$.
SO_3^{2-}	All sulfites are insoluble except those of the IA elements and $(NH_4)_2SO_3$.
PO_4^{3-}	All phosphates are insoluble except those of the IA elements and $(NH_4)_3PO_4$.
OH^-	All hydroxides are insoluble except those of the IA elements and NH_4OH. $Ba(OH)_2$, $Sr(OH)_2$, and $Ca(OH)_2$ are sparingly soluble.

TABLE 0.4.   GASES FORMED IN METATHESIS REACTIONS

Gas	Reaction
H_2S	Any sulfide (salt of S^{2-}) and any acid form H_2S and a salt.
CO_2	Any carbonate (salt of CO_3^{2-}) and any acid form CO_2, H_2O, and a salt.
SO_2	Any sulfite (salt of SO_3^{2-}) and any acid form SO_2, H_2O, and a salt.
NH_3	Any ammonium salt (salt of NH_4^+) and any soluble strong hydroxide react upon heating to form NH_3, H_2O, and a salt.

TABLE 0.5.   ELECTROLYTES

General Classes	Example Compounds
Strong electrolytes (complete ionization):	$HClO_4$, H_2SO_4, HNO_3, HCl, HBr, and practically all salts, including soluble hydroxides.
Intermediate electrolytes ($K = 10^{-2}$ to 10^{-4}):	H_2SO_3, H_3PO_4, HF, HNO_2
Weak electrolytes ($K = 10^{-5}$ to 10^{-7}):	$HC_2H_3O_2$, H_2CO_3, NH_3, $Pb(C_2H_3O_2)_2$
Feeble electrolytes ($K = 10^{-8}$ to 10^{-11}):	H_2S, $HClO$, NH_4^+, H_3BO_3, HCN, $HgCl_2$
Very feeble electrolytes ($K = 10^{-12}$ to 10^{-16}):	H_2O_2, H_2O

TABLE 0.6.   NAMES AND COMMON VALENCES OF SOME ELEMENTS AND GROUPS

Name	Valence	Example Compounds Ionic	Example Compounds Nonionic	Name	Valence	Example Compounds Ionic	Example Compounds Nonionic
Metals; cations				*Nonmetals; anions*			
Ammonium	1	NH_4Cl	—	Acetate	1	$NaC_2H_3O_2$	$HC_2H_3O_2$
Copper(I) (cuprous)	1		Cu_2O	Arsenite	1	$KAsO_2$	—
Copper(II) (cupric)	2	$CuCl_2$	—	Hydrogen carbonate	1	$NaHCO_3$	—
Hydrogen	1	—	H_2O	Bromide	1	$NaBr$	HBr
Mercury(I) (mercurous)	1	—	Hg_2S	Bromate	1	$NaBrO_3$	$HBrO_3$
Mercury(II) (mercuric)	2	HgF_2	$HgCl_2$	Chloride	1	$NaCl$	HCl
Potassium	1	K_2SO_4	—	Chlorate	1	$KClO_3$	$HClO_3$
Silver	1	$AgNO_3$	AgI	Cyanide	1	KCN	HCN
Sodium	1	$NaCl$	—	Dihydrogen phosphate	1	NaH_2PO_4	—
Barium	2	$BaCO_3$	—	Fluoride	1	KF	HF
Cadmium	2	CdF_2	—	Hydride	1	LiH	CH_4
Calcium	2	CaO	—	Hydroxide	1	$NaOH$	HOH
Cobalt	2	$Co(NO_3)_2$	$CoCl_2$	Iodide	1	KI	HI
Iron(II) (ferrous)	2	$FeSO_4$	$FeBr_2$	Nitrate	1	$NaNO_3$	HNO_3
Iron(III) (ferric)	3	$Fe_2(SO_4)_3$	$FeCl_3$	Nitrite	1	$NaNO_2$	HNO_2
Lead	2	PbF_2	$Pb(C_2H_3O_2)_2$	Permanganate	1	$KMnO_4$	$HMnO_4$
Magnesium	2	$MgSO_4$	—	Carbonate	2	$CaCO_3$	—
Nickel	2	NiO	$NiCl_2$	Chromate	2	K_2CrO_4	—
Strontium	2	$SrCl_2$	—	Dichromate	2	$K_2Cr_2O_7$	—
Tin(II) (stannous)	2	$SnSO_4$	$SnCl_2$	Hydrogen phosphate	2	Na_2HPO_4	—
Tin(IV) (stannic)	4	SnO_2	$SnCl_4$	Oxygen (oxide)	2	Na_2O	HgO
Zinc	2	ZnF_2	$ZnBr_2$	Oxygen (peroxide)	2	Na_2O_2	H_2O_2
Aluminum	3	Al_2O_3	$AlBr_3$	Sulfate	2	Na_2SO_4	H_2SO_4
Antimony	3	—	SbI_3	Sulfide	2	K_2S	CdS
Bismuth	3	BiF_3	$BiCl_3$	Sulfite	2	Na_2SO_3	—
Chromium	3	$Cr_2(SO_4)_3$	$CrCl_3$	Arsenate	3	K_3AsO_4	H_3AsO_4
				Phosphate	3	Na_3PO_4	H_3PO_4

TABLE 0.7.   TYPICAL MEASUREMENT PRECISION OF VARIOUS DEVICES

Device	Approximate Precision
Trip balance	±0.5 g
Triple beam balance	±0.01 g
Analytical balance	±0.0001 g
1 L graduated cylinder	±5 mL
100 mL graduated cylinder	±0.2 mL
10 mL graduated cylinder	±0.1 mL
50 mL buret	±0.02 mL
25 mL buret	±0.02 mL
25 mL pipet	±0.02 mL
10mL pipet	±0.01 mL
1 mL pipet	±0.01 mL
Thermometer (-10°C to 110°C)	±0.2°C
Barometer (mercury)	±0.5 torr
Meter stick	±0.5 mm

USEFUL APPROXIMATION:

$$1 \text{ mL} = 1 \text{ g} \approx 18 \text{ drops from a medicine dropper}$$

TREATMENT OF EXPERIMENTAL DATA

I. PRECISION AND ACCURACY

The only kind of physical quantity that can be measured with perfect accuracy is a tally of discrete objects, for example, dollars and cents or the number of objects in a museum case. In measuring a quantity capable of continuous variation such as mass or length there always is some uncertainty because the answer, like an irrational number such as π, cannot be expressed by any finite number of digits. The volume of liquid in a buret, for example, is capable of continuous variation and can only be estimated because obtaining a measurement requires guessing just where the liquid level lies between marked volume divisions on the buret wall. The precision of the volume estimate depends on the quality of the buret and the skill of the experimenter. In addition to errors which result from difficulties of constructing and using measuring devices, other errors over which the experimenter has no control are inherent in measurements. Therefore, at least two, preferably three or more, determinations of any quantity should be made for comparison purposes. After making several measurements of some quantity, the best value to use generally is the average of all the measurements. The "true" value—more correctly, the "accepted" value— of important quantities such as physical constants listed in a handbook (for example, the velocity of light in a vacuum), is chosen by some competent group of experts who critically examine all the available data to select the most probable value.

It is important to distinguish between the **precision** and the **accuracy** of a series of measurements.

The precision indicates how reproducible the measurements are.
Measurements whose values scatter widely are less precise than measurements whose values lie close together, even though the average value of the two sets of data might be exactly the same. Measurements of the diameter of a solid cylinder made with a micrometer will be more precise than the same measurements made with a meter stick because the meter stick is more coarsely graduated.

The accuracy indicates how well the measurement agrees with an accepted value.
The accuracy of a measurement is unrelated to its precision. Accuracy depends on how well the measuring device is calibrated with respect to some accepted standard, such as the international reference meter length at the National Bureau of Standards. If a precise micrometer reads 1.50 mm when it actually should measure 1.75 mm, then the average value of a very precise series of measurements will be in error by 0.25 mm, inaccurate by: $(0.25/1.75)(100\%) = 14.3\%$.

One can only judge the accuracy of a measurement by comparing it with an accepted value. If no accepted value can be found, the accuracy cannot be ascertained.

II. DETERMINING PRECISION:

The precision of an experimental determination may be taken to be a statement about how widely the individual values of a series of measurements deviate from the average value. The arithmetic average of a series of measurements is usually taken to be the "best" value, but the average value gives no information about the precision, or "scatter", of the separate measurements. The most common ways to express experimental precision are by the **average deviation, relative average deviation,** and **standard deviation.**

1. Average Deviation

The simplest measure of precision is the **average deviation**, which is determined as follows:
1. Calculate the average value of all the of measurements
2. Subtract the average value from each individually measured value; this quantity is called the **deviation,**
3. Sum the deviations (treat each deviation, whether positive or negative, as a positive quantity) and calculate their average.

Written as an equation, the average deviation $\bar{d}$ is:

$$\bar{d} = \frac{\displaystyle\sum_{i=1}^{n} \left| (x_i - \bar{x}) \right|}{n}$$

where n is the total number of measurements, the summation goes from i=1 (the value of x for the first measurement) to i=n (the value of x for the nth measurement), x_i is the value of the ith measurement, and $\bar{x}$ is the average value of all the measurements. The vertical lines on each side of the parenthesis in the numerator are an **absolute value** symbol. They indicate that the quantity between them is to be regarded as a positive quantity.

  Using the absolute value of a number simply means always treating the number as a positive quantity, regardless of whether the true value is positive or negative.

The numerator, then, is the sum of the absolute values of all the deviations.

EXAMPLE 1:

  In a series of measurements, the following values for the molarity of a potassium permanganate solution were obtained: 0.1010, 0.1020, 0.1012, 0.1015 mol L^{-1} (moles per liter). Calculate the average deviation.

ANSWER:

Individual measurements	Individual deviations from the average (absolute value)
0.1010	0.1014 - 0.1010 = 0.0004
0.1020	0.1014 - 0.1020 = 0.0006
0.1012	0.1014 - 0.1012 = 0.0002
<u>0.1015</u>	0.1014 - 0.1015 = <u>0.0001</u>
Σx_i = 0.4057	Σd_i = 0.0013

$$\text{Average } x = \bar{x} = \frac{0.4057}{4} = 0.1014 \text{ mol/L} \qquad \text{Average deviation} = \bar{d} = \frac{0.0013}{4} = 0.0003 \text{ mol/L}$$

These results would be reported as: **0.1014 ± 0.0003 mol/L**.

2. Relative Average Deviation

  Precision also may be expressed as the **relative average deviation (r.a.d.)**, defined as *the average deviation divided by the average value.*

EXAMPLE 2:

  The r.a.d. for the measurements in Example 1 is:

$$\text{r.a.d.} = \frac{\text{av. deviation}}{\text{av. measured value}} = \frac{\bar{d}}{\bar{x}} = \frac{0.0003}{0.1014} = 0.003 \text{ (dimensionless)}$$

(Only one significant figure is valid.) The percent r.a.d. is an often used variation. It is obtained by multiplying the r.a.d. by 100%:

$$\text{r.a.d.} = \frac{0.0003}{0.1014} \times 100\% = 0.3\%$$

  The precision of an experiment varies with the method and apparatus used. An experienced chemist using equipment commonly available for routine analytical work should be able to determine the chloride concentration in a solution with a precision of 0.1% r.a.d. The average inexperienced student is more likely to obtain a precision around 1.0% r.a.d.

3. Standard Deviation

  The **standard deviation** has greater theoretical validity than the average deviation. The average deviation is popular because of its simplicity but is reliable only if the number of measurements is very large, around 10 or more. For smaller sets of data, the standard deviation gives a much better indication of measuring precision. Both methods indicate the same precision for random errors in a very large number of measurements.

The standard deviation is determined as follows:
1. Calculate the average of a series of measurements.
2. Determine the deviation of each measurement from the average.
3. Square the deviations and add up their squares.
4. Divide the sum of the squares of the deviations by (n - 1), where n is the total number of measurements.
5. Take the square root of the result from step 4.

Let x_1 be the value of the first measurement, x_2 the value of the second, and so forth. Let $\bar{x}$ be the average value of all the measurements. Then the deviations in step 2 above are found as in the average deviation, by subtracting the average value from each measured value. Let the deviation of measurement 1 be d_1, of measurement 2 be d_2, etc. Then:

$$d_1 = x_1 - \bar{x}$$
$$d_2 = x_2 - \bar{x}$$
$$d_3 = x_3 - \bar{x}$$
etc.

Written as an equation, the standard deviation σ is:

$$\sigma = \left(\frac{d_1^2 + d_2^2 + d_3^2 + \ldots + d_n^2}{n - 1} \right)^{1/2} = \left(\frac{\sum\limits_{i=1}^{n} (x_i - \bar{x})^2}{n - 1} \right)^{1/2}$$

where n is the total number of measurements, the summation goes from i = 1 (the value of x for the first measurement) to i = n (the value of x for the nth measurement), x_i is the value of the ith measurement, and $\bar{x}$ is the average value of all the measurements. Example 3 illustrates how to determine the standard deviation of a series of measurements.

EXAMPLE 3:

Calculate the standard deviation of the measurements in Example 1 of the molarity of a potassium permanganate solution.

ANSWER:

Measurement number	Measured molarity (mol/L)	Individual deviations from the average	Deviations squared
1	0.1010	0.1010 - 0.1014 = -0.0004	$(-0.0004)^2$ = 1.6×10^{-7}
2	0.1020	0.1020 - 0.1014 = +0.0006	$(+0.0006)^2$ = 3.6×10^{-7}
3	0.1012	0.1012 - 0.1014 = -0.0002	$(-0.0002)^2$ = 0.4×10^{-7}
4	<u>0.1015</u>	0.1015 - 0.1014 = +0.0001	$(+0.0001)^2$ = $\underline{0.1 \times 10^{-7}}$
$\sum x_i$ =	0.4057		$\sum d_i^2$ = 6.6×10^{-7}

$$\text{Average molarity} = \frac{0.4057}{4} = 0.1014 \text{ mol/L} \qquad \text{std. dev.} = \sigma = \left(\frac{6.6 \times 10^{-7}}{4 - 1} \right)^{1/2} = 0.00047 \text{ mol/L}$$

These results would be reported as:

molarity of potassium permanganate solution = 0.1014 ± 0.0005 mol/L (std. dev.).

Notice that in this case, with only a few measurements, the standard deviation more realistically indicates less precision in the measurements than did the average deviation.

The standard deviation has a precise theoretical meaning.
If the deviations from one measurement to another are perfectly random, the magnitude of the standard deviation gives the range of spread from the average value within which 68% of all repeated measurements are expected to fall.
For example, if you make 100 duplicate measurements of the mass of a sample, 68 of these measurements should fall within plus or minus one standard deviation of the average value. Ninety-five percent of all duplicate measurements should lie within two standard deviations[1] (2σ: std. dev. multiplied by two). Two standard deviations (2s) is said to represent the 95% confidence level. Testing these statements on an actual sample can be a useful way of determining whether or not the errors in a given measurement are truly random or not. If some systematic error is present, such as a steady drift in an instrument calibration due to a changing temperature, the precision predictions of the standard deviation will not hold true.

Like the average deviation, use of the standard deviation is strictly valid only for an infinite number of measurements. When the total number of replicate measurements is quite small, perhaps four or five, the standard deviation should be regarded only as a rough estimate, but it is better than the average deviation and is the most useful indication of measurement uncertainty for finite data sets.

III. DETERMINING ACCURACY:

Precise measurements are not necessarily accurate. The **accuracy** expresses the agreement of the measurement with an accepted value for the quantity. If no accepted value is known, the accuracy cannot be ascertained. When a quantity is measured for which a "true" or accepted value is known, it is usual to express the accuracy in terms of the **absolute error** and **relative error,** both of which compare the measured value with the accepted value.

The absolute error (also called just the *error*), is the experimentally determined value minus the accepted value.

The relative error is the absolute error divided by the accepted value.

EXAMPLE 4:

Suppose that the accepted value for the normality of the permanganate solution in Example 1 is 0.1024 mol/L, as measured by the instructor of the course. What are the absolute error and relative error for the determination of the normality in Example 1?

ANSWER:

$$
\begin{array}{ll}
0.1014 \text{ mol/L} & \text{the determined average value} \\
\underline{- 0.1024 \text{ mol/L}} & \text{the accepted value} \\
-0.0010 \text{ mol/L} & \text{the absolute error}
\end{array}
$$

from which the relative error is -0.0098, obtained as follows:

$$\frac{-0.00010}{0.1024} = -0.0098$$

Notice that there are no units for the relative error.

Other ways to express relative error are by **percent relative error (%), parts per thousand (ppt),** and **parts per million (ppm).** The relative error from above may also be expressed as:

$$
\begin{array}{ll}
-0.0098 \times 100\% & = -0.98\% \\
-0.0098 \times 1000 \text{ ppt} & = -9.8 \text{ ppt} \\
-0.0098 \times 1{,}000{,}000 \text{ ppm} & = -9800 \text{ ppm}
\end{array}
$$

IV. PROPAGATION OF ERRORS:

Often, several different measured quantities are used to calculate another quantity, as when the density of an object is found by dividing its measured mass by its measured volume. Uncertainties in the measured quantities naturally will result in an uncertainty in the calculated quantity. If the uncertainties in the measured quantities have been determined, the most probable, or **statistical uncertainty**, in a calculated quantity can be found by using the following rules.

---

[1] Actually, 95% of the measurements should lie within 1.96 standard deviations. Using a value of 2σ is close enough and, in any case, it errs on the safe side.

1. *Statistical uncertainty in sums and differences*
Suppose a calculated quantity is $F = x \pm y$. Let U_F, U_x, and U_y be the statistical uncertainties in F, x, and y, respectively.

The statistical uncertainty in F equals the square root of the sum of the squares of the uncertainties in x and y.

In equation form: $U_F = (U_x{}^2 + U_y{}^2)^{1/2}$.

EXAMPLE 5:
The mass of water in a beaker is found by weighing the dry beaker empty and then weighing it with the water in it. The mass of the water is the difference between these two masses. Each mass is weighed four times. The results, with the calculated average deviations, are:

mass of beaker when empty (m_b): 9.8264 ± 0.0005 g

mass of beaker with water in it (m_{b+w}): 16.7193 ± 0.0005 g

mass of water (without uncertainties): $m_w = m_{b+w} + m_b = 16.7193 - 9.8264 = 6.8929$ g

statistical uncertainty in mass of water: $U_w = (0.0005^2 + 0.0005^2)^{1/2} = (5 \times 10^{-7})^{1/2}$

$U_w = \pm 0.0007$ g (rounded to one significant figure)

The measured mass of water is expressed correctly as: $m_w = 6.8929 \pm 0.0007$ g

2. *Statistical uncertainty in products and quotients*
Suppose a calculated quantity is $F = (xy)/z$. Let U_F, U_x, and U_y be the statistical uncertainties in F, x, and y, respectively. The statistical uncertainty in F is:

$$U_F = F\left\{\left(\frac{U_x}{x}\right)^2 + \left(\frac{U_y}{y}\right)^2 + \left(\frac{U_z}{z}\right)^2\right\}^{1/2}$$

EXAMPLE 6:
The density of an object is found by dividing its measured mass by its measured volume. The volume and mass each are measured seven times, the mass on an analytical balance and the volume by displacing water in a graduated cylinder. The results, with their calculated standard deviations, are:

mass of object: $m = 9.2152 \pm 0.0003$ g

volume of object: $V = 8.74 \pm 0.07$ mL

density of object (without uncertainties): $r = 9.2152/8.74 = 1.05$ g/mL

$$U_d = (1.05)\left\{\left(\frac{0.0003}{9.2152}\right)^2 + \left(\frac{0.07}{8.74}\right)^2\right\}^{1/2} = \pm 0.008 \text{ g/mL}$$

statistical uncertainty in density:

The measured density is correctly expressed as: $r = 1.05 \pm 0.01$ g/mL (rounded to 2 decimal places)

This example illustrates a very important point.

The precision of a result calculated by multiplication and/or division can be no greater than the precision of the least precise quantity used in the calculation.

When different measurements are combined in a calculation to obtain a new quantity, such as dividing mass by volume to get density, the precisions of the different measurements will, in general, be different. It can be seen that the uncertainty in the density measurement is dominated by the uncertainty of the volume measurement, which is known only to three significant figures. A less precise balance could have been used without decreasing the precision of the measurement at all. If a more precise density measurement is needed, the precision of the volume measurement must be improved first. There is no point in improving the mass measurement alone because it would make no difference in the final uncertainty of the density value.

V. SIGNIFICANT FIGURES:

Whenever a measured value is given, the number should be expressed in a manner that makes the degree of uncertainty perfectly clear. This is done by writing a value so that it contains only those digits that are known with certainty (have not been estimated), plus one more figure (the first estimated figure). The last figure in the number is always one that requires some degree of estimation and, therefore, it serves as an indication of the precision of the value.

Significant figures are the number of digits necessary to express a measurement or calculation to the precision with which it was made.

It is important to remember that the number of significant figures is unrelated to the position of the decimal point. The numbers 1056, 105.6, 1.056, 0.1056, 0.001056, and 1.056×10^7 all are written with just four significant figures. Zeroes which serve only to locate the decimal point, as the first two zeroes in 0.001056, and powers of ten which are needed to express the magnitude of a value, do not count as significant figures.

Usually, there is no problem in reporting the correct number of significant figures to express the result of a direct measurement because the measuring instrument simply does not give extra meaningless figures. However, when calculations are performed to obtain some desired quantity, extra numbers often appear that have no experimental validity. This is especially true when using an electronic calculator. Extra meaningless figures always should be eliminated by rounding off the calculated value to the correct number of significant figures.

Rules for Rounding Off to the Correct Significant Figures

1. **When the first digit after the last significant figure is less than 5**, simply cut off all extra figures without changing the last significant figure.

Example:

A calculation gives the number 3.9634263 for a value that should have four significant figures. The fifth digit (which is the first nonsignificant figure) is 4, which is less than 5. Therefore, the value is correctly given as **3.963.**

2. **When the first digit after the last significant figure is greater than 5**, add 1 to the last significant figure and drop all following digits.

Example:

A calculation gives the number 0.0462733 for a value that should have three significant figures. The first nonsignificant figure is 7 (the zeroes in front of the 4 serve only to locate the decimal point and do not count as significant figures), which is greater than 5. Therefore, the value is correctly given as **0.0463.**

3. **When the first digit after the last significant figure is equal to 5**, the situation is ambiguous and there is no universally accepted way to handle this case. A good guide to follow is:

    a. Look at the *second* digit after the last significant figure. If it is larger than 5, round the last significant figure upward by adding 1 to it. If it is 5 or smaller, or if there is no second non-significant figure, go to step **b.**

    b. When the first digit after the last significant figure is equal to 5, and the second digit after the last significant figure offers no basis for rounding upward, as explained in **a,** always round the last significant figure to the nearest *even* number. This offsets any psychological bias for rounding in a non-random manner, since there is an equal probability that choosing the next nearest even number results in rounding up or down.

Example:

a. A calculation gives the number 123.6257 for a value that should have five significant figures. The first non-significant figure is 5, which is ambiguous with respect to rounding up or down. Because the second non- significant figure is 7, which is larger than 5, the last significant figure should be rounded upward by adding 1. The value is correctly given as **123.63.**

b. A calculation gives the number 3.775 for a value that should have three significant figures. The correct value could be either 3.77 or 3.78, since 3.775 is exactly in the middle between them. By choosing the closest *even* number, the value is correctly given as **3.78.**

Rules for Finding the Correct Number of Significant Figures

An exact treatment for finding the correct number of significant figures is more complicated than necessary for most situations. If you remember that **no mathematical operation can increase the precision of an experimental result**, you can use three simple rules for determining the correct number of significant figures.

1.   When the precision of a number has been determined, as the standard or average deviation, the average value of a series of measurements should have the same number of decimal places as the ± value of the deviation.

Example:

The molarity of permanganate solution in Example 1 has an average deviation of ±0.0003. Therefore, the average value of the molarity should contain four figures after the decimal point. In calculating the average, the calculator result of

$$0.4057/4 = 0.101425000 \text{ mol/L}$$

must be rounded to **0.1014 mol/L** in order to use significant figures correctly.

2.   If a value is calculated using multiplication and/or division, the value should be rounded off to have the same number of significant figures (regardless of the position of the decimal point) as the quantity used in the calculation that has the least number of significant figures.

Example:

The velocity of a rolling ball is determined by observing that it rolls 135.6 cm in 12.1 seconds. Using a calculator, the velocity found to be:

$$(135.6 \text{ cm})/(12.1 \text{ s}) = 11.20661157 \text{ cm/s}$$

Because the time measurement, 12.1 seconds, has only three significant figures, the answer is limited to three significant figures. The answer is expressed correctly as **11.2 cm/s**.

3.   If a value is calculated using addition and/or subtraction, the value should be rounded off to have the same number of *digits after the decimal point* as the quantity used in the calculation that has the least number of digits after its decimal point.

Example:

<table>
<tr><td>

```
  265.3
   33.67
    1.0983
 ─────────
  300.0683
```

</td><td>Because 265.3 has only one digit after the decimal point, the answer must be rounded off to just one decimal place, as indicated by the vertical line. The answer is correctly given as **300.1.**</td></tr>
<tr><td>

```
   34.694
  -34.63
 ─────────
    0.064
```

</td><td>Because 34.63 has only two digits after the decimal point, the answer must be rounded off to two decimal places. The answer is correctly given as **0.06.**</td></tr>
</table>

CHEMICAL LABORATORY METHODS

Familiarize yourself with the equipment and glassware illustrated in Figures 0.1 and 0.2.

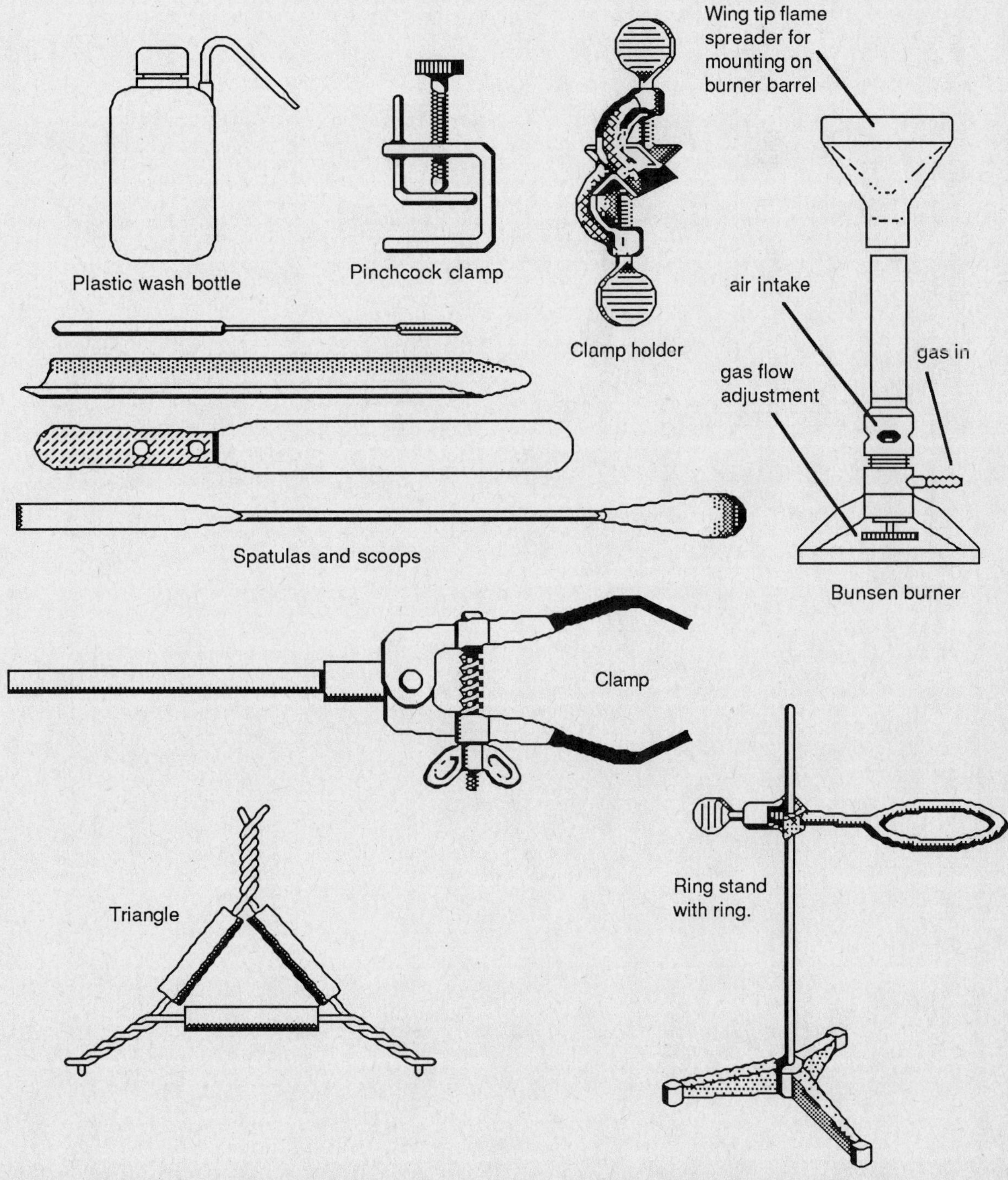

Figure 0.1. Some laboratory equipment commonly included in your bench drawer or locker.

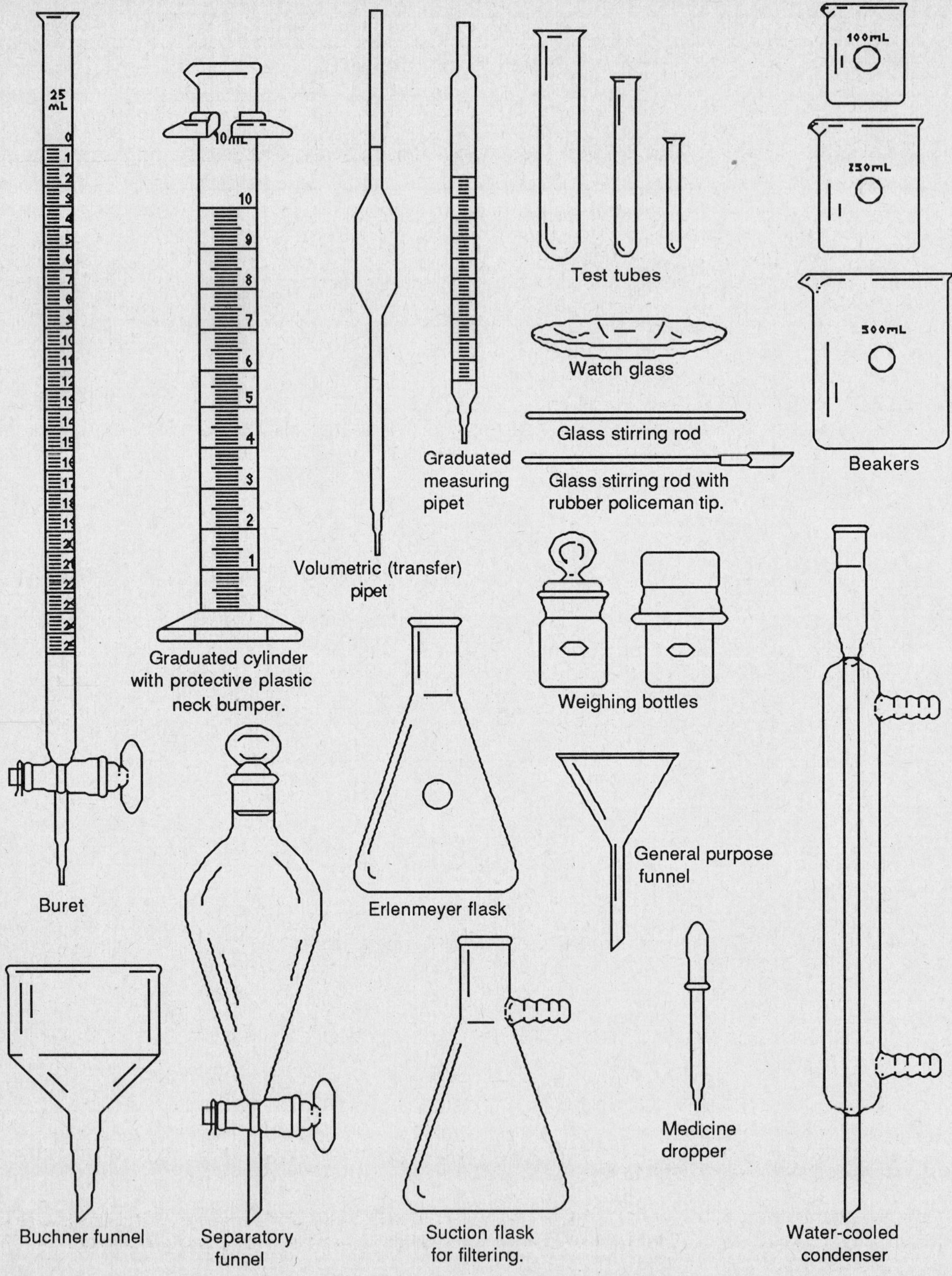

Figure 0.2. Some laboratory glassware commonly included in your bench drawer or locker.

I. POURING LIQUIDS

SAFETY

1. Check whether bottles are wet on the outside. If so, clean with wet sponge before handling.

2. Keep fingers out of path of flowing liquid. Rinse hands with water after operation.

Methods

Pour from a spout when possible. A funnel or glass rod (see Figure xx) may be used as a pouring aid. Lay only flat-top stoppers on the table with the sealing surface pointing up to avoid accidental contamination from contact with the table surface. Hold "pennyhead" stoppers and other types between the index and middle fingers, and hold the bottle with all fingers of the same hand, as illustrated in Figure 0.3. When removing these stoppers it is best to approach them with your palm up in order to grasp the stopper between your fingers from the back of your hand.

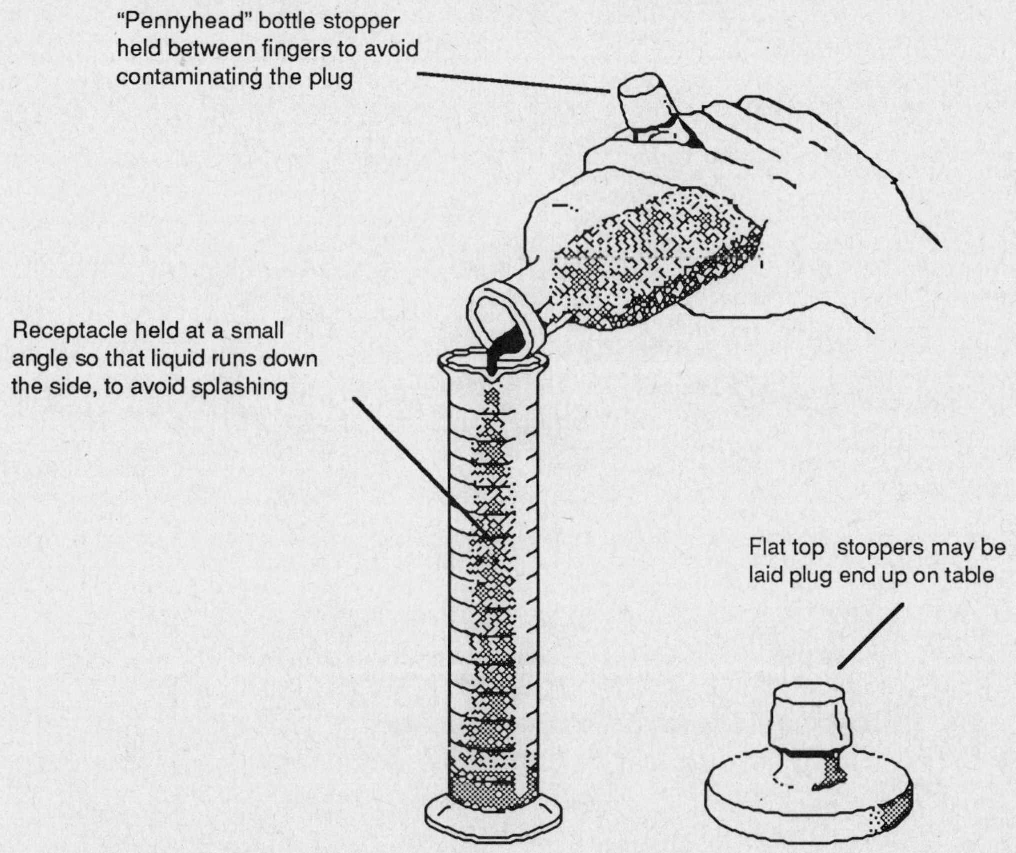

Figure 0.3. Pouring liquids.

Never pour directly from a 2 L or larger container into a narrow-mouthed vessel; pour into a beaker and then from the beaker into the final container.

Pour concentrated sulfuric acid into cold water (*never water into the acid*) slowly, with stirring. Use a vessel designed to withstand thermal shock (for example, a beaker or flask), not a bottle or graduated cylinder.

If too much is taken, discard the excess; never put anything back into a reagent bottle.

II. TRANSFERRING POWDERS

SAFETY

Many solid chemicals are corrosive and will damage skin and clothing. Many solids are poisonous.

Do not touch solid chemicals.

Methods

Use a disposable plastic weighing dish or place the powder on weighing paper or other smooth paper. Curl the paper into a chute and use it as a funnel for pouring. Do not try to pour powder directly from a container into anything that does not have a very wide mouth. With small-mouthed containers, use a **clean** spoon or spatula. If too much is taken from the reagent bottle, discard the excess. Reagent chemicals must be kept pure.

Never put anything back into a reagent bottle.

III. WEIGHING:

The Laboratory Triple-Beam Balance

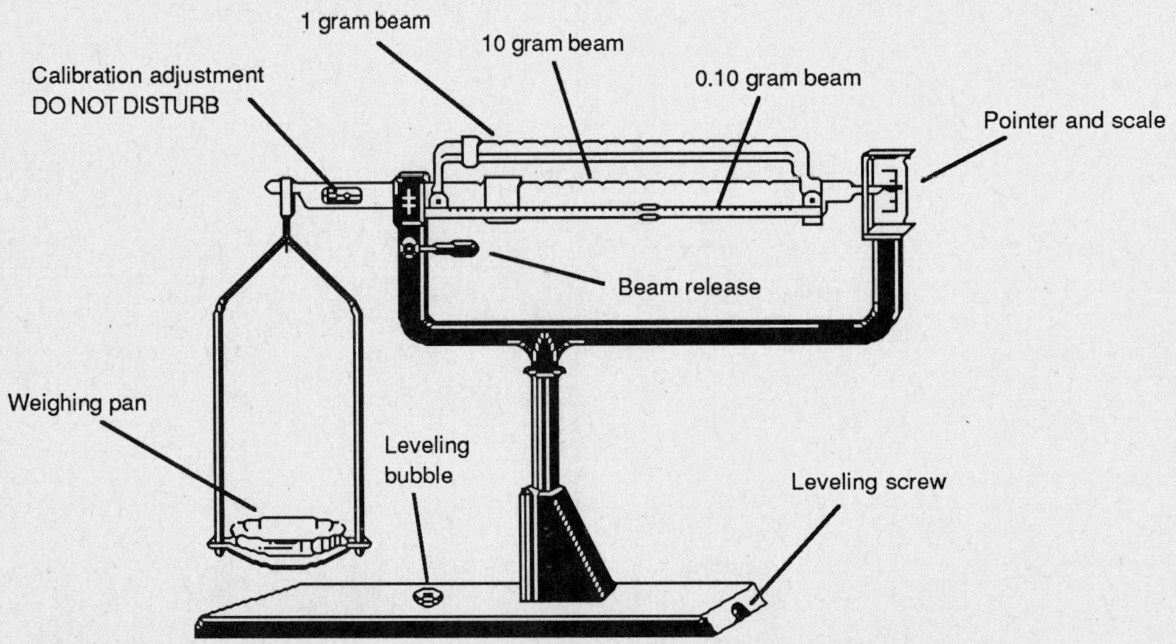

Figure 0.4. An agate or stainless steel knife-edge triple-beam balance (accuracy 0.01g).

Methods

This is the balance to use when a sensitivity of ±0.1 g is satisfactory. Set all rider weights to zero. Avoid drafts. Set balance swinging slightly (pointer should move two to five divisions away from center). Average one reading above center with one reading below center. If average (called "zero point") falls outside the range of -1 to +1, move the small screw at the calibration adjustment (ask instructor first) until the average falls within this range.

If the material to be weighed is a clean solid (metal bar, dry beaker, etc.) at room temperature, set it on the platform. If the material is a powder or liquid, use an appropriate **previously weighed (tared)** container (weighing paper, weighing bottle, watch glass, beaker, etc.). Never place a powder directly on the platform. Now move the rider weights until the swings of the pointer again average between -1 and +1. This is the "rest point." Read the values where the rider indices point and record the total weight.

Protect the balance from corrosive chemicals by keeping it clean at all times. Dust the pan before and after use.

The Single-Pan Analytical Balance

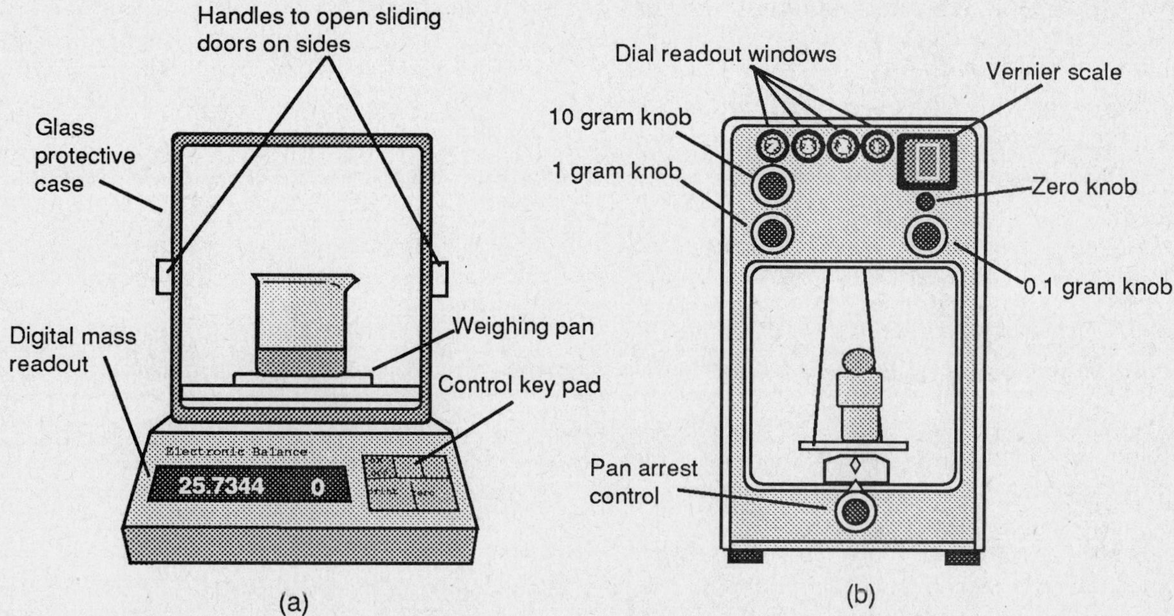

Figure 0.5. Typical single-pan analytical balances with weighing accuracy of ±0.0001 g:
   (a) An electronic balance detects the small downward motion caused by a mass on the weighing pan by its disturbance of a magnetic field.
   (b) A mechanical balance weighs by substituting weights on a single beam arm to balance the mass on the weighing pan.

Methods

The single-pan analytical balance is one of the chemist's most important tools. The analytical balance you will use is a sensitive instrument that has a precision of approximately ±0.0001g. You must treat it with courtesy and respect, for it does endure abuse very well. There are too many different models in use to give detailed operating instructions here. Two commonly used types, electronic and mechanical, are illustrated in Figure 0.5.

With an electronic balance, a mass on the weighing pan causes the pan to press a small metal mass downward, into a magnetic field. The disturbance of the field is sensed by an electronic circuit that activates an opposing electromagnet and measures the current required to force the pan back to its original position.

With mechanical balances, a mass on the balance pan causes the deflection of the beam from which it hangs. The deflection movement is optically measured with a system of mirrors and lenses. By turning the weight knobs, you can place appropriate counter-weights on the beam to balance out the mass on the weighing pan and restore the beam to its original position.

Your instructor will demonstrate how to weigh samples on your particular models.

Do not attempt to operate an analytical balance until your instructor has demonstrated the correct procedure and has approved your use of the instrument.

GENERAL PRECAUTIONS FOR USING AN ANALYTICAL BALANCE

1. An analytical balance is a delicate instrument and should not be treated roughly. Handle all controls gently. With mechanical balances, rapid turning of the knobs will throw the weights off of their rests. Never force a knob to turn. Protect the balance from corrosive chemicals by dusting before and after use. **If the balance does not work, notify the instructor. Do not try to fix it.**

2. Keep the door to the weighing pan closed at all times except when placing or removing objects on the pan. Air drafts will change your mass readings, and dust and fumes should be prevented from entering at all times.

3. The balance pan and floor must be kept very clean. If you find them dirty, clean them up. If you spill anything, clean it up right away. Always brush off the pan with the small brush provided just before you load anything onto it.

4. Before weighing any glassware, wipe it off with a towel to remove moisture and grease. Do not handle anything to be weighed with your fingers to avoid moisture and grease. Use tweezers, forceps, clean gloves, or toweling to handle samples. Never handle standard weights with your fingers, for the subsequent corrosion can ruin them.

5. **With mechanical balances, never load or unload the pan unless the balance is "arrested."** When "arrested," a support protects the delicate pivots against excessive loads, like accidentally dropping something on the pan. (This is not a problem with electronic balances.) A good practice is never to open the balance door without arresting the balance and never release the arrest unless the door is closed.

6. Never weigh anything that is not at room temperature. The analytical balance is much more sensitive than the triple beam balance and is affected by air currents created by thermal convection from hot objects. Atmospheric moisture may condense on cold objects.

7. Only hard bulky objects, such as an aluminum bar, may be placed directly on the pan for weighing. Liquids and powdered, granulated, or pelleted solids must be held in a previously weighed (tared) dry container at room temperature. If the material may interact with the atmosphere (evaporate, fume, adsorb moisture, or oxidize) during weighing, the container must be closed. For solids that do not require protection from the atmosphere, it is permissible to use chemical weighing paper. Weigh the empty paper first and subtract its weight from the total weight of sample and paper.

8. During the final part of the weighing, keep the balance door closed.

9. Keep your hands, arms, and feet off of the balance table to avoid vibration.

10. When you are finished weighing, leave the balance leveled, zeroed, clean, arrested (for mechanical balances), and closed.

The Top-Loading Balance

Another type of balance that might be available is the top-loading kind. These are intermediate in sensitivity between the analytical single-pan balance and the triple-beam balance, usually having a sensitivity of ±0.001g. They are very convenient and easy to use. Like the more sensitive analytical balance, both electronic and mechanical versions are available, although electronic top loaders have largely replaced the mechanical ones. The operating principles are similar to the analytical balance.

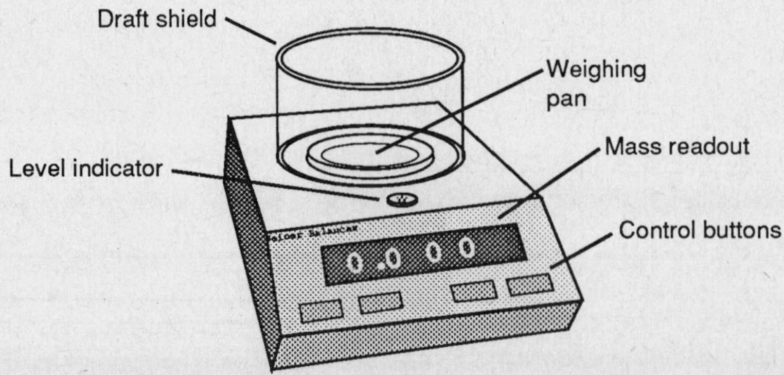

Figure 0.6.   An electronic top-loading single-pan balance (accuracy ±0.001g).

GENERAL PROCEDURES FOR WEIGHING SAMPLES ON ALL BALANCES

1. **Know the mass limit of your balance.**
2. Be sure the weighing pan is clean and empty.
3. Check the balance level and adjust if necessary.
4. Check readout zero and adjust if necessary.
5. Place sample in center of weighing pan. Always use a container or weighing paper.
6. Operate the controls as instructed to obtain an accurate measurement.
7. Read the mass value and write it down.
8. After weighing, arrest the pan if necessary, remove your sample, and return all controls to zero.
9. **Leave the balance in a clean condition.**

Weighing a Predetermined Quantity of Solid Sample

First, weigh the empty container. This is called the **tare mass**. This tare mass is recorded and then subtracted from the final mass to obtain the mass of the sample.

Electronic balances generally subtract the tare mass automatically and read the sample mass directly.

Transfer the sample with utmost care to the container in which it is to be used, using a camel's hair brush for the last few particles. A powder funnel (short, wide neck) or chute made of weighing paper may be used to transfer a sample into a flask. If you plan to dissolve the weighed sample, the last portion should be transferred by rinsing all containers and funnels several times with the solvent, adding the rinse liquid to the sample container. If the material to be weighed interacts with the atmosphere, put a little more than the desired amount quickly into a closed weighing bottle. Portions are removed in rapid operations, with minimal openings of the container, until the amount of sample remaining is close to the desired quantity.

One can never obtain a predetermined mass with maximum accuracy and the use of predetermined masses should be avoided whenever possible.

Weighing a Sample Whose Mass Is Not Decided in Advance

This is more accurate than trying to obtain a predetermined mass. In this case, one makes a rough estimate of a suitable quantity of sample, weighs it accurately, and then works with whatever accurately determined mass is obtained. A solid sample is often weighed out from a small test tube or weighing bottle into a larger beaker or flask. In such cases, it will be preferable to weigh the sample in its lighter container, then pour a portion into the beaker or flask and reweigh the original container with the remaining sample. When a sample is issued in a corked, labeled tube, transfer it before weighing to a clean, dry weighing bottle with no paper label, to avoid gross errors such as the transfer of cork particles and weight changes from variations of moisture content in the label and its adhesive.

Heating to a Constant Mass

To get a reproducible mass free from error caused by unknown amounts of moisture and other variable volatile impurities in the sample, it is necessary that empty crucibles and crucibles with contents be heated to constant mass. To accomplish this, heat the crucible for a minimum of 15 minutes at the required temperature. Cool the crucible to room temperature, in a desiccator if necessary to keep it dry, and then weigh it.

Allow 15-20 min for cooling; the crucible must be at room temperature before weighing (see item 2 under "General Precautions for Using an Analytical Balance," page 15.

Reheat for about 5-10 min, cool, and reweigh. Repeat the process until the mass remains constant within 0.1 to 0.3 mg. depending upon the precision required. The use of crucible tongs is recommended.

A desiccator is a container that provides a comparatively dry atmosphere in which crucibles and other materials may be stored. The lower portion of the desiccator contains a desiccant, usually anhydrous calcium chloride or silica gel.

IV. LIQUID MEASURE
The Graduated Cylinder

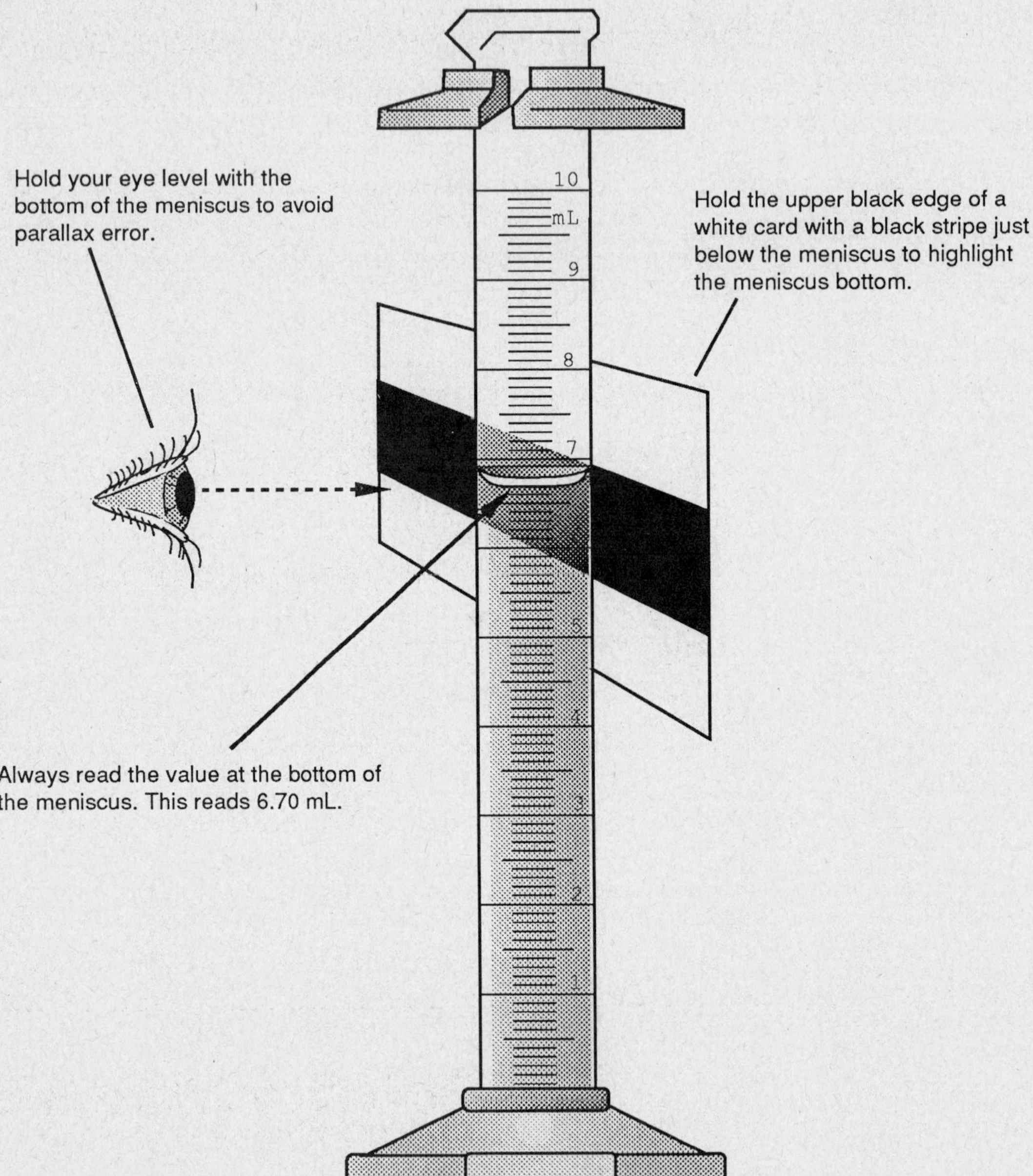

Figure 0.7. Reading a graduated cylinder. A white card with a black rectangle held with the upper black edge just below the meniscus will highlight the meniscus bottom and aid reading. Pipets, burets, and volumetric flasks are read in a similar manner.

SAFETY

1. See Section I., page 11, for filling a graduated cylinder .

2. Do not put hot liquid into a graduated cylinder.

Methods

Read the **bottom** of the meniscus (see Figure 0.7; instructor will demonstrate). Be certain your eye is level with the meniscus bottom to avoid parallax. Note that the lowest division on the graduate cylinder may be missing because of the curvature of the bottom of the cylinder.

The Pipet

SAFETY

1. Always use a rubber-bulb pipeting device to draw the liquid up into the pipet. Your instructor will demonstrate how to adjust the liquid level by controlling the entrance of air.

    Do not pipet *any* liquids with your mouth.

2. Be careful to always keep the pipet tip well below the surface of the liquid from which you are filling. If the surface drops below the pipet tip during the filling operation, air will enter the pipet rapidly, bubbling up through the liquid and pushing liquid high into the pipet. This can contaminate the liquid by bringing it into contact with the rubber bulb.

3.   **Never insert the pipet into a reagent bottle.**

Instead, pour some of the liquid into your own container before pipetting it.

Methods

Practice filling the pipet with water. Submerge the tip well below the liquid level, then draw up the water until the level in the pipet rises above the upper graduation mark. Control delivery with your forefinger. Practice control. Note and understand the calibration marks on your pipet before you use it. Now draw in some of the liquid to be measured, rinse the pipet, and discard rinsing liquid (see page18 for pipet rinsing technique). The pipet is now ready for use. Rinse it well with water when you are finished.

The Buret

SAFETY

1. Use a funnel or small beaker for filling the buret.

Methods

To fill the buret, support it vertically in a buret holder with a small funnel in its top and pour directly from the reagent bottle through the funnel.

    The funnel must be rinsed with water and dried before being used again for refilling the buret.

If the bottle is too large to be held comfortably or its mouth is constructed so that liquid cannot safely be poured from it into the funnel, the buret may be filled from a small beaker. However, any solution left in the beaker will dry out, and the beaker must be rinsed with distilled water and dried before each use.

Before using it, rinse the buret with water. Remove the stopcock plug (remove retaining ring first if it is present). Clean the plug and the inside of the barrel by wiping and drying carefully with cloth or soft paper. Make sure the hole in the plug is clean. If the plug is glass, smear one thin line of grease along the length of the plug between (and away from) the hole openings. (Teflon plugs do not require grease.) Smear another thin line on the other side. Insert the plug into the barrel in the open position (hole aligned vertically) and press firmly for a few moments while the grease spreads under the pressure. Do not twist. The grease should spread sufficiently to seal without clogging the passage hole. Save the retaining ring; do not replace it until you have finished with the buret. Now close the stopcock and support the buret with a ringstand and buret clamp.

Fill the buret to about one-fifth of its volume with the liquid to be used. Rinse the stopcock by letting liquid flow through it into the sink or a waste beaker; repeat the rinsing. Then, fill the buret above the zero mark and discharge the first portion of solution rapidly to remove any bubbles from the tip.

Close the stopcock and fill the buret above the zero mark again. Allow liquid to be discarded through the stopcock until the upper level reaches the zero mark or slightly below. Check the following:

a. none of the liquid must be on the **outside** of the buret due to gross carelessness in filling;

b. the stopcock and tip must be full of liquid (no air bubbles);

c. when handling the stopcock, always maintain a slight positive (inward) pressure on it to avoid the possibility of a leak.

The buret is now ready to deliver the required liquid volume.

Adding a measured volume of one solution to another solution to complete a reaction is known as *titration*.

When you have finished with the buret, drain the liquid completely, then rinse three times with water.

Summary of Buret and Pipet Cleaning Technique

Before a buret or pipet is used, it must be rinsed with several small portions of the solution that it will contain, so that this solution will not be diluted by the water remaining on the walls.

A buret can be rinsed by pouring the solution slowly down the walls, rotating to insure that no part of the surface is missed.

This rinsing must be performed at least three times.

To rinse a pipet, draw a small amount of the solution into the glass bulb portion and, by shaking and tilting, bring it into contact with the entire glass surface of the bulb portion and of the upper stem, as far up as the mark. Rinse three or more times.

A buret or pipet measures the volume of liquid that is to be delivered to another container.

To insure reproducible delivery, the inner wall of the buret or pipet should be clean enough to leave an *unbroken* film of water after drainage.

Sufficient time should be allowed for drainage, about 10-20 seconds, before a buret reading is made or a pipet is considered to be empty.

The tip of a buret or pipet should be touched to the inner wall of the receiving vessel to transfer any hanging partial droplet. Any liquid that remains in the pipet tip should *not* be blown out.

Most pipets are not "calibrated for blowout." All measurements are made by reading the liquid level at the lowest point of the meniscus. Your eye should be level with the meniscus to avoid parallax error. The meniscus can be highlighted conveniently by a white card containing a black rectangle. The card is held in position with the top edge of the black rectangle directly below the meniscus (see Figure 0.7, page 16).

The Volumetric Flask

<table><tr><td>

SAFETY

1. Fill the flask with a funnel or from a small beaker.

2. Never put a hot liquid into a volumetric flask.

</td></tr></table>

Methods

The volumetric flask is made so that it contains an accurately known volume of liquid, at a certain temperature, when it is filled to a calibration mark on the neck.

The neck must be clean so that water drains in an unbroken film.

To prepare a solution containing a known weight of a solid solute in a known volume of solvent, either of the following two procedures should be used:

1. Weigh the solid into a beaker and add enough water (or other solvent) to dissolve it. Stir as needed. If heating is necessary, be careful not to lose any material by spattering; if the solution must be boiled, keep the beaker covered with a watch glass and then rinse the condensate on the underside of the watch glass into the beaker.

Let the solution cool to room temperature before adding it to the volumetric flask.

Introduce the solution into the volumetric flask with the aid of a funnel, using a stirring rod to guide the flow. Rinse the beaker (including the outside of the lip), the stirring rod, and the funnel (including the outside of the stem) several times with water from a wash bottle. When all of the sample is in the flask, add distilled water nearly up to the bottom of the stem.

2.   If the solid is finely divided (no lumps), flows freely, and dissolves easily without pronounced evolution of heat or gas, it may be weighed directly into the volumetric flask, with the aid of a **dry** powder funnel (a funnel with a **short, wide** stem). After all the solid has passed through the funnel, rinse the funnel (including the outside of the stem) with water from a wash bottle. When all of the sample is in the flask, add water nearly up to the bottom of the stem and let stand, with occasional swirling, until the solid is dissolved.

Now add water carefully until the **bottom** of the meniscus is just at the mark on the neck. **Use a medicine dropper to add the last few drops.**
Let the water run down the neck and allow time for drainage after each drop. Stopper the flask with a rubber or a glass stopper (not a cork) and mix by inverting at least twenty times.

V. CLEANING VOLUMETRIC GLASSWARE

SAFETY

1. Some cleaning solutions for volumetric glassware are extremely corrosive. Handle them with great care.

2. Draw cleaning solution up into a buret by suction from an aspirator *through a safety bottle.*

The standard of cleanliness demanded for burets, pipets, and the necks of volumetric flasks is much higher than for other glassware because water hangs in drops on even a slightly greasy surface. **Solution that sticks to the glass surface without draining is not measured properly and is not reproducible, causing measuring errors.**
After washing with soap and water and thorough rinsing, the last traces of grease can usually be removed with a strong laboratory detergent. Sometimes, however, it is necessary to use a strong oxidizing solution of $K_2Cr_2O_7$ in H_2SO_4, commonly called "cleaning solution."
This solution is extremely corrosive, especially when hot, and must be handled with great care. DO NOT USE THIS CLEANING SOLUTION WITHOUT PERMISSION FROM YOUR INSTRUCTOR.

VI. BORING CORKS AND RUBBER STOPPERS

SAFETY

1 . The boring tool used for making holes through corks and rubber stoppers must be regarded as a

dangerous cutting tool, like a knife or ice pick.

Never use your hand as the support for a stopper being bored.

Methods
Holes often must be made in corks and stoppers so that glass tubes and thermometers can be passed through the stopper into a container.
For corks, first soften them by rolling in a cork softener, a wheel that turns in an off-center track so that the cork is squeezed into an ever-smaller space. If a softening wheel is not available, roll the cork on a tabletop under pressure, using a board or heavy book as the rolling device.
Remember that rolling to soften a cork also will reduce its diameter.
A borer tube is selected so that its outside diameter is a trifle less than that of the glass tubing to be inserted.
1. Sharpen the borer with a borer sharpener (a knife blade resting in a metal cone; your instructor will demonstrate its use).
2. Place the cork or stopper on a solid, relatively soft surface that you can afford to damage (e.g., wooden board, heavy cardboard, or paper pad), and proceed to to bore through from one side by simultaneously pushing and twisting the borer into it. After penetrating about half-way, withdraw the borer and press out any material in the tube with the metal rod that is provided.
3. For small corks and stoppers, continue cutting until the borer emerges from the other side. For large corks and stoppers, invert the plug and complete the hole from the other side.

VII. FITTING GLASS TUBING INTO CORKS, RUBBER STOPPERS, AND RUBBER TUBING

> **SAFETY**
>
> Jamming the jagged end of a piece of fractured glass tubing into your hand while trying to force the tubing through a rubber stopper may be the most common injury-causing accident in student chemistry laboratories. The strong pressures exerted just before the tubing breaks make it almost impossible to avoid hurting yourself. Such accidents can be avoided, however, by following the recommended procedure.

ALWAYS follow these rules:

1. Be certain the glass tip is fire polished.

2. Use a lubricant (glycerin, soapy water, or stopcock grease) in the hole of the stopper AND on the glass tip.

3. Grip the glass with your finger tips very close to the entry into the stopper so that the the glass has less tendency to flex when pushed.

4. Use a towel to protect your hands.

5. In the early stage, before the glass has come through the other side of the stopper, DO NOT HOLD YOUR HAND FLAT AGAINST THE STOPPER SURFACE AS A SUPPORT.

6. *Rotate* the stopper onto the glass . Do not force the glass into the stopper with straight pushing.

7. When removing glass from stoppers, observe precaution 3. **A cork borer lubricated with glycerin is a handy tool for removing glass from a stopper.**

8. If the glass becomes stuck at any time, do not try to remove it. See your instructor.

VIII. DECANTATION AND GRAVITY FILTRATION
Methods

Decantation and filtration are two common methods for separating solid material from a liquid.

Decantation: Allow the sample to stand undisturbed for a time so that the solid material can settle to the bottom. Then, carefully pour off the liquid leaving the settled solid undisturbed. Use a glass rod as a pouring aid (see Figure 0.8). A **centrifuge** is an instrument for speeding the settling of solids in liquids by rapidly spinning the sample container so that centrifugal force acts to drive the solid toward the container bottom.

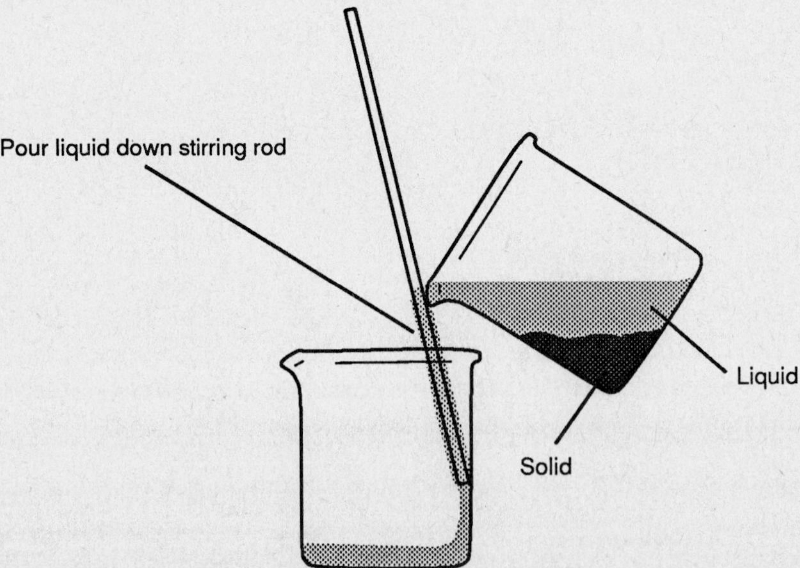

Figure 0.8. Decantation, using a glass rod as a pouring aid

Filtration: Fold a circle of filter paper sharply in half to give a half-circle. Then fold it in half again to give a quarter-circle. Shape the paper into a cone by separating one thickness from the other three (see Figure 0.9). Insert this filter paper cone into a funnel cone. Use a little water[2] to make the cone fit snugly in the funnel: wet the paper thoroughly and press the paper against the funnel wall with you finger or a glass rod, carefully pressing out any air pockets between the paper and the glass.
     A glass rod or any other tool used to press the filter paper into the funnel must be blunt and smooth, to avoid tearing the wet paper.

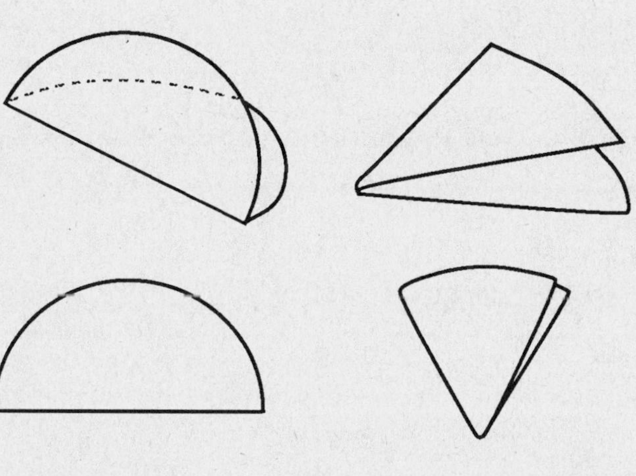

Tear off this outer corner so that the torn edges cannot press together. This torn corner allows both sides of the fold to seal against the funnel wall to prevent air from entering and breaking the suction created by flowing liquid.

Fold filter paper circle evenly in half.

Fold again, leaving the top quarter section a little short.

Open out the larger quarter section and insert into funnel. Moisten with distilled water and and seal against funnel wall with finger pressure.

Figure 0.9.   Preparation of a filter paper cone.

Pour the mixture to be filtered directly into the filter cone, not down the side of the funnel. Use a glass rod as a pouring aid (see Figure 0.10).
     The liquid level in the funnel must never rise above the top of the filter paper.
     The tip of the funnel stem should touch the side wall of the receiving container, to avoid splashing.

The funnel should be supported in a ring attached to a ring stand or in an arm support for funnels.
The liquid part of the sample is called the **supernatant**. The liquid collected after it has passed through the filter is called the **filtrate**. The solid that is retained on the filter is called the **residue**.

---

[2] If the sample contains a liquid other than water, that liquid should be substituted for water in this procedure.

Washing a Solid by Decanting through a Filter

Decant the supernatant liquid through the filter. Add about 25 mL of water or wash solution to the solid, stir with a clean rod to thoroughly mix the solid and wash water, and allow the solid to settle. Then decant the wash water. Repeat the washing three or four times.

Figure 0.10. Gravity filtering, using a glass rod as a pouring aid

Transferring the Precipitate

After washing, it frequently is necessary to completely transfer all of the solid to the filter. Most of the solid can be transferred during the last decantation. To complete the transfer, hold the stirring rod across the top of the beaker firmly with your forefinger, holding the beaker with your remaining fingers (see Figure 0.10). Tilt the beaker so that its bottom is above its pouring spout and the glass rod runs from the spout into the filter. Use a wash bottle to direct a stream of water to flush the solid with the water down the glass rod into the filter. Then, place the beaker on the desk and flush any particles adhering to the wall back into the bottom of the beaker, using as little water as possible. Then flush the solid into the filter as before.

A **policeman**, which is a rubber tip fitted onto one end of a glass rod, may be used to scrub the remaining traces of solid from the beaker walls. Use a wash bottle to rinse the solid from the policeman directly into the filter.

IX. VACUUM (SUCTION) FILTRATION

| **SAFETY** |
| Use thick-walled glassware to prevent implosion. |

Methods

A water aspirator or vacuum line is used to provide the suction. A Buchner funnel is inserted through a rubber stopper into the filter flask and a safety bottle should be inserted in the suction line between the vacuum source and filter flask (see Figure 0.11a). All connections are made with thick-walled rubber or plastic vacuum tubing and rubber stoppers. The filter paper should be just large enough to cover al the holes in the flat-bottomed portion of the funnel, but not so large that it has to be wrinkled to make it fit. Wet the paper with the same solvent as is in the sample to be filtered.

Support the flask and filter with a ring or clamp to prevent the apparatus from tipping over.
Porcelain filter crucibles with permanent porous beds also are commonly used. These require a special rubber holder to attach the crucible to the filter flask (see Figure 0.11b).

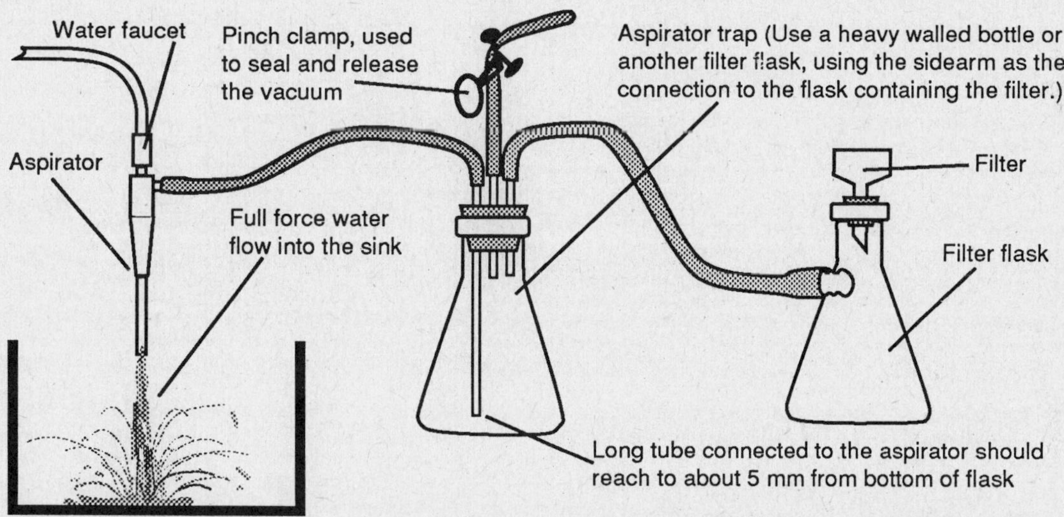

Figure 0.11a.   Suction filtering setup with aspirator, aspirator trap (safety bottle), and filter in filter flask.

Keep the aspirator faucet turned on to its fullest extent to obtain maximum suction and to prevent aspirator water from backing-up into your filtering apparatus.

Never not turn off the water during a filtering or washing operation. Turn the suction off and on by opening and closing the pinch clamp.

Always add sample to the filter with the pinch clamp *open*. Then close the pinch clamp to draw the solvent through the filter.
This procedure is especially important when washing the product. Adding the washing solvent to the filter with the suction *off* insures that the washing solvent remains in contact with the solid sample long enough for efficient washing. Close the pinch clamp to turn on the suction after insuring that all the solid has been covered with washing solvent.

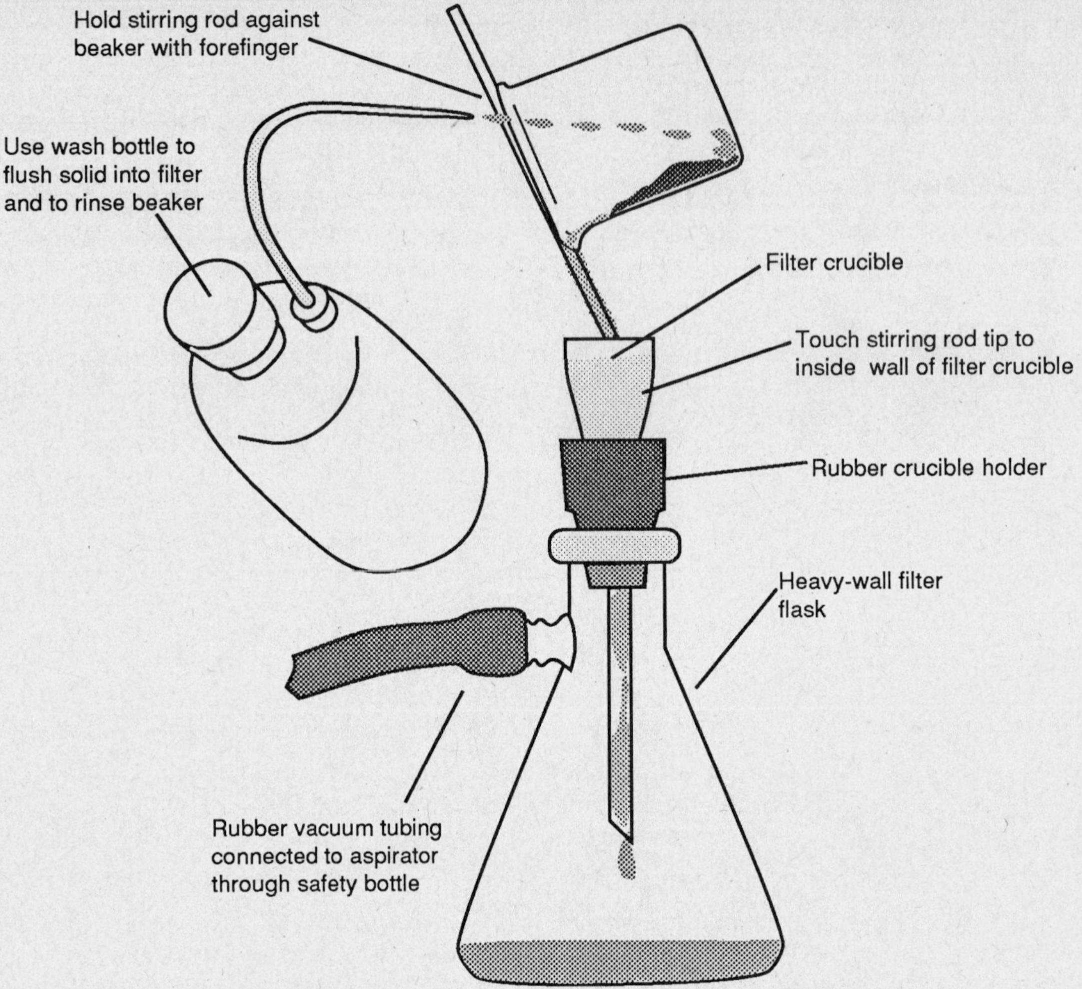

Figure 0.11b. Transferring solid to a porcelain filter in a suction filtering apparatus

X. SETTING UP AND USING GLASS EQUIPMENT

SAFETY

1.   Clamp glass parts very carefully to avoid causing stress and strains.

2.   Never heat a "closed" system — one which is tightly sealed from the atmosphere.

3.   Never heat flammable liquids with an open flame.

4.   Never point the open end of a test tube being heated toward anyone, including yourself.

5.   Understand the functions of all the parts of your equipment.

Methods

General principles to keep in mind:

1.   Glassware larger than test tubes usually should be supported from below in addition to being clamped. A ring or tripod with a wire gauze square makes a good support for heating beakers and flasks.

2.   When heating, observe the following precautions:
   a.   Use a wire gauze to distribute the heat more evenly when heating beakers, flasks and
        evaporating dishes.

b. Test tubes may be heated directly in a flame.
> **Never heat a test tube strongly above the liquid level nor at the very bottom of the tube. Heat the tube at one side near the bottom.**
Heating above the liquid level may overheat the glass and crack it. Heating at the bottom may cause a large vapor bubble which could blow the contents right out of the mouth of the tube.

c. Crucibles should be supported in a clay or wire triangle and heated directly with a flame.

d. Clamp flasks and test tubes very close to their open ends. Clamp just firmly enough to prevent motion; excessive pressure may break the glass.

e. The wire-spring test tube holder should be used only to hold small test tubes. It will not hold heavy objects securely.

f. Never heat graduated cylinders or other volume calibrated glassware with a flame. Thes items can, however, be dried in an electric oven.

g. Do not use a compressed air jet for drying. The air often is contaminated with oil and water from the compresser.

h. Do not heat equipment at or near the point where they are supported by a clamp or near cork or rubber parts.

i. When heating a solid that may melt, be sure the container tilts upward enough to insure the melt does not run out.

XI. DISTILLATION

> **SAFETY**
>
> When setting up and using a distilling apparatus, observe all the safety precautions listed above in Section X.

Liquid solutions and mixtures often are separated and purified by **distillation**, a process which vaporizes the most volatile components and condenses them into a separate container. Condensation of hot vapors is accomplished with a flowing-water condenser. The vapor system makes contact with the atmosphere at the receiving flask after all vapor has condensed so that evaporation losses are minimal. The liquid formed by condensation is called the **distillate** and the liquid remaining in the distilling flask is called the **residue**. A typical apparatus is illustrated in Figure 0.12. For safe and efficient operation of the distilling apparatus, you must observe the following precautions:

1. Cooling water must flow through the condenser from bottom to top, to insure uniform contact with the inner cooling wall.

2. The condenser must not be sealed to the receiving flask, to prevent any pressure build-up.

3. The distilling flask should not be more than half-full at the start of distillation, to prevent any liquid splashing over into the condenser during boiling.

4. Add several boiling chips at the beginning to prevent liquid "bumping". NEVER add boiling chips to a hot liquid sample.

5. Never heat the distilling flask or residue to dryness. With no liquid in the flask, the temperature can rise high enough to crack the glass or decompose the residue, sometimes with violent results.

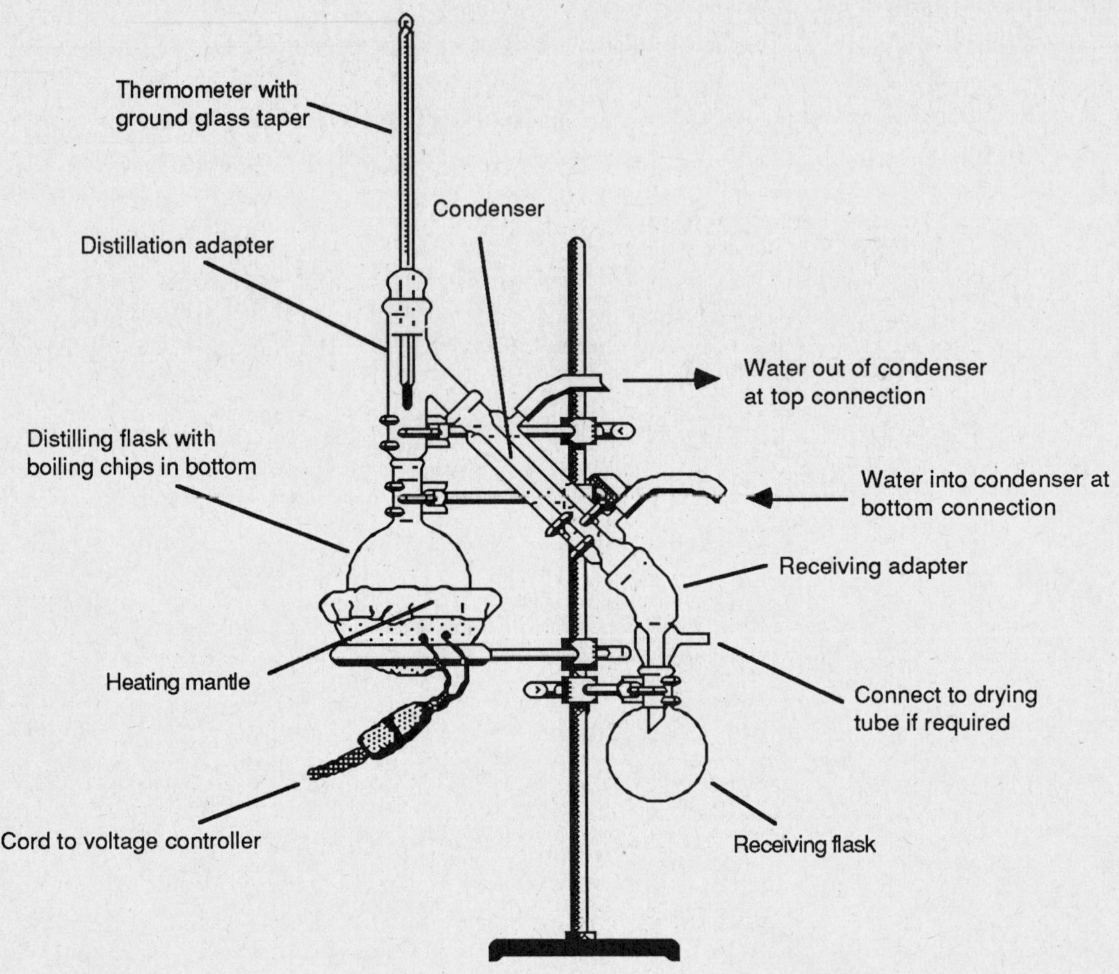

Figure 0.12.   Distillation apparatus.

XII. THE pH METER

A pH meter is an electronic instrument for directly and continuously measuring the hydrogen ion concentration of solutions. A sensing electrode and a reference electrode are immersed in the sample solution, generating a voltage that is proportional to the solution pH. The results are displayed on a meter or chart recorder as the pH value. The sensing and reference electrodes may be two separated units or they may be combined into a single electrode structure. A typical pH meter is illustrated in Figure 0.13.

pH electrodes are very delicate and easily damaged. They must be handled with care. Avoid bumping them against any solid surfaces, such as beaker walls or bottoms.

Procedure for Using a pH Meter

1.   The electrodes must be hydrated. They should have been soaking in distilled water overnight.

2.   Some meters require one-half to one hour warm-up time for maximum stability. Be certain warm-up has been sufficient and do not turn the meter off between measurements. Use the "standby" switch position during warm-up and between measurements.

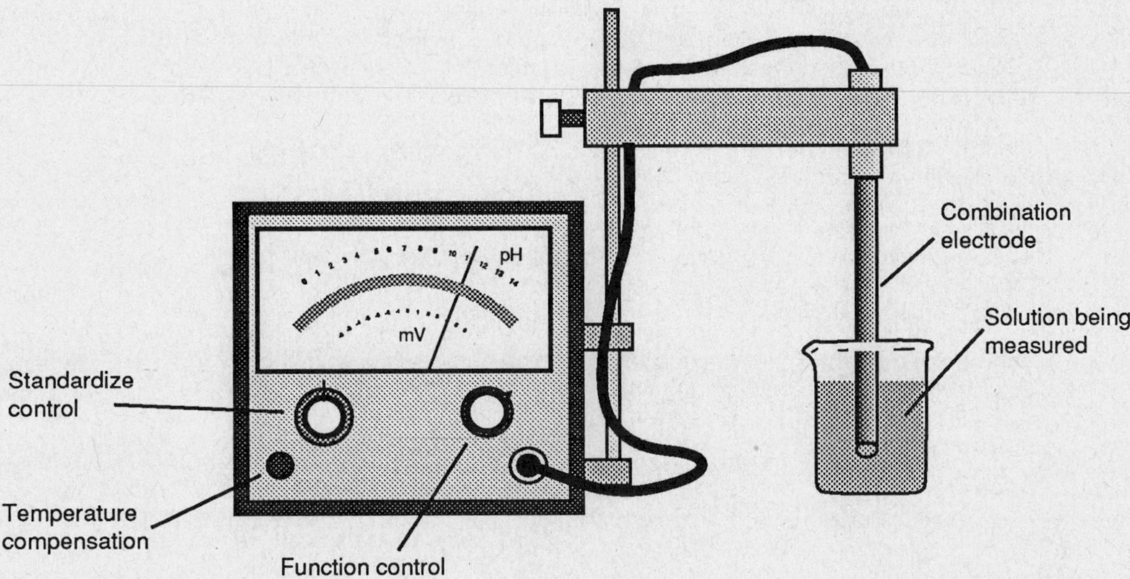

Figure 0.14. pH meter with combination electrode that combines both the sensing and reference electrodes.

3. Raise the electrodes out of their soaking bath, rinse them thoroughly with distilled water from a squeeze bottle, and dry them carefully with soft tissue.

   Well-soaked electrodes can be left in the air for 10-20 minutes without losing their water of hydration.

4. If your meter does not have automatic temperature correction, you must set the proper control to correct for the temperature of the solution you are measuring. pH electrodes are temperature sensitive. Measure the solution temperature with a thermometer.

   Samples may be measured conveniently in 150-250 mL beakers.

5. Lower the electrodes into the buffer solutions to be used for calibration. Wait about two minutes before reading to allow the electrodes to reach thermal equilibrium. It is best to use at least two calibration buffers, about 3 pH units apart. If only one buffer is used, its pH should be near the pH to be measured.

6. Switch from "standby" to the "measure pH" position and adjust the meter to read the known buffer pH.

7. Turn back to "standby," raise the electrodes from the buffer solution, and rinse and dry them carefully as before.

8. Measure and record the sample volume.

9. Immerse the electrodes in the sample solution. The electrodes should be covered with solution but should not be in danger of bumping into the beaker bottom.

10. Read and record the sample pH. After the measurement, switch the meter to "standby," raise the electrode carefully out of the sample solution, and rinse them well. When not in use for longer than 10-15 minutes, leave the electrodes soaking in distilled water.

    Always rinse and wipe dry the electrodes between immersions in different solutions.

XIII. CHEMICAL ANALYSIS WITH AN ABSORPTION SPECTROMETER

Absorption spectrometry can be used to identify components of a solution and measure their concentrations. The amount of light absorbed by a compound at different wavelengths is a characteristic of the compound and the pattern of light absorption, called a **spectrum**, can serve as an identifying "fingerprint," The higher the concentration of the compound, the stronger the absorptions at each wavelength, allowing the magnitude of absorption to be used to measure the sample concentration (see Figure 0.15a, b).

A compound is identified by comparing its spectral "fingerprint" with spectra of known compounds, using a dictionary of standard spectra.

To find the concentration of a compound:

1. **Calibrate the sample by measuring the amount of light absorbed at a particular wavelength by the compound at several known concentrations.**

2. **Make a graph of absorbance vs. sample concentration (see Figure 0.15b)**

3. **Measure the absorbance of the unknown sample at the same wavelength and read the sample concentration from the calibration graph.**

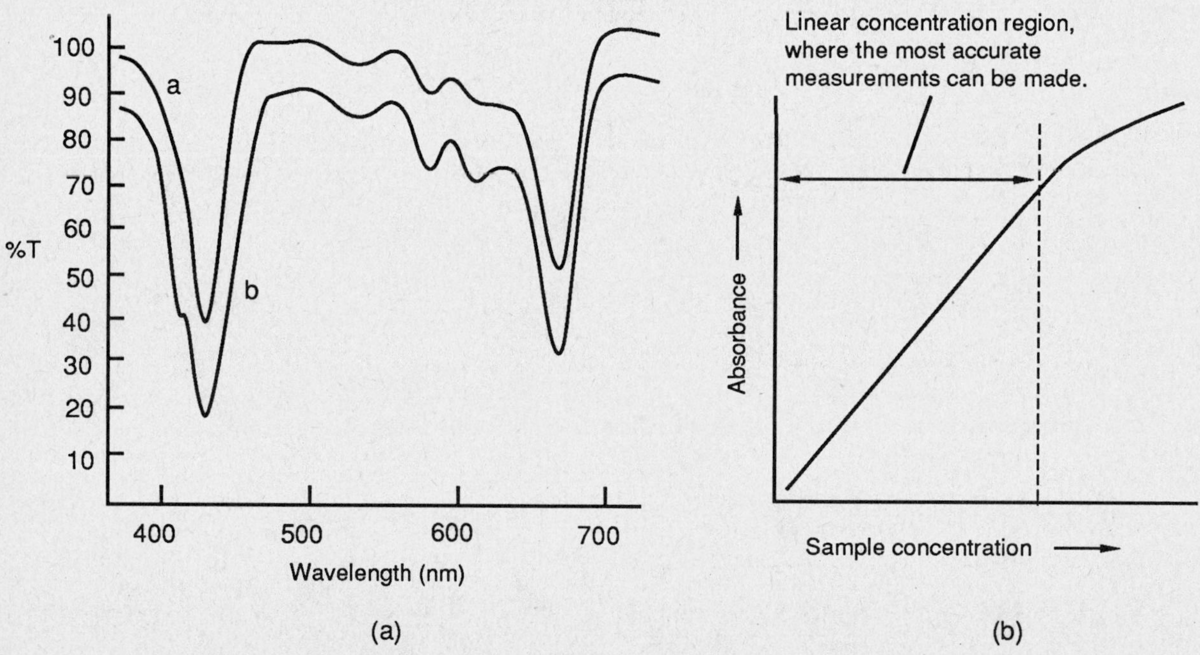

Figure 0.15.   Chlorophyll-*a* spectra and a characteristic absorbance graph, showing how sample absorbance increases with concentration.

(a)   Transmission spectrum of a solution of chlorophyll. Curve *b* is at a higher concentration than curve *a,* and the spectrum shows that the *b* sample absorbs more light than the *a* sample.

(b)   Typical plot of sample absorbance at a single wavelength vs. sample concentration. If the sample concentration is too high, the plot becomes nonlinear.

The spectrometer contains a light source, optical components to disperse the light and pass the separated wavelengths through the sample, and a detector that measures how much light passes through the sample at each wavelength. Your instructor will explain the specific operating procedure for the particular instruments that are in your laboratory. Some general principles are presented below.

Principles and Procedure

For accurate analyses, it is necessary to distinguish between light absorbed by the solute sample and light lost in unrelated ways, such as by reflections and scattering from the sample holder (called a **cell** or **cuvette**) and absorption by the solvent. In addition, the light source and detector have their own characteristic spectral output and response. For these reasons, it is necessary to make two measurements, one using a cell containing only the solvent with no sample dissolved in it (called the **blank**), and another using a different cell with both solvent and sample. It is important that the two cells be as identical as possible, so that their light-loss characteristics are the same. When the spectra from the blank and sample are compared, their differences are due only to light absorption by the sample, since other light losses are the same in both measurements.

Some instruments, called **double-beam spectrometers**, can accomodate both the blank and sample cells simultaneously and measure both at the same time, presenting only the desired difference signal. With a **single-beam spectrometer**, two separate measurements must be made of the blank and sample and the difference signal calculated by hand.

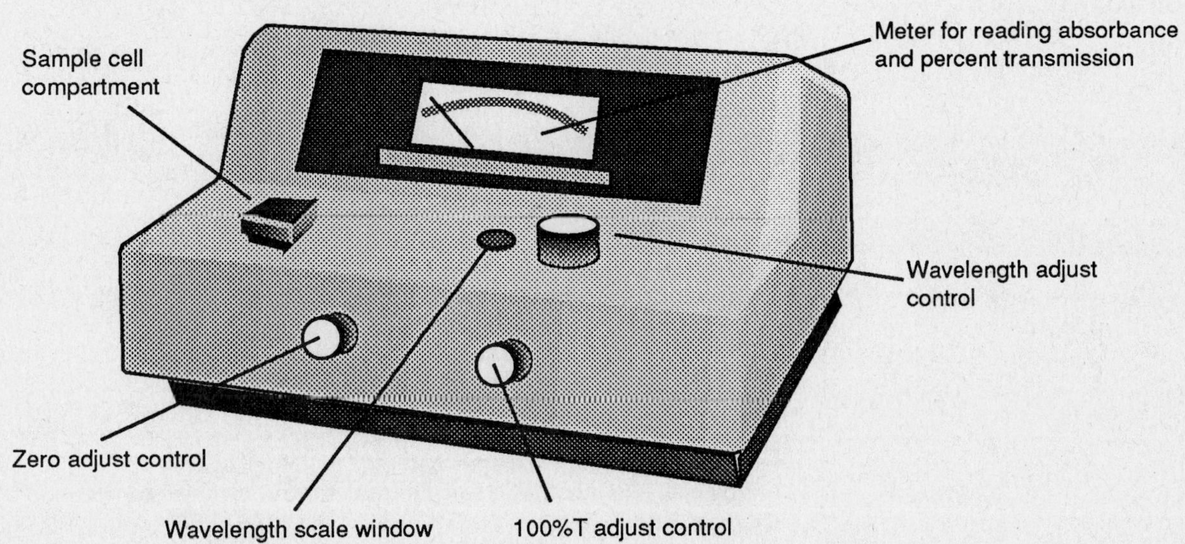

Figure 0.16.   Typical spectrometer for routine analysis.

Light Measurements

Spectrometers measure the light reaching the detector in terms of the light *transmitted* by the sample or the light *absorbed* by the sample.

The **percent transmission** through the sample is defined as:

$$\%T = \text{percent transmission at a particular wavelength} = \frac{\text{light intensity measured with sample cell}}{\text{light intensity measured with no cell}} \times 100$$

The **absorbance** of the sample is defined as:

$$A = \text{absorbance} = \log_{10}\frac{100}{(\%T)} = 2 - \log_{10}(\%T)$$

Each of the measurement quantities has different advantages. *%T* has a linear scale on the spectrophotometer meter, making it easier to read than *absorbance,* which has a logarithmic scale. However, the relationship between sample concentration and *absorbance* is frequently linear, whereas sample concentration and *%T* are related in a non-linear manner. A graph of sample concentration versus absorbance is frequently a straight line and allows easy interpolation between calibration point when measuring unknowns.

Calibration graphs should always be made by plotting *absorbance* on the vertical axis and *sample concentration* on the horizontal axis.

Stepwise Procedure for Spectrometer Analysis

1. Most instruments require one-half to one hour warm-up time to insure light source and detector stability. Be certain warm-up has been sufficient and do not turn the instrument off between measurements.

2. Clean the outside of your cells by wiping them carefully with lab tissue.
   Do not wipe the cells with paper towels, which might cause tiny scratches.
   Any lint or scratches will scatter light and cause errors. If the inside of the cell is dirty, ask your instructor for guidance. Always use the same cell for the blank and the same cell for the sample.
   Do not touch the lower three-fourths of the cell with your fingers.
   The light passes through this part and any smudges or oil on the surface will alter the absorbance.

3. Rinse each cell with distilled water to remove any traces of previous samples. Shake out the rinse water as completely as possible.

4. Rinse each cell three or four times with small amounts of the solution to be measured, to prevent any dilution of the solution by residual water.
   Use pure solvent in the blank cell and sample solution in the sample cell.
   Then dry the outside of the cell carefully with lab tissue.

5. Fill the cells to the proper level with the liquids to be measured (usually one-half to two-thirds full, sometimes indicated by an index mark).
   Be sure there are no small air bubbles attached to the inner surface and no dirt or lint on the outer surface.

6. Adjust the spectrometer:
   a. **Turn the wavelength control to the desired setting.**
   b. Adjust the zero control with the sample compartment empty and its cover closed. Under these conditions, a shutter blocks the light beam from the detector, so that the correct reading on the meter is 0.0% transmission (infinite absorbance).
   Turn the *zero adjustment* until the meter reads 0.0% transmission.
   This adjustment should be checked and reset from time to time.
   c. Insert the cell containing the blank.
   Adjust the *100% T* control until the meter reads 100% transmission (or zero absorbance).
   Inserting the cell automatically opens the light shutter and allows the light beam to pass through the cell to the detector. Adjusting the blank cell to 100% T compensates for all instrument factors and light losses other than sample absorption. Remove the blank.
   d. Insert the sample cell.
   Read the transmission and absorbance values from the meter.

7. If necessary, run a series of calibration measurements all at the same wavelength, obtaining the absorbance of several samples with known concentrations. You can use a graph of absorbance versus known concentrations to find unknown concentrations from their measured absorbances. It is wise to check the instrument zero and 100% T setting occasionally.

EXPERIMENT 1
Observations on Popcorn

PRELABORATORY PREPARATION
This experiment introduces several laboratory techniques you will use often in this course:
1. heating samples with a hot plate or gas burner,
2. weighing with an analytical balance,
3. calculating the standard deviation of a series of measurements.

In addition, you will make qualitative and quantitative observations of a commonplace phenomenon, the popping of popcorn, and you will try to explain what happens.

Before beginning the experiment, you should:
1. Do the Prelaboratory Exercise and turn it in at the beginning of your laboratory period.
2. Read the section titled "Treatment of Experimental Data" in the front of the manual, particularly the part about the standard deviation .
3. Read Section III, Methods, page 12, on weighing with laboratory balances.

INTRODUCTION
    How can we determine the characteristic properties of a sample of popcorn when each kernel is slightly different and when any two measurements on the same kernel are unlikely to give exactly the same value? This might not seem to be a very important matter, at least in the case of popcorn. Popcorn, however, is typical of many other substances in its variability from sample to sample; coal, oil and bacterial cultures are some other examples.
    For such materials, we can only describe their *average* properties.
The only exact measurement is a count of discrete objects, such as coins. When we measure properties that can vary continuously, such as mass, values for duplicate measurements will be different, even when we repeat the same measurement on the same sample, because of limitations of the measuring instruments and uncontrolled changes in experimental conditions (such as temperature).

To illustrate the concept, we can ask three questions about popcorn:
    1. How much does a particular popcorn kernel weigh?
    2. How much does an average popcorn kernel weigh?
    3. If I select a popcorn kernel at random, how closely can I predict its mass, having previously made several measurements on other similar kernels?
We will see that exact answers are not possible and that the desired answers to questions 1 and 2 can be said only to lie within a certain range of values. The answer to question 3 requires that we have information about the variability in mass from kernel to kernel.
    The most useful statement of a quantitative value expresses two characteristics of a series of measurements:
        1. the average value.
        2. the reproducibility of repeated measurements.
Techniques for determining average values and reproducibility are described in Section I, Methods, where average deviation, relative average deviation and standard deviation are discussed. You must read this material before doing this experiment.

PLAN OF EXPERIMENT
    You will try to determine some of the general properties of popcorn, deducing them from observations on a small sample of kernels. Your standard deviation from a small sample can be compared with that from a much larger sample by using data from your classmates. By comparing the results from small and large samples, you can judge whether or not your small sample size was large enough to give reliable average properties.

> **SAFETY**
> 1. Wear approved eye protection.
> 2. Contamination of your hands, table surfaces, and glassware with toxic substances is very likely in a chemistry laboratory.
>
> **DO NOT EAT ANY OF YOUR POPCORN.**

PROCEDURE

1.   Your laboratory instructor will provide a package of popcorn. Enter the total net "weight" (mass) written on the popcorn package in the designated space at the top of the Data sheet. Assume the mass given on the package label is accurate.

2.   **Handle kernels with tweezers to avoid transferring moisture and oil from your fingers.**
Place 22 kernels of popcorn in a small beaker. You will use 5 for practice popping, 5 for individual kernel measurements, 10 for popping all together as a group, and 2 as extras, in case you have trouble.

3.   Devise a convenient way to identify and keep separate at least 7 kernels (labelled as kernel 1, kernel 2, kernel...). You will have to carry your kernels to and from the analytical balance without mixing them up. Convenient methods might be to label the depressions in a spot testing plate with a grease pencil or crayon, to number some small paper baking cups, or to number 7 small test tubes.
Identify 7 kernels for weighing by placing them in the numbered locations.

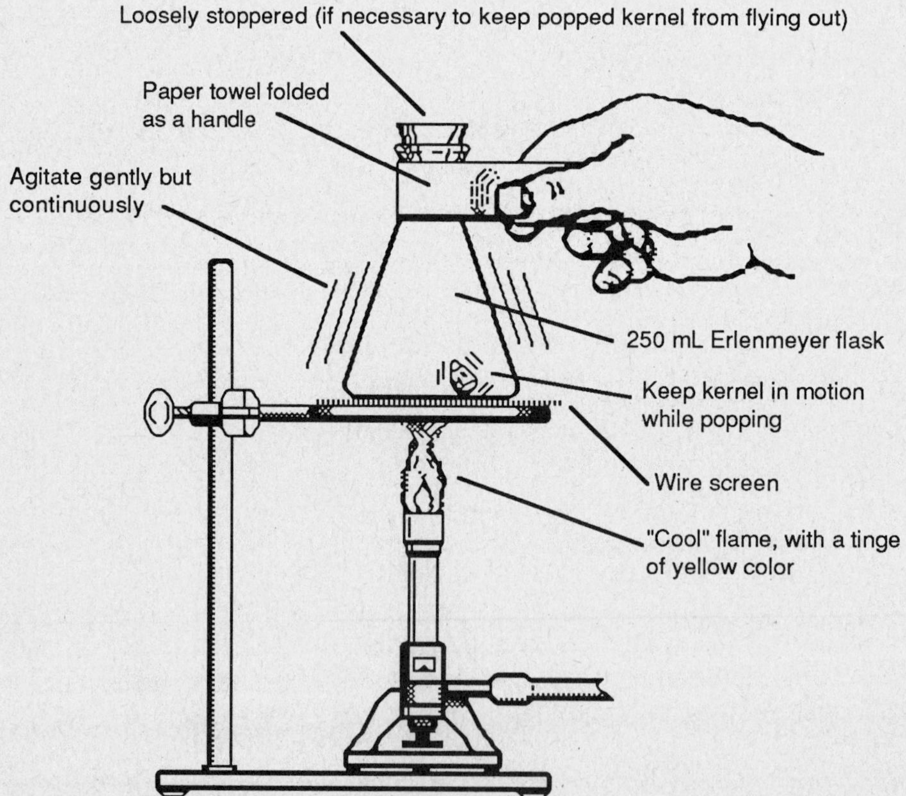

Figure 1.1. Method for popping a corn kernel.

4.   Your instructor will demonstrate the use of the analytical balance. While you are waiting to use the balance, practice your corn popping technique, as described in step 7.

5.   When you have access to a balance, weigh separately, on a weighing dish or watchglass, 7 identified kernels to the nearest 0.001 g. If you have never used an analytical balance before, your instructor may want you to make each weighing twice as a check on your technique. Record the masses on the data pages. Then weigh (all together) a *different* group of 10 kernels. Keep this group of 10 separated in a small, clean beaker. Record their total mass.

6.   You will use either a gas burner or a hot plate to heat the kernels.
        If you use a gas burner, adjust the air for a "cool" flame by partially closing the air intake to reduce the air flow until there is just a tinge of yellow color in the flame. Place a ring stand over the burner and, using tongs to handle it, place a wire gauze square, preferably with a ceramic center, on the stand over the burner flame. The gauze will diffuse the heat more evenly over a broader area.
        If you use a hot plate, turn it on the "high" setting.

7.   Obtain a cork (not a rubber stopper) that can be fitted loosely in the neck of a 250 mL Erlenmeyer flask. You will use it to loosely close your flask if the corn begins to jump out when popped.
        Fold a paper towel to make a thick paper strip about 2-3 cm wide. This will serve as a handle for holding the hot flask when it is wrapped around the flask neck as shown in Figure 1.1. Practice popping your extra unweighed kernels. Avoid scorching the kernel or the popped "flower."

Popping technique
        Have a glass rod handy for quickly unsticking the popped flower from the flask, if necessary. Preheat the 250 mL flask for about 10 seconds, and then, with tweezers, gently drop in an unweighed kernel while keeping the flask moving briskly. Continue heating the flask, constantly agitating the kernel, until it pops (1-3 min).
        If you are using a hot plate, keep the flask bottom flat on the hot plate while sliding the flask back and forth to shake the kernel around.

        Remove the flask from the heat as soon as the kernel has popped and immediately pour the flower out of the flask onto a paper towel, taking care not to break the flower. This must be done quickly to prevent scorching the flower, which might then stick to the bottom of the flask. Use the glass rod to quickly dislodge the flower from the flask wall if it sticks.
        Do not allow the flower to touch any water that might have condensed in the flask neck.

If you have trouble with the popped flowers jumping out of the flask, keep the flask **loosely** corked during the popping operation. When you are satisfied that you can pop the corn successfully without scorching it, go to Procedure step 8.

8.   Pop 5 of the singly weighed kernels, one at a time. You have 2 extra kernels already weighed in case you have trouble. After each kernel has popped, return the corn flower to its identified location or container to cool.

9.   Place the group of 10 weighed kernels into the Erlenmeyer flask, cork it *lightly,* and proceed to pop them all at once. Observe closely any changes in the system that might help explain a mass change that occurs in corn upon popping. Watch the upper, inner wall surfaces of the flask closely. Record your observations on the data pages. When all of the kernels have popped, return them to their storage beaker to cool.

If you didn't see anything obvious happening that could explain a mass change, redo this step as follows:
        a. Place 15 unweighed kernels in the flask and stopper it a bit more tightly.
        b. Pop the corn and watch the wall of the flask carefully. Record your observations.
Discard these flowers. (Be sure you have saved the 10 you popped earlier.)

10.   Cut 2 or 3 kernels in half with a sharp knife or a pair of wire cutters and try to pop them. Record the results of this effort on the Data sheet.

11.   The 5 single corn flowers should be cool by now. Weigh each of them, record their masses, and calculate the mass lost by each kernel upon popping. Then, calculate the percent mass loss for each kernel.

12.   When the group of 10 corn flowers are cool, weigh them all together and record their total mass. Calculate the total mass lost by the 10 kernels. Then, calculate the percent mass loss upon popping for the group of 10 kernels. Enter the results on the Data sheet.

13.   For the 5 single kernels, calculate their average initial mass, average mass loss, and average percent mass loss. Then calculate the standard deviation of the initial masses and the mass losses. Enter these results on the Data sheet.

14.   From data for the group of 10 kernels, calculate the average initial mass and average mass loss of *one* kernel. Record the result on the Data sheet.

15.   Compare the average initial masses and average mass losses determined from the 5 individual kernels and from the group of 10 kernels. Do they agree within one standard deviation? What does this tell you about your sample size for the standard deviation calculation?

16.   Use the average initial mass calculated from the group of 10 kernels to calculate the total number of kernels originally in the package of popcorn. Assume the net mass marked on the package is exact. Enter this result on the Data sheet.

17.   Each student should post on the blackboard her or his value for the total number of kernels originally in the package. Copy the results from the entire class on your Data sheet and calculate an average value and standard deviation for the number of kernels initially in the popcorn package.

18.   Be sure to answer the questions on the Question sheet.

Name ____   Date ________________

EXPERIMENT 1
PRELABORATORY EXERCISE

First, read Section I, Methods. The calculations in this exercise are the same as those you will use for analyzing your data from the experiment. By working through this exercise you will learn how to express the uncertainty and reproducibility in your experimental measurements. Be sure to finish it before coming to the laboratory to do Experiment 1.

The table below gives sample mass measurements made on 5 kernels of corn, before and after popping. Complete the table by calculating the percent mass loss for each kernel. Then calculate the quantities to be entered below the table.

Table 1.1: Sample mass measurements on 5 kernels of corn

Kernel number	m_i Initial mass before popping (grams)	m_f Final mass after popping (grams)	Δm Mass loss $(m_i - m_f)$ (grams)	$\%\Delta m$ Percent mass loss (mass loss x 100/m_i)
1	0.1434	0.1297	0.0137	(0.0137 x 100)/0.1434 = __9.55%__
2	0.1033	0.0928	0.0105	__________
3	0.0820	0.0742	0.0078	__________
4	0.1123	0.1035	0.0088	__________
5	0.1223	0.1100	0.0123	__________

total initial mass, m_i = __________   total $\%\Delta m$ = __________

average initial mass, $\overline{m_i}$ = __________   average $\%\Delta m = \overline{\%\Delta m}$ = __________

The usefulness of the average values determined above is increased greatly if you include the standard deviation as an indication of the reproducibility, or **precision,** of the measurements.
There is a worked-out example of a standard deviation calculation in Section I, Methods.

Use Table 1.2 to calculate the standard deviation of the initial masses, from the data above.

$$\text{standard deviation} = s = \sqrt{\frac{\sum_{n=1}^{N}(x_n - \overline{x})^2}{N-1}} \tag{1.1}$$

where x_n is the *nth* value of a set of N total measurements, $\overline{x}$ is the average value of all the measurements in the set, and the summation is from $n = 1$ to $n = N$ (i. e., x_1 is the value of x for the first measurement and x_N is the value of x for the last measurement). Substitute the measured m_i values for the x_n values in Equation 1.1

PRELABORATORY EXERCISE

$\bar{m}_i$ = _______________ (Enter the value from the results below Table 1.1)

Table 1.2.

Kernel number, (n)	m_n	$(m_n - \bar{m}_i)$	$\overline{(m_n - m_i)^2}$
1	_____________	_____________	_____________
2	_____________	_____________	_____________
3	_____________	_____________	_____________
4	_____________	_____________	_____________
5	_____________	_____________	_____________

$\sum (m_n - \bar{m}_i)^2$ = _____________   s = _______________

The final result of the measurement should be expressed by giving the average initial mass and the standard deviation:

The av. mass before popping = $\bar{m}_i \pm s$ = _____________ ± _________

Use Table 1.3 to find the standard deviation for the percent mass loss. Substitute measured $\overline{\%\Delta m}$ values for the x_n values in Equation 1.1 .

$\overline{\%\Delta m}$ = _____________ (Enter the value from the results below Table 1.1.)

Table 1.3

Kernel number, (n)	$\%\Delta m$	$(\overline{\%\Delta m} - \%\Delta m)$	$\overline{(\%\Delta m - \%\Delta m)^2}$
1	9.55%	_____________	_____________
2	_____________	_____________	_____________
3	_____________	_____________	_____________
4	_____________	_____________	_____________
5	_____________	_____________	_____________

$\overline{(\%\Delta m)^2}$ = _____________   s = _________

The average percent mass loss of a kernel = _____________ ± _________

Calculations:

Name ___________________________________   Date ________________

EXPERIMENT 1
DATA
(Observe significant figures in all calculations.)

1. Net "weight" (mass) of popcorn package. ____________________

2. Mass data of kernel samples:

Kernel No. ⟶	1	2	3	4	5	group of 10
mass before popping	_____	_____	_____	_____	_____	_____
mass after popping	_____	_____	_____	_____	_____	_____
mass loss	_____	_____	_____	_____	_____	_____
percent mass loss	_____	_____	_____	_____	_____	_____

3. What happened when you tried to pop some kernels that had been cut in half?

4. Record any observations on the popping of the group of 10 kernels that might be related to a change in mass of the corn when popped. Observe the flask walls near the stopper carefully.

Use measurements on the 5 single kernels to calculate entries 5-10.

5. Average initial mass: ____________________

6. Average mass loss: ____________________

7. Average % mass loss: ____________________

8. Median value of initial mass: ____________________

9. Standard deviation of initial mass: ____________________

10. Standard deviation of % mass loss: ____________________

Use measurements on the group of 10 kernels to calculate entries 11-13.

11. Average initial mass of a single kernel in the group of 10:   __________

12. Average mass loss per kernel in the group of 10:   __________

13. Average % mass loss per kernel in the group of 10:   __________

14. Do the average initial masses in **5** and **11** agree within one standard deviation?   __________

Calculations:

DATA

15. Do the average mass losses in 6 and 12 agree within one standard deviation?   _________

16. Do the average % mass losses in 6 and 13 agree within one standard deviation?   _________

17.   Based on your answers to 14, 15, and 16, were your standard deviation calculations based on a sufficiently large sample size? Explain your answer.

18. From entries 1 and 11, calculate the total number of kernels originally in the popcorn package.

____

Post this number on the blackboard where the instructor designates.

19. In the space below, copy the posted values of the total number of kernels originally in the popcorn package that were calculated by your classmates.

20. Average total number of kernels originally in package (based on class results):   _________

21. Standard deviation of average total number of kernels (based on class results):   _________

Calculations:

Name _______________________________________   Date _________________

EXPERIMENT 1
QUESTIONS

1. The fact that mass was lost from the kernels implies that matter was lost. Based on your observations, what is the nature of the material lost when corn is popped?

2. Based on what happened when you tried to pop some kernels that were cut in half, suggest one or more possible reasons why corn pops when it is heated.

3. Would you expect the standard deviation to be larger for the average initial mass of the kernels or for the average % mass loss? Explain your answer.

4. In Procedure steps 11 and 12, you weighed popped corn flowers after they had cooled. Why must you wait for them to cool before weighing? (See Section III, Methods.)

EXPERIMENT 2
Separation and Purification
by Physical Methods

PRELABORATORY PREPARATION
1. Do the Prelaboratory Exercise and turn it in at the beginning of your laboratory period.
2. Read distillation procedure, Section XI, Methods, page 25.
3. Read decantation and filtration procedures, Section VIII, Methods, page 20.

INTRODUCTION
The techniques you use in this experiment are commonly used methods for water purification. They also find applications in many other processes, such as separating mineral ores, making coffee, and pouring wine without serving the sediment. This experiment is divided into two parts. Each part will take one laboratory period.

    I.   Apply and practice each of the methods of decantation, sublimation, filtration, extraction, coagulation, adsorption, and distillation in separate experiments.

    II.   Combine the above techniques to obtain pure potable water from a sample of simulated oil-contaminated ocean water. You can use any series of methods you think will bring success. Your goal is to purify the simulated ocean water, which will contain sand, oil, salt and possibly other things, **as economically and efficiently as possible.**

After learning to use the methods of Part I, you might be tempted to do Part II by simply distilling the water from all the unwanted contaminants. This would not satisfy the economical and efficient requirements of the goal of Part II, since distillation is, by far, the most expensive of the separation techniques because of its heat energy requirements. Furthermore, your reaction vessel would soon fill up with sand, salt, oily scum, and plant and fish residues. You then would be forced to stop in order to clean out your apparatus before finishing to purification. You might even have to discard the "gunked-up" reaction vessel. The distillation step should be performed on as small and clean a sample as possible, to save energy and to avoid troublesome dirtying of your most complicated and expensive piece of apparatus.

    You must try to remove as many contaminants as possible *before* **using distillation. To do this efficiently, you must choose an optimum sequence of separation methods prior to distillation.**

As you plan your purification procedure, it will be helpful to have in mind the two kinds of mixtures, **heterogeneous** and **homogeneous.** Separation techniques usually are different for the two.

    Homogeneous mixtures result when one substance dissolves in another. They appear uniform throughout and cannot be separated by sifting, decanting or filtering. **Distillation, extraction,** and **precipitation** are methods often used to separate homogeneous mixtures. Salt water is a homogeneous mixture.

    Heterogeneous mixtures are not uniform throughout, and the different components can be seen. They often can be separated manually, by **sorting, sifting, decanting,** or **filtering.** Sand in water, oil in water, and mixed salt and pepper are heterogeneous mixtures.

PLAN OF EXPERIMENT
1.   **1st laboratory period:** Practice different separation techniques to learn their uses.

2.   **Before the 2nd laboratory period:** Plan the best sequence of techniques for purifying contaminated ocean water and draw a **flow chart** for your plan on the answer sheet, similar to the example in Figure 2.1.

3.   **2nd laboratory period:** Follow your flow chart to purify a polluted water sample. If you make any changes in your plan during the experiment, be sure to note them on your flow chart and explain your reasons.

Figure 2.1. A flow chart showing a sequence of separation steps that will purify raw ocean water, when the correct separation methods are chosen from the table at the bottom.

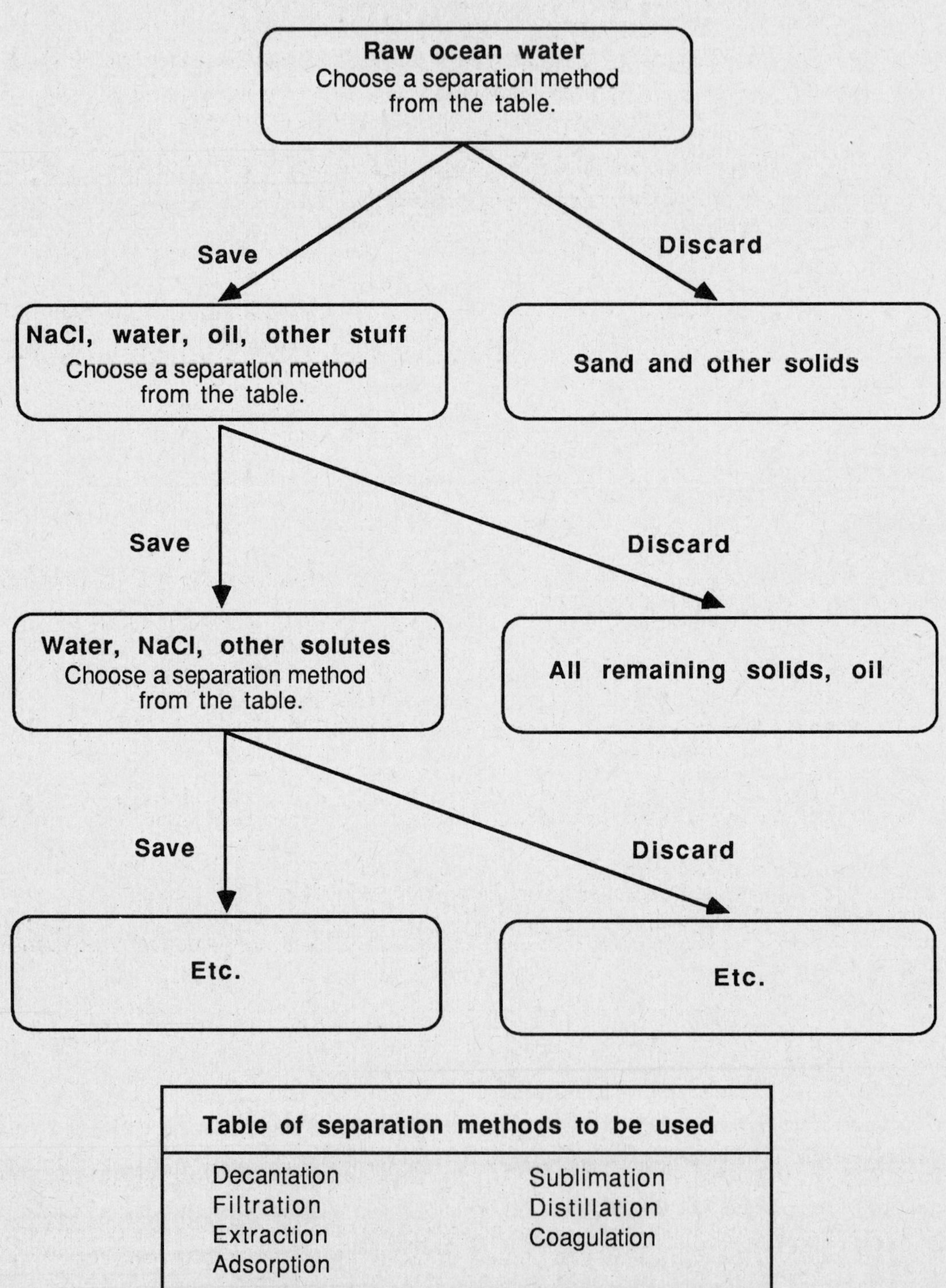

SAFETY

1. Wear approved eye protection.
2. Iodine and its vapor are very corrosive and toxic. Do not allow contact with skin or clothing and avoid inhalation.
3. Carbon tetrachloride is very toxic if inhaled or absorbed through the skin.
    USE IT IN THE HOOD ONLY.
    Wear plastic gloves and avoid inhalation and contact. Its toxicity is enhanced markedly by the ingestion of alcohol. Avoid all alcoholic beverages for 24 hours before or after working with carbon tetrachloride. (If available, a liquid freon may be substituted for carbon tetrachloride in section D of this experiment.)
4. In Section C, you filter an ethanol-sugar mixture. If you are heating your distillation flask with an open flame, keep the filter apparatus well away from any flames, for ethanol is flammable.

PROCEDURE

PART I

A. Distillation

Distillation is used to separate the components of a solution. The component with the lowest boiling point distills off first and can be separated from the components with higher boiling points. **Start the distillation first**, for this takes the most time but does not require your entire attention. You can practice other techniques at the same time, while the distillation proceeds.

1. Set up the apparatus as shown in Figure 0.12, Section XI, Methods, page 25.

2. Obtain 30-40 mL of colored solution (water containing dye) for distillation and transfer it to the distillation flask using a funnel. Add 2 or 3 boiling chips and insert the thermometer.
    The distillation flask should not be more than half full to reduce the possibility of liquid splashing over into the condenser during boiling.

3. Be sure that water is flowing through your condenser in the proper direction (from the lower to the higher end) and begin heating the flask gently. The entire distillation should take about 1 hour. Distillation is complete when nearly all the liquid has passed over to the receiving flask.
    Stop heating when there still is a little liquid left in the boiling flask to prevent contamination of your distillate and to simplify cleaning the flask.

4. Do the following sections of the experiment while the distillation proceeds, but watch the distillation process closely to make certain the boiling proceeds smoothly and the sample does not not evaporate to dryness.

B. Decantation

Decantation is used to separate liquid from a solid that will settle to the bottom of the container in a "reasonably" short time.

1. Shake up the reagent bottle marked "sand in water" and immediately pour about 25 mL of the mixture into a 100-mL beaker, before the sand settles. Wait about 10 min. for the sand to settle as completely as possible.

2. Hold a stirring rod so that it touches both the inside of the receiving beaker and the lip of the pouring beaker (see Figure 0.8, Section VII, Methods, page 20). Carefully pour the liquid down the rod so that all of the solid remains behind.
    You will have to leave a little liquid in the pouring beaker in order to avoid transferring any of the solid.
You now have separated *most* of the liquid from the solid. Decantation usually is only a preliminary step in a complete separation. Other methods are necessary to remove the remaining water from the sand.

3. After decantation, pour the decanted water back into the sand, shake the mixture well, and return it to the reagent bottle. Answer question 1 on the Question sheet.

C. Filtration

Filtration is one method for separating liquid from a solid that is too finely divided to settle out in a "reasonably" short time. Some very fine solids may never settle out at all. Your laboratory instructor will demonstrate the proper way to fold filter paper (see Figure 0.9, Section VII, Methods, page 21).

1.   Set up a gravity filter as in Figure 0.10, Section VII, Methods, page 22.
Be sure that the tip of the funnel stem touches the beaker wall. Keep your filtering setup well away from any distilling apparatus using an open flame, because ethanol is flammable.

2.   For your practice mixture, take about 20 mL of a well-shaken mixture of ethanol and sugar. You are to try to obtain ethanol-free sugar.

3.   Decant most of the liquid through the filter, using a glass rod to guide the flow (See Figure 0.8).

4.   Allow the ethanol to evaporate from the sugar on the filter paper. You have obtained ethanol-free sugar, but not sugar-free ethanol. Sugar is slightly soluble in ethanol, and the ethanol **filtrate** (the liquid that has passed through the filter) is not entirely sugar-free. Answer question 2 on the Question sheet.

D. Extraction

Extraction is a method for separating a substance that has very different solubilities in two different solvents. You will be given a homogeneous mixture of iodine in water. Because iodine is much more soluble in carbon tetrachloride, CCl_4, than in water and because water is insoluble in CCl_4, the iodine can be transferred to the CCl_4 by the following extraction procedure.

1.   Obtain about 50 mL of the iodine-water solution and transfer it to a 250-mL separatory funnel (see Figure 2.2). Add about 40 mL of CCl_4.
The fumes of CCl_4 are poisonous; avoid inhalation. Do not drink any alcoholic beverages (even beer) for at least 24 hours before or after using CCl_4.
DO THE EXTRACTION EXPERIMENT IN A HOOD.

2.   Stopper the funnel and shake well to mix the solutions.
Hold the stopper in place while shaking.
Shaking will warm the solutions slightly and cause a pressure increase in the funnel because of an increase in vapor pressure. Release the pressure by opening the stopcock as in Fig. 2.2d.

3.   Close the stopcock and place the funnel in an iron ring attached to a ringstand. Allow the immiscible liquids to separate into two layers. Carbon tetrachloride is more dense than water and will sink to the bottom. Most of the iodine will have been extracted from the water into the CCl_4. Take careful note of the colors of the solutions.

4.   Remove the stopper from the funnel and open the stopcock to drain the bottom layer into a beaker.
If the stopper is not removed before draining, the pressure in the vapor above the liquids will decrease, causing bubbles to come up through the stopcock and mix up the liquids in the separatory funnel.
Stop draining before the water comes through. The remaining water could be further purified by repeating the extraction with pure CCl_4, but you need not do this now. Answer question 3 on the Question sheet.

E. Sublimation

Direct vaporization of a solid is called **sublimation.** It can be used, like liquid distillation, to purify some substances. Sublimation is similar to distillation in that one substance is separated from a mixture by vaporizing it and collecting the vapor by condensation in a different place. Certain solids, such as iodine and carbon dioxide, will vaporize directly without first melting to a liquid when they are heated at atmospheric pressure. A mixture of iodine and sodium chloride can be separated by subliming the iodine and collecting it on a cool surface away from the sodium chloride.

1.   Obtain a small sample (about 1/2 g or 1/2 of a small spatula) of a sodium chloride-iodine mixture or impure iodine crystals. Place it in a 100-mL beaker and cover with a watch glass.
Do not touch iodine with your fingers. Avoid inhalation of its vapors.

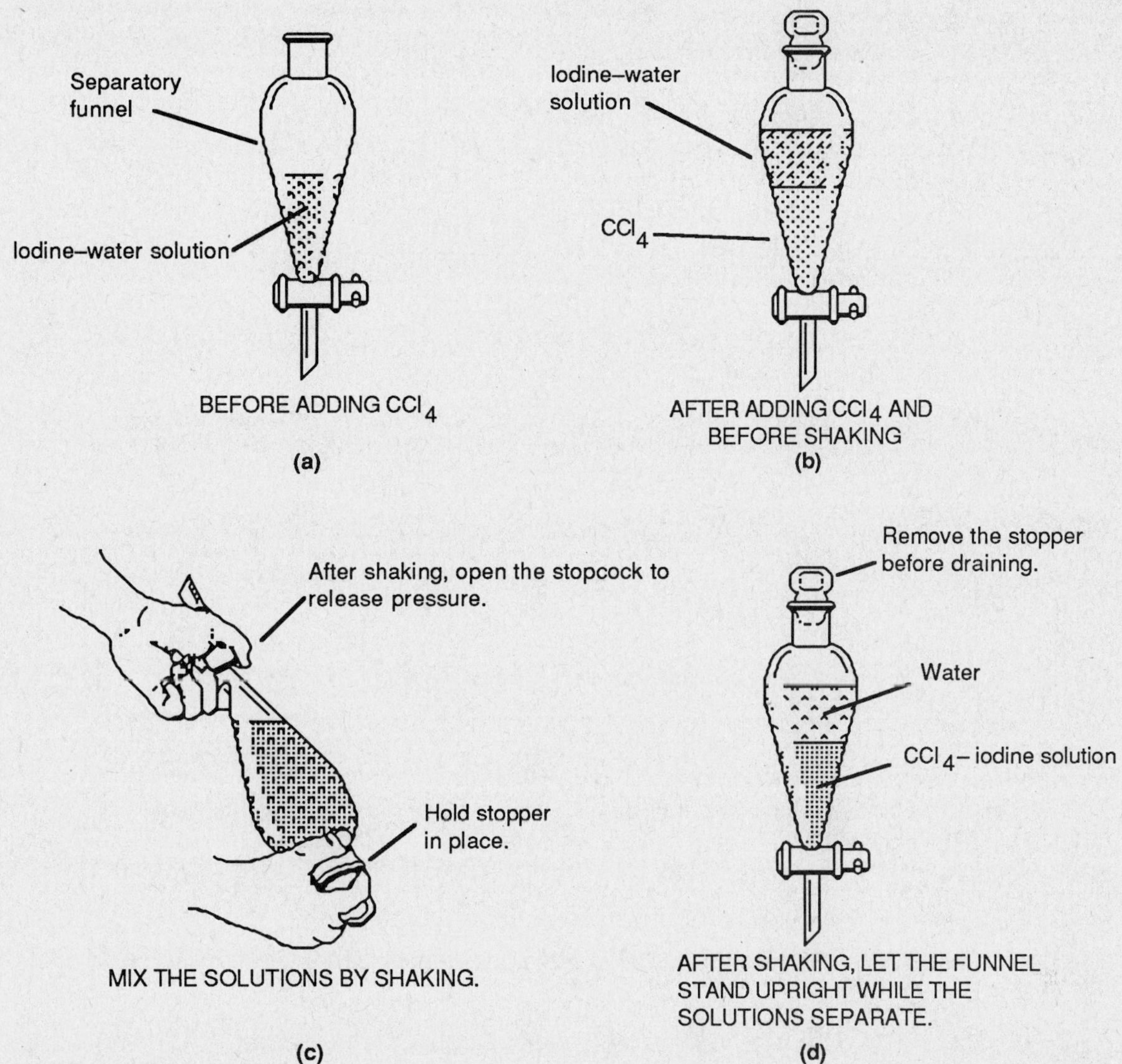

Figure 2.2.   Extraction of iodine from water.

2.   Place a few pieces of ice in the watch glass cover to cool it, and warm the beaker gently on a hotplate or on a wire screen over a burner.
   Iodine will vaporize from the bottom of the beaker and condense as crystals on the cool bottom side of the watch glass.

3.   When sublimation is complete, remove the watch glass and collect the purified crystals onto a clean watch glass. Answer question 4 on the Question sheet.

F. Coagulation

   When small particles bind together or bind to larger particles, the process is called coagulation. Settling out the solids, which was necessary before you could decant the liquid in Section B, can be speeded up by introducing coagulating reagents to the water. Aluminum sulfate, $Al_2(SO_4)_3$, and calcium hydroxide, $Ca(OH)_2$, react to form a precipitate of aluminum hydroxide, $Al(OH)_3$:

$$Al_2(SO_4)_3 + Ca(OH)_2 \longrightarrow Al(OH)_3 + CaSO_4 \quad \text{(Unbalanced - see question 5 on Question sheet.)}$$

The precipitate is produced as a light, fluffy mass (a **flocculant**) with an extremely large surface area that attracts and traps small suspended particles and carries them to the bottom of the container as the precipitate settles. This method is much more common in water treatment plants and industry than it is in chemical laboratories, where centrifuging is usually feasible to speed particle settling.

1.   Place 20 mL of muddy water into each of two test tubes. Shake them well.

2.   Add 8 mL of saturated $Ca(OH)_2$ solution [about 1 g $Ca(OH)_2$ per 500 mL water] and 2 mL of 0.1 M $Al_2(SO_4)_3$ solution to one of the tubes. Shake up both tubes again.

3.   Let the tubes stand and compare their rates of clearing. Answer question 5 on the Question sheet.

G. Physical Adsorption

Activated charcoal is used in water treatment to remove undesirable colors, odors, and tastes by adsorption.

1.   Place 1 or 2 drops of a food coloring dye in a test tube with 10 mL of water, and add one or two drops of some aromatic food flavoring, such as wintergreen oil, peppermint oil, vanilla, or rum. You now have a colored, smelly solution.

2.   Add about 0.3 g of activated charcoal; stopper and shake the test tube vigorously. Allow the charcoal to settle and record your observations. If the charcoal is finely divided, it will not settle very rapidly and you will have to centrifuge or filter it. Another way to speed up the settling would be to add the coagulating reagents used in section F. If you try this, you also should add the coagulating reagents to the dye solution alone, with no charcoal added, to see if the $Al(OH)_3$ precipitate will remove the dye color and odor by itself. Answer question 6 on the Question sheet.

PART II
(Requires one laboratory period and is to be done in the next period. Read the procedure before coming to lab.)

H. Purification of Ocean Water

Before starting Part II, use your experience from Part I to plan an efficient and economical sequence of separations for purifying contaminated ocean water. Draw a flow chart for your plan on the Data sheet. Be certain that your plan addresses the concerns about efficiency and economy, described in the Introduction. **The *last* separation step should be a distillation.**

1.   Weigh a clean, empty 200 or 250 mL beaker on a triple beam balance and record the mass.

2.   Put into the weighed beaker about 100 mL of simulated ocean water, made of approximately 12 parts water, 3 parts oil, 1 part NaCl, 3 parts sand, and 1 part miscellaneous substances.

3.   Weigh your sample in the beaker on a triple beam balance and record the mass.

4.   You already should have devised an efficient and economical purification scheme and drawn the flow chart of your scheme on the data sheet.

5.   Use your scheme to purify your sample. If you make any changes to your original flow chart as you do the experiment, be sure to note and explain them on the data sheet.

6.   Place about 5 mL of your final purified distillate in a test tube and test it for the presence of NaCl by first adding about 3 drops of 6 M HNO_3 and then adding a few drops of silver nitrate, $AgNO_3$, to the distillate. A white precipitate indicates that some NaCl remains in the water. The reaction is:

$$AgNO_3 + NaCl \longrightarrow AgCl(s) + NaNO_3$$

The purpose for acidifying the solution with HNO_3 is to prevent the precipitation of Ag_2O, which could give a false indication that Cl^- is in the solution.

Just in case your procedure did not work perfectly,
DO NOT TASTE THE WATER!

Name ____   Date ____________

EXPERIMENT 2
PRELABORATORY EXERCISE

1. How can you tell a homogeneous mixture from a heterogeneous mixture?

2. Which separation technique in this experiment is the most expensive to perform?

3. In a sequence of separation steps, why should distillation be the last step? Give at least two reasons.

4. Why should you avoid drinking any alcohol before or after exposure to carbon tetrachloride?

5. Which separation techniques would you use to separate sand from water?

6. In the extraction procedure, section D, why does shaking the separatory funnel to mix the solution, cause an increase in pressure?

    What should you do to release the pressure?

7. In which direction should the condenser water flow in the distillation apparatus? ________________

8. A sample of simulated ocean water weighs 135.205 g before purification. After purification, the volume of pure water collected was 76.3 cm^3. The density of water at room temperature (25°C) is 1.00 g cm^{-3}. What weight percent of the original sample was collected as pure water?

Weight % collected: ________________________________

Calculations:

Name ____   Date ____________

EXPERIMENT 2

QUESTIONS ON PART I, SECTIONS A and B

1. In section B, the sand remaining after decantation was still wet with some retained water. How could you remove the rest of the water to obtain dry sand and, also, to save the water?

2.   **a.** In section C, is the ethanol-sugar *filtrate* a heterogeneous or homogeneous mixture?

   b. How could you complete the separation and obtain pure ethanol?

3. In section D, why is it necessary to remove the stopper of the separatory funnel before draining the bottom layer into a beaker?

4. In section E, note any differences in appearance of the iodine crystals before and after sublimation.

5. Write the balanced equation for the reaction, in section F, producing a flocculant.

What was the effect of the flocculant on the settling rate of muddy water?

6. What happened, in section G, when activated charcoal was added to the colored, odorous solution?

DATA FOR PART II
(Only data from Section H, the purification of ocean water, is to be entered here.)

1.   Draw a preliminary flow chart (after you have practiced the separation techniques) in the space below for the efficient and economical purification of ocean water. Correct it, if necessary, as you proceed with Section H of the experiment.

2.   Mass of empty beaker for holding simulated ocean water: _____________

Mass of beaker + simulated ocean water sample: _____________

Mass of simulated ocean water sample: _____________

3.   Mass of purified water collected after last separation step (distillation): _____________

Mass percent of original sample collected as pure water: _____________

4.   Result of $AgNO_3$ test for residual NaCl in purified water: _____________

Calculations:

EXPERIMENT 3
Paper Chromatography:
Qualitative Analysis of Amino Acids

PRELABORATORY PREPARATION
Do the Prelaboratory Exercise and turn it in at the beginning of your laboratory period.

INTRODUCTION
The first step in the chemical analysis of an unknown material often is the separation of the different components that are present in the unknown. Once separated, identification of the components is generally much easier to accomplish. Every separation method has its own limitations and advantages, and a particular analytical sample will respond differently to different methods of separation. It is helpful to be familiar with many different separation techniques so that, if one method fails or is difficult, you can try another. All chromatographic techniques separate the components of a mixture by making them move through an absorbing medium that causes them to move at different speeds. Thus, the components become separated spatially and can be isolated for species identification. Paper chromatography uses paper fibers as the absorbing medium. It is a method of separation for homogeneous mixtures that is easy to use, and is well suited to separating and purifying small quantities of chemicals that can be dissolved into the same solvent.

In this experiment, the sample to be analyzed is placed near one end of a strip of paper. This end of the paper strip is immersed in a solvent that then is drawn upward through the paper fibers by capillary action. When the solvent reaches the sample, components of the sample are dissolved and carried along with the solvent as it moves through the paper. Spatial separation of the components is based on the properties of cellulose fibers, from which paper is made. Cellulose fibers contain many polar hydroxyl groups (-OH) that exert attractive forces on other molecules. The attractions cause molecules to be adsorbed from solution onto the fiber surface. Water is adsorbed especially strongly onto cellulose, and even "dry" paper generally has about 15% water by mass adsorbed from the atmosphere to the fiber surfaces. This water is so tightly bound to the paper fibers that it is immobilized and is not displaced by the solvent or sample components.

The separation process depends on two phenomena, one involving the cellulose fibers and one involving the adsorbed water:

1. Different kinds of molecules generally have different attractive strengths of adsorption to the cellulose fibers.
2. If solute molecules are dissolved in a solvent that is immiscible with water, and the solution is brought into contact with water, some of the solute will leave the immiscible solvent and dissolve into the water. The solute molecules **partition** themselves between the water and the immiscible solvent in a manner dependent on their relative solubilities in the water and in the solvent.

In practice, the solvent may be partly miscible with water, so that both effects play a role in the separation process. Different components of the sample are attracted out of the moving solvent and become bound to the fibers or partitioned into the adsorbed water. The partition effect often is the more important of the two phenomena, and so the adsorbed water, which is immobilized on the cellulose fibers, is an essential factor in the separation process.

- The sample dissolved in the moving solvent is called the *moving phase.*
- The paper fibers and the immobilized adsorbed water are called the *stationary phase.*
- The solvent, which often is made from a recipe involving several compounds, is called the *developer solution.*
- The process of separating the sample components is often called *developing the chromatogram.*

During a separation, the moving phase is drawn in one direction over the stationary phase by capillary action. Those sample molecules that are carried in the moving phase and are most strongly attracted to the stationary phase move more slowly than than more weakly attracted molecules. As an example, the behavior of two different compounds, **A** and **B**, is described below as they are separated by paper chromatography analysis.

Example of separation process

1. The mixture of **A** and **B** is dissolved in a solvent.

2. A small droplet of the sample solution (1 to 2 μL) is placed near one end of a strip of filter paper and dried by allowing the solvent to evaporate. This step is called **spotting** the chromatogram.

3. The end of the paper strip that carries the dried sample is placed in contact with a reservoir of the developer solution, which is drawn by capillary action through the paper (see Figures 3.1 and 3.3).

4. When fresh solvent reaches the spot of dried sample, it tends to dissolve **A** and **B** and carry them along with the moving solvent.

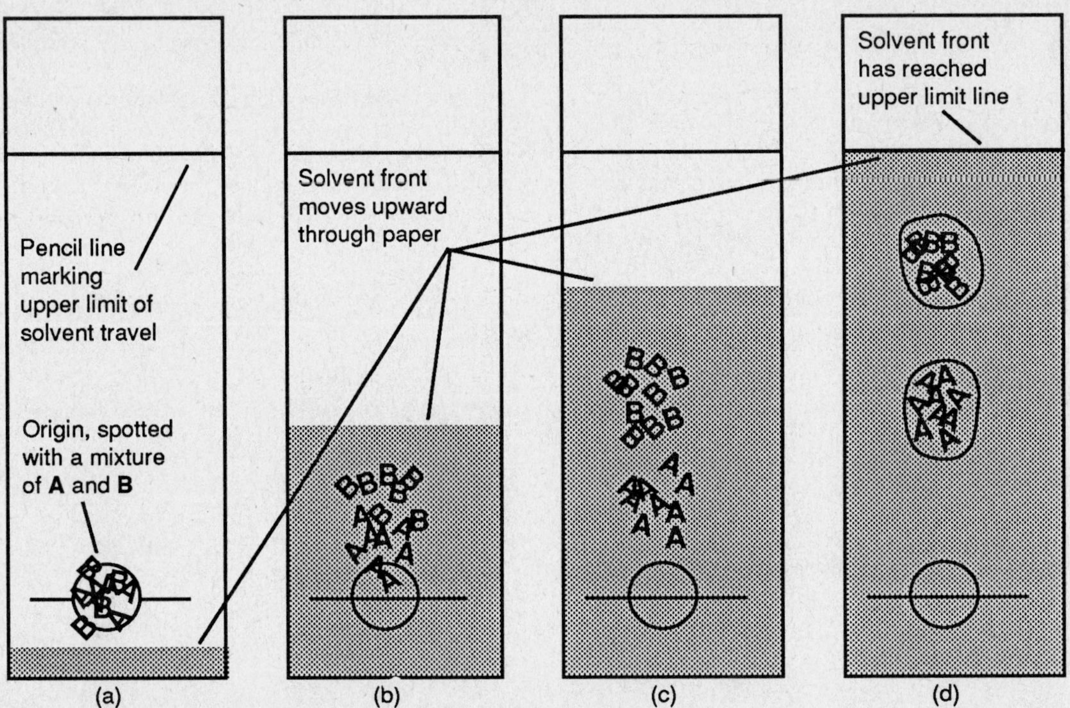

Figure 3.1. Illustration of the separation of a 2-component mixture of **A** and **B** by paper chromatography. The shaded area represents the solvent as it moves upward through the paper strip.

(a) The solvent front has not yet reached the origin, where a mixture of **A** and **B** are spotted.

(b) As the solvent front passes through the origin, it carries with it molecules of **A** and **B**. The **A** molecules, which bind more strongly to the stationary phase than do **B** molecules, move upward more slowly than **B**.

(c) As the the solvent continues upward, the separation of **A** and **B** becomes more complete.

(d) When the solvent reaches the line marking its upper limit, **A** and **B** have separated sufficiently to allow their identification by measuring their **retention factors** (see below and Figure 3.4).

5.   This tendency is offset by the fact that **A** and **B** are attracted to the stationary phase, which holds them and acts to prevent their motion.
> **The ease with which A and B can be detached from the stationary phase and dissolved into the moving solvent depends on the relative strengths of the attractive forces to the stationary and moving phases.**

Let **A** be more strongly bound to the paper than **B**. Then **A** will dissolve into the solvent more slowly than **B**, and **A** will progress along the paper with the moving phase more slowly than **B**.

6.   As the solvent moves through the paper, molecules of **A** and **B** are dissolved from the dried sample spot and carried along a little way, until they chance to become bound again to the stationary phase. In this way, **A** and **B** move along the paper in the direction of the solvent flow, alternatively binding to the stationary phase and then redissolving into the moving solvent.

7.   **A** is held more strongly than **B** to the stationary phase and moves with the solvent more slowly than **B**; it is released from the stationary phase into the solvent more slowly, and is quicker to reattach to the stationary phase.

8.   As **A** and **B** are carried along by the moving phase, molecules of **A** lag farther and farther behind the molecules of **B**. Eventually, **A** becomes completely separated from **B**.

9.   After **A** and **B** are separated, the paper strip is removed from the solvent reservoir, stopping the solvent motion in the paper. If **A** and **B** are colored, they will appear as two separated colored spots on the paper. Some colorless compounds become visible by fluorescing under ultraviolet light. Compounds that are not naturally colored or fluorescent must have their locations on the paper determined with a chemical treatment that makes them either colored or fluorescent.
        The final distance that an unknown compound has moved along the paper is compared with the movement of known standard compounds to help identify the unknown.

This comparison is made by means of the **retention factor,** described below.

RETENTION FACTOR
        After the separation is complete, the **retention factor** is determined as follows:

1. Mark the position of the **solvent front**, the maximum height attained by the solvent, with a pencil. The solvent front is the border between the portion of the paper wetted by solvent and the dry part not yet reached by solvent (see Figure 3.4).

2. Mark the positions of the separated compounds on the paper strip by outlining each spot with a pencil. Then, mark the center of each outlined spot (see Figure 3.4).

3. Retention factors are obtained by measuring the distances travelled by the different compounds and by the solvent. The measurements are made from the starting spot of dried sample to the centers of the separated spots and to the solvent front, as in Figure 3.4.

$$\text{retention factor} = R_f = \frac{\text{distance travelled by the compound}}{\text{distance travelled by the solvent}}$$

Each different compound will have a characteristic retention factor for a particular set of chromatographic conditions.

EXAMPLE
        After stopping the solvent flow, component **A** has moved 3.3 cm, component **B** has moved 4.1 cm, and the solvent has moved 5.4 cm. The **R$_f$** values characteristic of **A** and **B** are:

$$R_f(A) = \frac{3.3 \text{ cm}}{5.4 \text{ cm}} = 0.61 \quad ; \qquad R_f(B) = \frac{4.1 \text{ cm}}{5.4 \text{ cm}} = 0.76$$

An unknown is identified by its R$_f$ value and sometimes the color of its spot, if it is distinctive.

The usual procedure is to run a chromatogram of the unknown substance and compare the results with chromatograms of known standards, that are obtained under identical conditions. Experimental conditions for the unknown and the standards must be exactly the same, because the **R$_f$** value obtained for a given substance depends upon many experimental factors, such as temperature and the particular paper and solvent used. Even the orientation of the paper fibers relative to solvent flow direction can affect the **R$_f$** factor.
        Unknowns and known standards often are measured simultaneously on the same paper strip, to insure identical experimental conditions.

AMINO ACIDS
        Paper chromatography is useful for separating and analyzing many biochemical molecules, such as amino acids. Amino acids are the building blocks that are assembled into protein molecules. A protein is a very large polymer made of amino acid monomer subunits. There are about twenty different amino acids that are assembled in various arrangements to form all the different kinds of proteins found in the human body. We obtain the amino acids we need for good health by eating foods containing protein. Digesting and metabolizing these foods breaks down the proteins into their amino acid monomer subunits. In our body

cells, the amino acids are then reassembled into whatever particular proteins are needed at that time. Different foods contain different kinds and amounts of amino acids. The ability to measure the amino acid content of foods is important to understanding how diet and nutrition affect our body processes and health.

PLAN OF EXPERIMENT

In this experiment, you will determine the R_f values for three amino acids and analyze an unknown mixture containing some of them. The moving phase is an alcohol-ammonium hydroxide solution, and the stationary phase is a strip of filter paper with its adsorbed water. Your unknown mixture may contain any or all of the three amino acids whose structures are shown below:

$$CH_3-CH-\overset{\displaystyle O}{\overset{\|}{C}}-OH \qquad CH_2-\overset{\displaystyle O}{\overset{\|}{C}}-OH \qquad CH_2-CH-\overset{\displaystyle O}{\overset{\|}{C}}-OH$$

Alanine   Glycine   Phenylalanine

These amino acids will separate according to their different structures, which cause them to bind with different strengths to the stationary phase. Amino acids are colorless, so their separated spots cannot be seen unless they are chemically treated to make them visible. The amino acid spots are located on the chromatogram by reacting them with ninhydrin reagent to form colored compounds. Ninhydrin reagent is applied, after separation has occurred, by spraying it onto the filter paper or by dipping the filter paper into ninhydrin solution.

SAFETY

1. Wear approved eye protection.
2. Avoid inhalation of any solvent vapors or of the ninhydrin spray.
3. Avoid skin contact with ninhydrin. If ninhydrin gets on your skin, wash immediately with soap and water. Ninhydrin reacts with amino acids in your skin to produce a purple color that lasts for days.
4. All the solvents and the ninhydrin reagent are flammable. Be sure there are no open flames in the laboratory.

PROCEDURE

1. Add 45 mL of developer solvent solution (butanol:glacial acetic acid:water, in a 12:3:5 ratio by volume) to each of two clean and dry 250 mL Erlenmeyer flasks. Stopper the flasks snugly so that solvent vapors do not escape. The flask volume must become saturated with solvent vapor, to insure uniform solvent motion through the filter paper.

2. Cut two strips of filter paper, 15 cm by 3 cm. Avoid touching the paper surfaces with your fingers, because oil on your skin contains amino acids that might interfere with your experiment.
   Handle the paper strips carefully by their edges only, or with tweezers, throughout the rest of the experiment, to avoid contamination.

3. Draw two parallel lines on each strip *with a pencil,* as in Figure 3.2. One line is 2 cm from the bottom of the strip and the other is 10 cm from the bottom.
   All markings on the filter paper must be made with pencil. Inks contain substances that will dissolve in the moving phase and be chromatographed along with the amino acids.

4. Lightly mark three evenly spaced spotting marks on the bottom line, as shown in Figure 3.2.

5. Label these spots, as indicated, to identify the different amino acids and unknowns.

Spotting the chromatogram

6. You now are ready to spot your filter paper strips. Follow the procedure below carefully.

 a. Lay the prepared filter paper on a clean paper towel, which serves as an absorbent pad.

 b. Carry your filter paper on the towel to where the amino acid samples are located. Each solution will have its own micro-capillary spotting tube (1-2 μL) in it.
 Do not mix up the spotting tubes. Each tube must remain in its own   solution and never contact any other solution.

 c. Just lift the end of the micro-capillary spotting tube out of the solution that you are going to spot. Liquid will be held in the tube by capillary action. Touch the tip to the side of the solution container to remove the hanging drop.

 d. Hold the capillary tube perpendicular to the filter paper and touch its tip lightly to the marked spotting location for that sample (see Figure 3.2). Allow the solution to run into the paper until the wet spot is about 5 mm in diameter. Allow the spot to dry and repeat the spotting procedure once more, with the same sample on the same location.
 Each sample is to be spotted 2 times on the same location.
 Drying may be speeded with a hot-air blower or by holding the strips over a warm hotplate.

 e. Repeat this procedure for each spot on both filter strips.

 f. Spot glycine and phenylalanine on the same spot, marked *mixture,* one after the other.
 Allow each application to dry before the next is added to the same   spot. Be careful not to mix up the capillary spotting tubes.
 These two amino acids will serve as standards for identifying the unknowns.

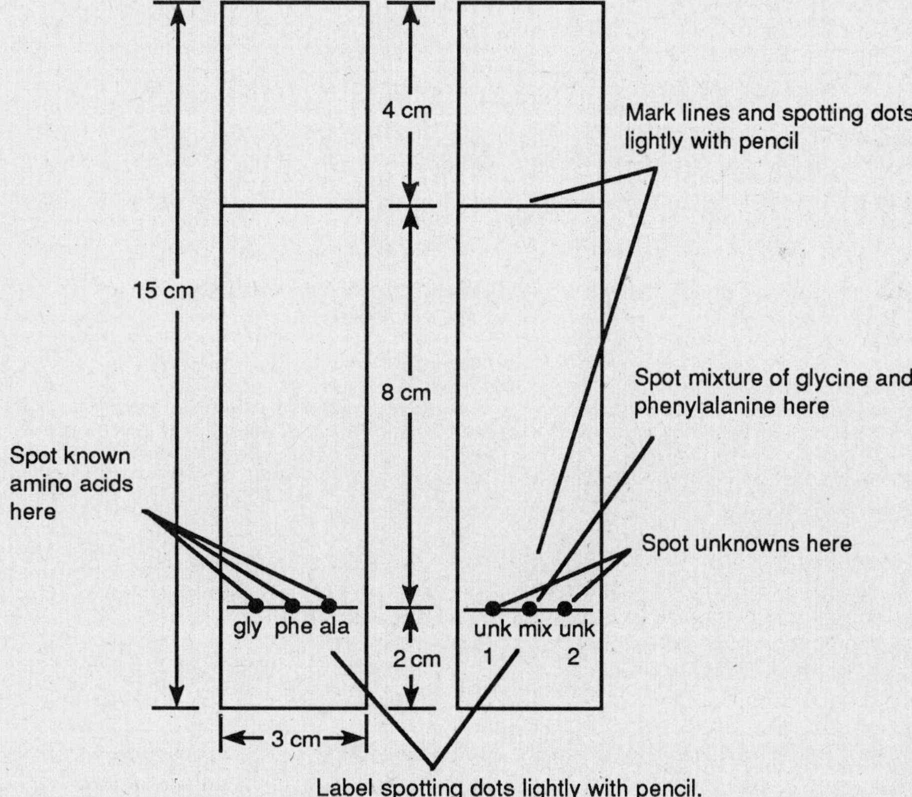

Figure 3.2. Preparation of chromatogram strips

7.   When all the spots are dry, prepare to place each strip into one of the Erlenmeyer flasks containing the developer solution.

First, swirl the flask gently before inserting the strips to wet the lower flask walls with solvent. This will speed evaporation and uniform vapor saturation.

Insert the strips quickly to keep the flask volume saturated with solvent vapor. Make certain the spots are not immersed in the solvent but are several mm above the liquid level.

Replace the stopper snugly into the flask neck, so that the filter paper fits between the stopper and the side of the flask (see Figure 3.3). The upper pencil line must be in the flask, below the stopper.

8.   Make both chromatograms simultaneously. Allow the solvent to rise to the upper pencil line, which will take about 1-2 hours. This step is called **developing** the chromatogram. Your instructor may want you to use the waiting time to prepare for the next experiment.

Do not disturb the flasks or move the stopper during development.
The space in the flask above the solvent must remain saturated with solvent vapor, or else there will be uneven evaporation loss of solvent from the filter paper strips. This would cause nonuniform motion of the solvent through the paper and cause unpredictable errors.

9.   Remove the strips as soon as the solvent reaches the upper pencil line and let them dry. If hot-air blowers are available, use them to hasten drying.

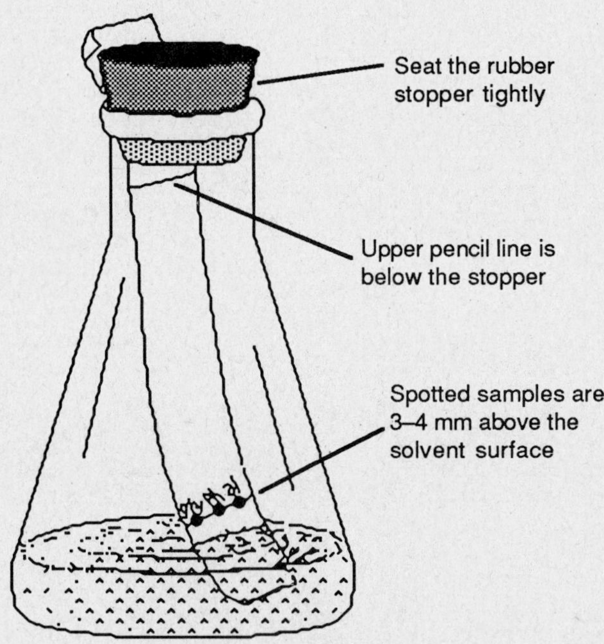

Figure 3.3.   Chromatogram strip in a solvent chamber made from an Erlenmeyer flask.

10.   Lay the dried chromatograph strips on a piece of paper towelling and take them to the laboratory fume hood. In the hood, you either will dip the strips into ninhydrin reagent or spray them with ninhydrin from a spray bottle.
Be careful to keep ninhydrin reagent off your skin.

11.   Let the strips dry, then place them in an oven at 95 °C (or warm them carefully over a medium hotplate) for 2 or 3 minutes. Colored spots should appear.

12.   Outline each spot with pencil, because they will fade in a day or two. Mark the center of each spot, as shown in Figure 3.4.

13.   Measure the distance between the lower spotting line and the center of each spot. Then measure between the lower spotting line and the upper line marking the solvent front position when development was stopped. Record the values on the data sheet.

14.   Calculate R_f values for the three known amino acids chromatographed singly, the two known amino acids in the mixture, and the components in your unknown solution. Record the values on the data sheet. Compare the R_f values to identify the unknowns.

15.   Present all of your results in the table on the Data sheet. Tape your chromatograms to the back of the sheet.

Name ____   Date ____________

EXPERIMENT 3
PRELABORATORY EXERCISE

Use Figure 3.4 to measure some hypothetical R_f values for glycine, alanine, and phenylalanine. Measure with a ruler and record the values below the figure.

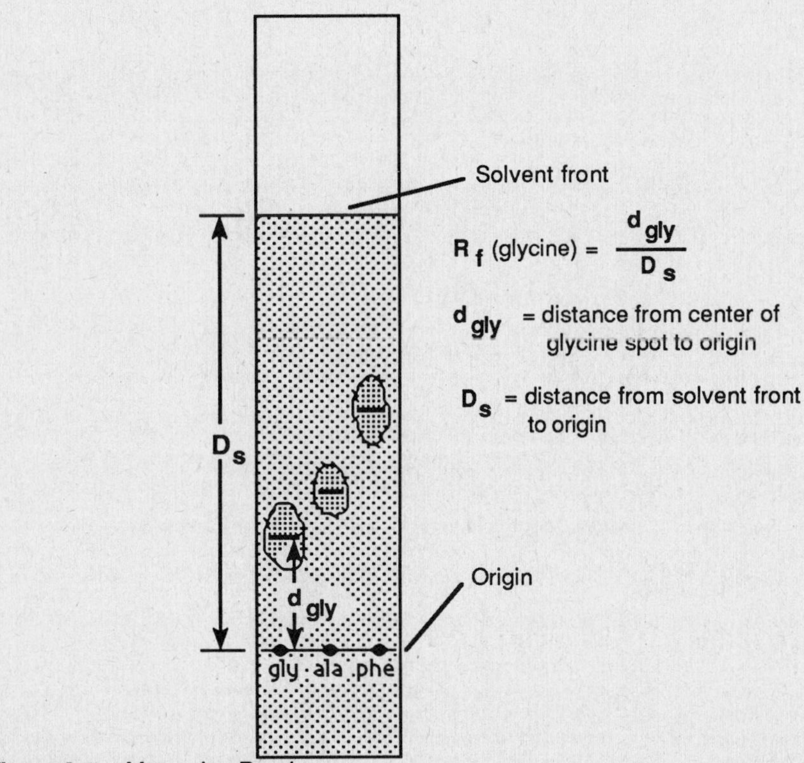

Figure 3.4.   Measuring R_f values.

a. R_f (glycine) =   ________________

b. R_f (alanine) =   ________________

c. R_f (phenylalanine) =   ________________

Name ____   Date ____________

EXPERIMENT 3
DATA

Distance travelled by solvent
(d_s)

Paper strip 1 with knowns: _______________________
Paper strip 2 with unknowns
 and known standard mixture: _______________________

	Distance travelled by amino acids	**R_f**
Glycine:	_______________________	_______________
Phenylalanine:	_______________________	_______________
Alanine:	_______________________	_______________
Glycine in mixture:	_______________________	_______________
Phenylalanine in mixture:	_______________________	_______________
Unknown No. 1:	_______________________	_______________
Unknown No. 2:	_______________________	_______________

Composition of Unknown No. 1:   ____

Composition of Unknown No. 2:   ____

DATA

Tape your chromatograms to this page.

Name ____   Date ____________

EXPERIMENT 3
QUESTIONS

1.   Based on your results, are the R_f values for the amino acids combined in a mixture exactly the same as for the amino acids measured separately?

      Give a brief explanation of your answer, in terms of the chromatographic separation mechanism.

2.   How might you isolate and collect a sample of one of the amino acids in your mixture?

3.   Based on the structures of the amino acids, suggest a reason why they separated in the order observed.

EXPERIMENT 4
Thin-Layer Chromatography: Separation of Plant Pigments

PRELABORATORY PREPARATION
1. Do the Prelaboratory Exercise and turn it in at the beginning of your laboratory period.
2. Read the Introduction to Experiment 3, especially definitions of **retention factor, stationary phase, moving phase,** and **developer solution.**
3. Review decantation procedure, Section VIII, Methods, page 20.

INTRODUCTION
Thin-layer chromatography, or **TLC** as it is commonly called, is quite similar to paper chromatography. It will be helpful to refer back to Experiment 3 for certain definitions and techniques. The stationary phase in TLC is a thin coating of finely powdered adsorbant material, such as cellulose, aluminum oxide, or silica gel, spread on a glass plate or plastic sheet. Its advantages over paper chromatography are that a wider variety of stationary phases are available and the coatings can be made very uniform. TLC can separate a greater variety of compounds than paper chromatography, and generally gives sharper separations in a shorter time. Very small quantities can be detected. Although analysts sometimes make their own TLC plates, they usually use commercially manufactured sheets that can give very reproducible results. The only disadvantages of TLC, with respect to paper chromatography, are a somewhat higher cost, a slightly more delicate experimental technique, and the need for specially prepared coatings and supporting materials.

PLAN OF EXPERIMENT
You first extract the pigments from some fresh flower petals and green leaves. The flower pigments are extracted into boiling water and the green leaf pigment into an acetone-petroleum ether solution. Then you use TLC to determine how many different pigments were extracted from each of the plant samples. There are three main groups of pigments found in flowers and plants:

1. **chlorophylls - green colors**
2. **carotenoids - red, orange, and yellow colors**
3. **flavenoids - red, blue, and yellow colors**

SPECIAL MATERIALS FOR THIS EXPERIMENT
1. Enough flowers of different colors to provide each student with 4 or 5 petals from each of two differently colored flowers. Roses, carnations, and chrysanthemums provide a wide range of colors. During the winter, flowers can be purchased from local greenhouses. Greenhouse "throwaways" are quite suitable. **Flowers must be fresh for good results.**

2. Green plant material from one source, such as pine needles, spinach, or green leaves of trees and bushes. Each student will need 1-2 g.

3. TLC plastic sheets to be cut with scissors or a paper cutter into strips about 1 cm wide and 8-10 cm long. Many types are available. Alumina sheets of the Chrom AR type are suitable.

4. Ultraviolet lamps of the long-wavelength type.

5. Two different TLC solvent solutions, one for flower petal extract and one for green plant extract.

SOLVENTS FOR CHROMATOGRAPHING
flower pigment extract: 1-butanol, glacial acetic acid, and water in a 60:15:25 ratio by volume.
green plant extract: petroleum ether and acetone in an 8:2 ratio by volume.

SAFETY

1. Wear approved eye protection.
2. Petroleum ether is very flammable. Allow no open flames in the laboratory while it is being used.
3. The high energy of ultraviolet light can damage your eyes. Do not look directly at the ultraviolet lamp when it is turned on.
4. Iodine is a corrosive chemical. If iodine is used, avoid contact with skin or clothing and do not inhale its vapors.
5. Most organic solvents have some toxicity. Avoid inhaling the vapors of acetone or petroleum ether.

PROCEDURE

A. Extraction of Flower Pigments

1. Extract the pigments from two different colors of flower petals. Do them separately. For each color, grind about 4 petals in a clean porcelain mortar and pestle. Place the ground material from each type of petal into separate small beakers, each containing 6 mL distilled water. Mark the 2 beakers to identify the flower petals used.

2. Boil to extract the pigments. Stir constantly to prevent the petal material from sticking to the side of the beaker and charring.
Do not let the solution go dry. Add more water, if necessary, to keep the petals in a boiling solution, but use as little water as possible.
The water will become colored as the pigments are extracted.

3. When you think most of the pigment has been extracted from the petals, remove any solid material from the solution with tweezers or a stirring rod and decant the solution into another small beaker, marked to identify the type of flower petal used.

4. Continue boiling to concentrate the solution further.
Boil off enough of the liquid to make the solution deeply colored and concentrated. You need only 1 or 2 drops for the separation.
If the solution is not sufficiently concentrated, the TLC separation will not work properly.

B. Extraction of Pigments from Green Plants

1. In a clean porcelain mortar, place about 1 g green plant material, 1 g clean sand, 5 mL acetone, and 5 mL petroleum ether.

2. Grind the plants in the solvent solution until the green pigment appears to be completely extracted. The sand helps break down the cell walls and release their contents.

3. Decant the solution into a clean, marked, small beaker. Cover the beaker with a watch glass to prevent evaporation.

C. Chromatography

1. Obtain 3 TLC strips from your instructor.
Handle them by their edges only. Avoid touching the coated surface.

2. Prepare them as in Figure 4.1a. Across the coated side of each strip, draw a light pencil line 1 cm from each end. With scissors, cut a point at one end of each strip, as shown. Then use the pencil to mark the spotting origin.

3. Using one 15 mm test tube for each of the flower petal extracts, place into each test tube enough **flower petal extract solvent** to fill the tube bottom to a depth not over 7 mm. Label the test tubes to identify the extracts.

4. Place just 7 mm of **green plant extract solvent** into a labelled 15 mm test tube.

5. Use micro-capillary tubes or pipets (1 - 2 µL) to spot your 3 TLC strips. Read the spotting procedure in Experiment 3. Use a very light touch for spotting.
TLC strips are much more delicate than paper strips. Do not exert any pressure against the TLC coating with the capillary tip, or you may chip or pit the surface.
Spot each sample twice at the origins of the TLC strips. Use a separate strip for each sample. Allow the spots to dry.

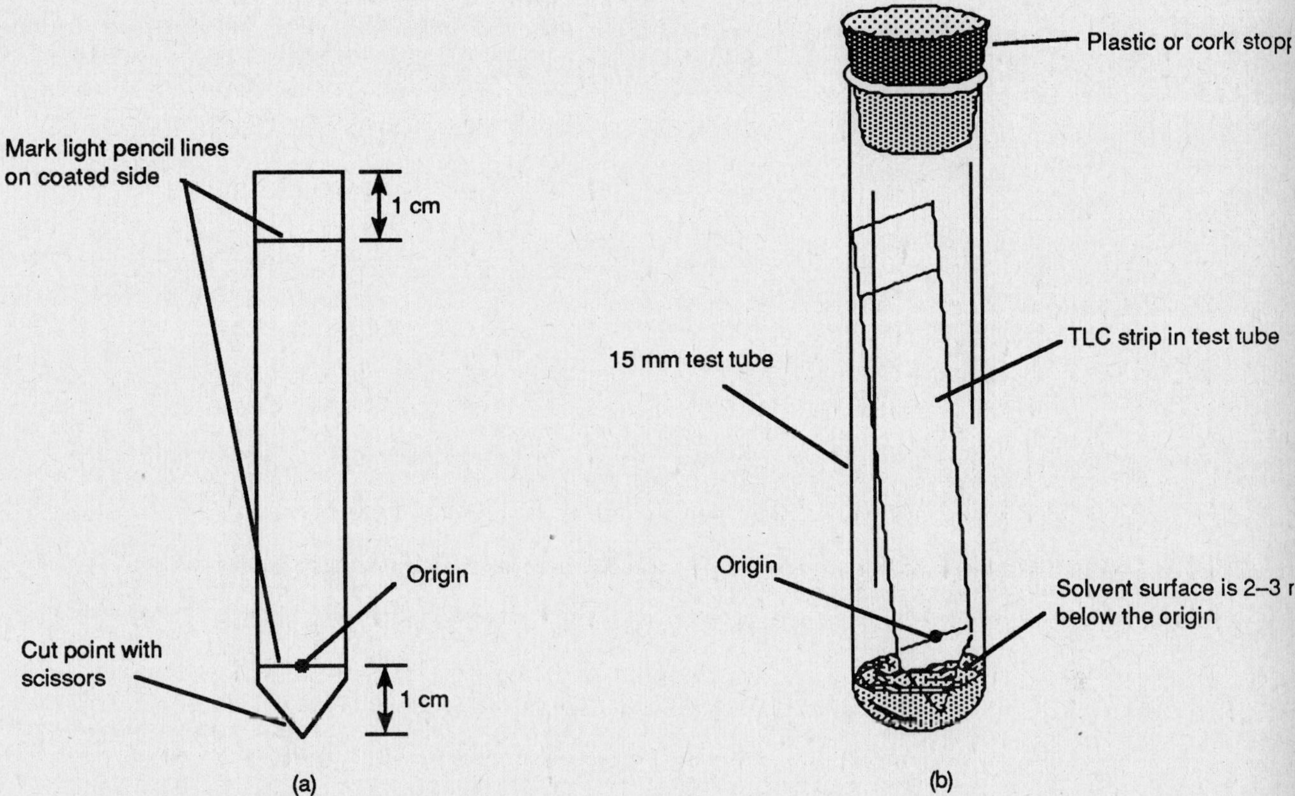

Figure 4.1. **(a)** Preparation of TLC strip; **(b)** TLC assembly

6.   Develop the chromatograms by putting each TLC strip into a test tube containing the proper TLC solvent. The pointed tips should be immersed in the solvent.
     It is important that the solvent surface lie below the TLC origin, see Figure 4.2b.

7.   Tightly stopper the test tubes and stand them in a rack where they will not be disturbed.
     Use plastic or cork stoppers — not rubber. Rubber may partly dissolve in the solvent and interfere with the analysis.

8.   Allow the solvent to ascend until it reaches the pencil line at the top of each TLC strip. Then remove the strips and air-dry them.

9.   Examine your developed strips under room lights and also under an ultraviolet lamp[1].
     Do not look directly at the ultraviolet lamp.
Use a pencil to outline the ultraviolet fluorescing spots on the TLC strips. Mark the center-line of each spot.

10.   Measure the distances from the origin to the spot centers and to the solvent fronts. Calculate and record the R_f values for each component separated. (See Experiment 3 for details.)

11.   Enter your results in a table prepared by your instructor on the blackboard, in which each type of flower and green plant is listed, so that you can compare the pigments from all the plants used in your laboratory.

12.   Before leaving, copy data from the blackboard into the table on your data sheet. Mark which results are yours.

---

[1]If ultraviolet lamps are not available, place TLC strips in a covered jar containing some crystals of solid iodine. The iodine vapor reacts with the pigments, making them visible.

Name ____   Date _____________

EXPERIMENT 4
PRELABORATORY EXERCISE

1. What are some of the advantages and disadvantages of thin-layer chromatography, compared to paper chromatography?

2. What part of the TLC strip serves as the stationary phase?

3. A TLC separation yields 3 pigment components. Component 1 is green, component 2 is yellow, and component 3 is blue. What pigment types correspond to the 3 components?

4. Why is sand used in part B?

Name ____ Date ____________

EXPERIMENT 4

DATA

Kind of plant	Color of TLC spot in ordinary light	Color of TLC spot in UV light	Possible pigment types	Distance of solvent boundary from origin	Distance of TLC spot center from origin	R_f

QUESTIONS

1. The table above contains your results and those of other students. With this data, answer the following questions:

a. Do any plants of different colors seem to have any pigments in common? Which ones?

b. List the plants having 1 pigment only, 2 pigments, 3 pigments, etc.

c. Do plants of similar color always contain the same pigments? _____

d. How many pigments would not have been detected without using the UV lamp? _____

2. How might you recover one of the pigments for further analysis?

DATA

Tape TLC strips to this page.

EXPERIMENT 5
Stoichiometry:
Finding the Equation of a Reaction and the Formula of a Product

PRELABORATORY PREPARATION
1. Do the Prelaboratory Exercise and turn it in at the beginning of your laboratory period.
2. Review use of the pipet, Section IV, Methods, page 17.
3. Review decantation and suction filtration, Section XIV, Methods, pages 20-22.

INTRODUCTION

A chemical molecular **formula** describes a molecule and tells which elements and how many atoms of each element are contained in that molecule.

For example, the formula $AgNO_3$ describes the silver nitrate molecule, which contains 1 atom of silver (Ag), 1 atom of nitrogen (N), and 3 atoms of oxygen (O).

A balanced chemical **equation** describes a reaction and tells what kinds of reactant and product molecules are involved and gives the relative numbers of each. The reaction between silver nitrate and sodium sulfide in aqueous solution is described by the balanced equation 5.1.

$$2\,AgNO_3(aq) + Na_2S(aq) \longrightarrow Ag_2S(s) + 2\,NaNO_3(aq) \tag{5.1}$$

It states that 2 molecules of silver nitrate ($AgNO_3$) will react, in aqueous solution, with one molecule of sodium sulfide (Na_2S), to form one molecule of solid silver sulfide (Ag_2S) and 2 molecules of sodium nitrate ($NaNO_3$). It implies that the moles of $Ag_2S(s)$ formed will be equal to exactly 1/2 the moles of $AgNO_3$ consumed. With such information, it is possible to make quantitative calculations about a reaction that tell how much reactant will be consumed and how much product will be formed. In particular, one can calculate the relative number of molecules, relative number of moles, and relative masses of all the reactant and product compounds. The term **stoichiometry** refers to these quantitative relationships and the calculations are called **stoichiometric calculations**.

Remember that the balanced equation only gives molar relations, not mass relations. All masses must be converted to moles to make a stoichiometric calculation. The answer in moles can be converted back to mass, if needed.

In this experiment, you will study a reaction, find the balanced equation that describes it, and determine the formula of one of its products. The technique used is similar to that described in the example below.

EXAMPLE:
Measuring the Yield of a Product to Find the Balanced Equation for a Reaction.

The reaction in aqueous solution between barium nitrate, $Ba(NO_3)_2$, and sodium chromate, Na_2CrO_4, produces a precipitate. The stoichiometry of this reaction and the nature of the precipitate can be investigated by varying the amount of Na_2CrO_4 added to a fixed amount of $Ba(NO_3)_2$ and weighing the precipitate obtained. Typical results of the experiment are illustrated graphically in Figure 5.1. As more and more Na_2CrO_4 is added, a point is reached where the mass of precipitate no longer increases, indicating that enough Na_2CrO_4 has been added to completely react with the fixed amount of $Ba(NO_3)_2$. At this point, the starting quantity of $Ba(NO_3)_2$ and the amount of Na_2CrO_4 added are in their exact stoichiometric ratios. Adding more Na_2CrO_4 after the stoichiometric point is passed causes no further reaction, because all of the $Ba(NO_3)_2$ is used up. The experiment was performed by dividing 120 mL of 1.00 M $Ba(NO_3)_2$ solution into 12 portions of 10 mL each. Each portion then contained:

$$1.00\,\frac{mol}{L} \times 10\ mL \times \frac{1\ L}{1000\ mL} = 0.0100\ mol\ Ba(NO_3)_2\ \text{per portion}$$

Different amounts of Na_2CrO_4 are added to each $Ba(NO_3)_2$ portion, starting with approximately two millimoles (0.002 mol) to the first portion, 4 millimoles to the second, and so on through all 12 portions. Figure 5.1 shows the mass of precipitate obtained from each portion in a typical experiment.

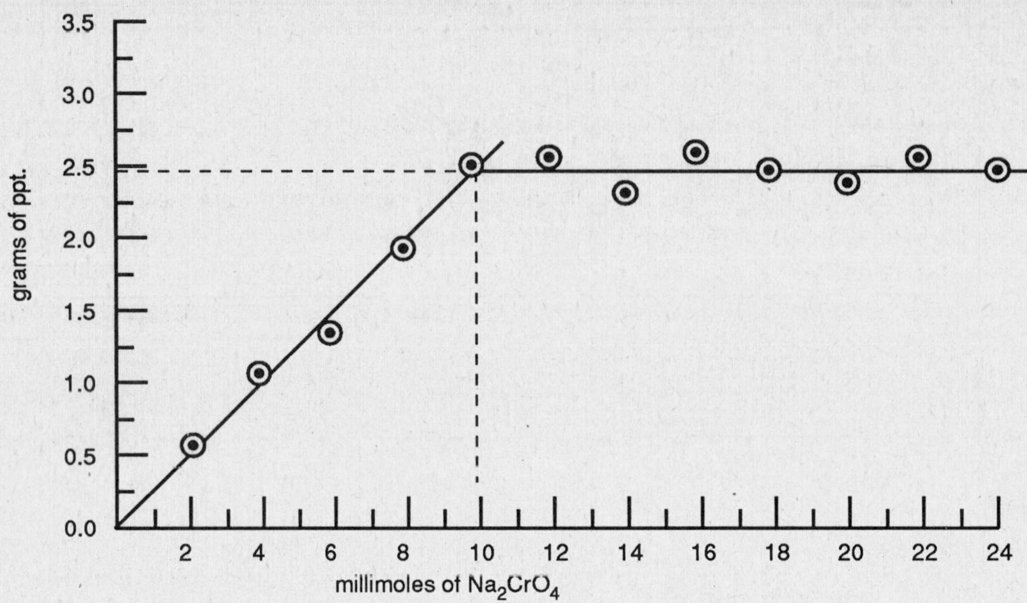

Figure 5.1. Grams of precipitate obtained vs. millimoles of Na$_2$CrO$_4$ added to 10 millimoles of Ba(NO$_3$)$_2$.

Each portion contained 10 millimoles (0.0100 mol) of Ba(NO$_3$)$_2$ and different amounts of Na$_2$CrO$_4$. It is clear from Figure 5.1 that adding more than 10 millimoles of Na$_2$CrO$_4$ to a solution containing 10 millimoles of Ba(NO$_3$)$_2$ does not cause any further increase in the mass of precipitate. Thus, 10 millimoles of Na$_2$CrO$_4$ react exactly with 10 millimoles of Ba(NO$_3$)$_2$ and the stoichiometric point is reached when the mole ratio of Na$_2$CrO$_4$ to Ba(NO$_3$)$_2$ is 1:1.

Figure 5.1 indicates that equal molar quantities of Ba(NO$_3$)$_2$ and Na$_2$CrO$_4$ react together, i.e., their mole ratio in the balanced reaction equation is 1:1.

By analogy with Equation 5.1, the likely products are BaCrO$_4$ and NaNO$_3$ Which product is the precipitate? Equation 5.1 indicates that NaNO$_3$ is soluble and data from chemical reference books indicate that nitrates are generally soluble while many chromates are insoluble. Therefore, BaCrO$_4$ is likely to be the precipitate. This conclusion will be confirmed later by weighing the precipitates. The balanced reaction equation must show the experimentally determined 1:1 mole ratio of the reactants and, also, must indicate the nature of the products and be balanced with respect to atoms, as in Equation 5.2, which assumes that BaCrO$_4$ was correctly chosen as the precipitate.

$$Ba(NO_3)_2(aq) + Na_2CrO_4(aq) \longrightarrow BaCrO_4(s) + 2\,NaNO_3(aq) \tag{5.2}$$

The balanced equation gives the mole ratios of the products and reactants.

Quantitative Verification that BaCrO4 is the Precipitate

Assuming that Equation 5.2 is correct, we can use it to calculate the masses of BaCrO$_4$ and NaNO$_3$ that are formed by complete reaction of 10 millimoles of Ba(NO$_3$)$_2$. By comparing the calculated masses with the measured mass of precipitate, from Figure 5.1, the correctness of our choice of BaCrO$_4$ as the precipitate can be checked.

From the reaction stoichiometry, 10 millimoles of Ba(NO$_3$)$_2$ should form 10 millimoles of BaCrO$_4$ and 20 millimoles of NaNO$_3$. Find the grams of BaCrO$_4$ and NaNO$_3$.

Atomic masses: Ba = 137 g;   Cr = 52.0 g;   O = 16.0 g;   Na = 23.0 g;   N = 14.0 g

Molar mass NaNO$_3$ = 23.0 + 14.0 + 3(16.0) = 85.0 g mol^{-1}

Molar mass BaCrO$_4$ = 137 + 52.0 + 4(16.0) = 253 g mol^{-1}

20 millimoles NaNO$_3$ weigh (20x10^{-3} mol) x 85.0 g mol-1 = 1.70 g

10 millimoles BaCrO$_4$ weigh (10x10^{-3} mol) x 253 g mol-1 = 2.53 g

From Figure 5.1, the average mass of precipitate, after it became constant, was about 2.4 g, confirming the choice of BaCrO$_4$ as the precipitate.

PLAN OF EXPERIMENT
A. Visual Determination of Stoichiometry

You will study the stoichiometry of the reaction between lead nitrate, $Pb(NO_3)_2$, and potassium chromate, K_2CrO_4, in a manner similar to the technique of the example above. As in the example, this reaction forms a precipitate.

You first prepare 6 identical aqueous solutions, each containing the same amount of K_2CrO_4. Then progressively larger amounts of $Pb(NO_3)_2$ are added to each sample. More $Pb(NO_3)_2$ added will result in more precipitate, as long as K_2CrO_4 is in excess. In samples where enough $Pb(NO_3)_2$ was added to completely use up the K_2CrO_4, extra $Pb(NO_3)_2$ does not form more precipitate.

As long as any K_2CrO_4 is present, the liquid above the precipitate is colored yellow. In solutions where all of the K_2CrO_4 has reacted, the liquid is colorless.

By noting how much $Pb(NO_3)_2$ is required to turn the solution colorless, you can determine approximately how much $Pb(NO_3)_2$ is needed to react stoichiometrically with the known amount of K_2CrO_4. Then you can write a balanced equation for the reaction, using your best judgement to identify the nature of the precipitate.

B. Gravimetric Determination of Stoichiometry

You will filter, dry, and weigh the precipitate from **one** of your samples, each student doing a different sample. Mass data from the entire class can be used to make a plot similar to Figure 5.1, which should give a more precise value for the stoichiometric point of the reaction. The stoichiometric mass of precipitate can be used to verify your identification of the precipitated compound, as was done in the example.

SAFETY

1. Wear approved eye protection.

2. All lead compounds are toxic. Handle all solutions carefully, avoiding skin contact. If contact occurs, wash immediately with soap and large amounts of water.

3. Be sure to use a rubber pipeting bulb — never pipet by mouth.

PROCEDURE
A. Visual Determination of Stoichiometry

1. Prepare a hot water bath, consisting of a 400 mL beaker about 2/3 full of water on a hot plate or on a ring stand with a wire gauze over a gas burner. Add one or two boiling chips and begin heating the water. **Once the water boils, adjust the burner to keep it at a low simmer.**

2. While the water is heating, obtain 50 mL of 0.09 M K_2CrO_4 and 50 mL of 0.10 M $Pb(NO_3)_2$ in two small, clean, dry beakers. Record the exact molarities on the Data sheet. Label the beakers to identify the solutions. Self-adhesive paper labels marked with pencil are satisfactory.

3. Obtain 6 small, clean test tubes (15-16 mm). The test tubes need not be dry. Number them from 1 to 6, using pencil on paper adhesive labels fastened around the **tops** of the tubes, and place the tubes in a test tube rack (Figure 5.2). Rinse a clean 5 or 10 mL graduated pipet once with the K_2CrO_4 solution and then pipet accurately 4 mL of the K_2CrO_4 solution into each of the tubes. Using the **accurate molarity** on the K_2CrO_4 solution bottle, calculate the number of moles of K_2CrO_4 added to each test tube. Record this value on the Data sheet.

4. Rinse a clean 5 or 10 mL graduated pipet once with $Pb(NO_3)_2$ solution and use it to transfer 1 mL of the $Pb(NO_3)_2$ solution to test tube no. 1, 2 mL to tube no. 2, 3 mL to tube no. 3, 4 mL to tube no. 4, 5 mL to tube no. 5, and 6 mL (deliver 3 mL twice if using 5 mL pipet) to tube no. 6. A precipitate will form in the tubes.

5. Use a wash bottle to add distilled water to tubes 1-5 until their levels are just a little lower than the liquid level in tube 6.

Each test tube contains 4 mL of 0.09 M K_2CrO_4. The amount of 0.1 M $Pb(NO_3)_2$ pipeted into each test tube is:

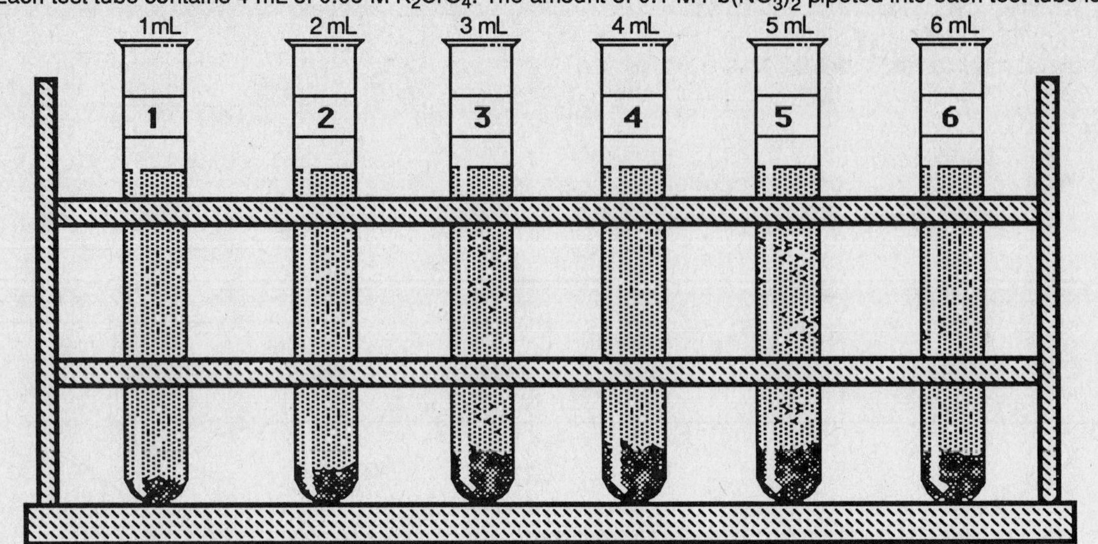

Figure 5.2. Preparation of sample test tubes.

6.   Shake each tube thoroughly, 15 to 20 seconds, closed with a rubber stopper.
Remove the stopper after shaking, rinse any precipitate adhering to it back into the tube with a wash bottle containing distilled water, and wipe it clean with a paper towel before moving it to a new tube for shaking. Then use the wash bottle to rinse down any precipitate adhering to the upper part of the tube. Use a minimum amount of water for these rinses.

Use the wash bottle to add distilled water to each tube until their levels are all approximately the same (Figure 5.2). Place all the tubes in the hot water bath and simmer them for 15-20 minutes to coagulate the precipitate into large crystals (Figure 5.3).

7.   Remove the tubes from the bath and arrange them in numerical order in the test tube rack. Let them sit undisturbed for about 10 minutes, or until the precipitate has completely settled.

8.   Hold a sheet of white paper behind the test tubes and examine the liquid above the precipitate for color. On the Data sheet, indicate whether the liquid in each tube is colorless or yellow.[1]

9.   Calculate the moles of $Pb(NO_3)_2$ in the highest numbered tube that contains yellow-colored liquid and the lowest numbered tube that contains colorless liquid. Record these values on the Data sheet.
The moles of $Pb(NO_3)_2$ that react stoichiometrically with 4 mL of 0.09 M K_2CrO_4 must lie between these two values.

10.   Determine the best value for the stoichiometric amount of $Pb(NO_3)_2$ by averaging the two values from step 9. Record this value.

B. Gravimetric Determination of Stoichiometry
1.   Your instructor will tell you which test tube to analyze. Record its number on the Data sheet.

2.   Set up a suction filtration apparatus. Weigh a filter paper disc to the nearest mg and record its mass. Place it in a Buchner funnel, and moisten it thoroughly with distilled water. Connect the apparatus to an aspirator and turn on the water to start the suction.

---

[1] Careful attention to the procedure usually will produce a precipitate that settles completely. If coagulation is not complete, however, it might be difficult to identify which solutions contain colorless liquid. In this case, it will be faster to centrifuge the tubes than to start over. If needed, your instructor will demonstrate proper use of the centrifuge.

3.   Use a clean glass rod as a guide to pour most of the liquid from your assigned test tube into the center of the filter. Then, with the rod, stir up the solid in the bottom of your assigned test tube. Using the rod as a guide, pour the remaining liquid and as much solid as flows easily, down the rod into the filter. Then rinse the glass rod into the filter with distilled water from a wash bottle.

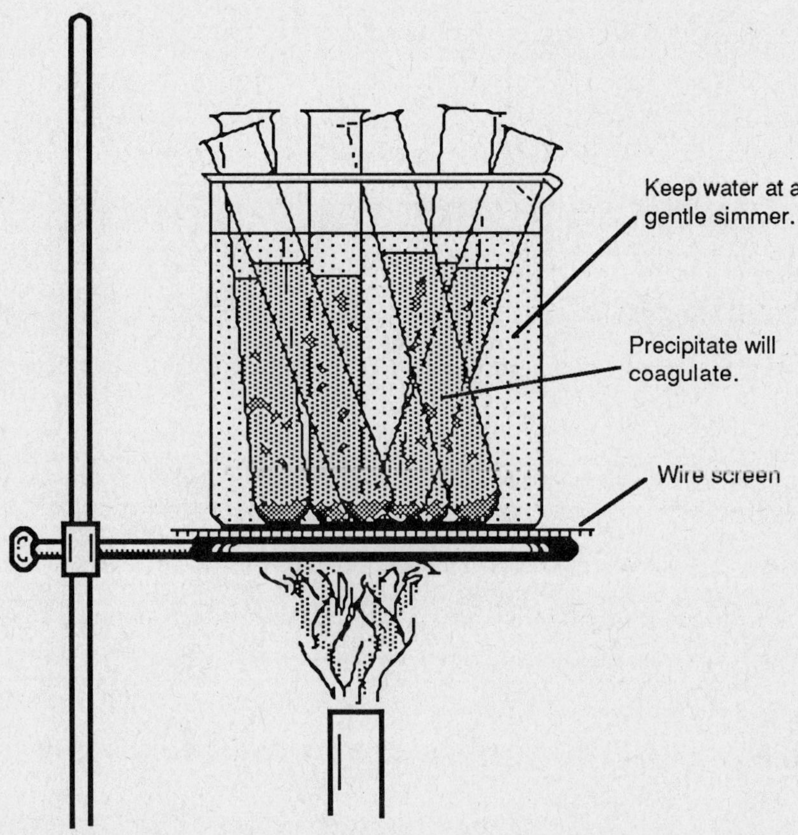

Figure 5.3. Coagulating the precipitate in a water bath.

4.   Use the wash bottle to rinse the remaining solid down the rod into the filter. Use the rod to further loosen the solid, if necessary. When all of the solid has been transferred, rinse the test tube into the filter with an additional 15 mL of water from the wash bottle and rinse off the glass rod into the filter again.
    After all the solid has been transferred to the filter paper, continue drawing air through the filter for about 5 more minutes, to speed the drying process.

5.   Carefully remove the filter paper from the funnel and place it on a piece of paper towel to absorb excess moisture. Dry it in the laboratory oven at 60°C (about 15 minutes) or, if directed, place it in your equipment drawer to dry until next laboratory period.
    Handle wet filter paper very carefully so that it is not accidentally torn.

6.   When the filter paper with precipitate is dry, weigh it to one milligram. Record the mass of paper plus precipitate and determine the mass of precipitate.

7.   Your instructor will have prepared a table on the blackboard where you and your classmates can record your test tube number and the mass of precipitate obtained from it. When everyone has entered their data, copy the table onto your Data sheet.

Name ____   Date ____________

EXPERIMENT 5
PRELABORATORY EXERCISE

1.   A flask contains 10 mL of 0.20 M Na_2S solution. Use Reaction 5.1 to answer the following questions:

$$2\,AgNO_3(aq) \;+\; Na_2S(aq) \longrightarrow Ag_2S(s) \;+\; 2\,NaNO_3(aq) \tag{5.1}$$

a. What mass of precipitate is formed if 3 mL of 0.15 M $AgNO_3$ is added?

b. What mass of precipitate is formed if 50 mL of 0.15 M $AgNO_3$ is added?

c. How many mL of 0.15 M $AgNO_3$ are needed to react stoichiometrically with the solution in the flask?

Name ____   Date _____________

EXPERIMENT 5
DATA
(Observe significant figures in all calculations.)

A. Visual Determination of Stoichiometry

1. Labelled molarity of $Pb(NO_3)_2$ solution: _________________
2. Labelled molarity of K_2CrO_4 solution: _________________
3. Volume of K_2CrO_4 added to each test tube: _________________
4. Calculated moles of K_2CrO_4 in each test tube: _________________

Test tube No.	1	2	3	4	5	6
mL $Pb(NO_3)_2$						
moles $Pb(NO_3)_2$						
color of liquid (yellow or colorless)						

3. Moles of $Pb(NO_3)_2$ in highest numbered test tube with yellow liquid: _________
4. Moles of $Pb(NO_3)_2$ in lowest numbered test tube with colorless liquid: _________
5. Stoichiometric moles of $Pb(NO_3)_2$ (average of 3 and 4):   _________

B. Gravimetric Determination of Stoichiometry

1. Number of test tube chosen for analysis: _________________
2. Initial mass of filter paper: _________________
3. Mass of dry filter paper + precipitate: _________________

4. Enter mass data from entire class in the table below. Put your own measurement in the first row and circle it.

Tube No.	1	2	3	4	5	6
Sum of masses for each tube:						
Av. mass for each tube:						

Use the average mass for each tube to make a graph similar to Figure 5.1 on the graph paper on the next page.

EXPERIMENT 5
QUESTIONS

1. Write a balanced equation for the reaction between $K_2CrO_4(aq)$ and $Pb(NO_3)_2(aq)$.

2. Why was the precipitate heated in a water bath before filtering?

3. Why must you rinse the stopper into the test tube after shaking the solution?

EXPERIMENT 6
Le Chatelier's Principle:
Influence of Concentration and Temperature on Reaction Equilibrium

PRELABORATORY PREPARATION
1. Do the Prelaboratory Exercise and turn it in at the beginning of your laboratory period.
2. Review the use of the graduated cylinder and buret, Section IV, Methods, pages 16-18.

INTRODUCTION

A chemical system is said to be in equilibrium when there are no measurable changes occurring. Reaction 6.1 describes a hypothetical system where **a** moles of reactant **A** and **b** moles of reactant **B** combine to form **c** moles of product **C** and **d** moles of product **D**.

$$aA + bB \rightleftharpoons cC + dD \qquad (6.1)$$

When **A** and **B** are first mixed, the system is not in equilibrium. As the reaction proceeds, the decrease in concentration of the reactants can be measured, as can the increase in concentration of the products. Since all reactions are reversible to some degree, as soon as any **C** and **D** are formed, they can react in reverse to form **A** and **B** again.

$$cC + dD \rightleftharpoons aA + bB \qquad (6.2)$$

The **rate** of a reaction generally varies with the concentration of the reagents, decreasing if the reagent concentrations decrease, and increasing if the reagent concentrations increase. As **A** and **B** are consumed, they react in the forward Reaction 6.1 more and more slowly. As **C** and **D** are formed and increase in concentration, they react in the reverse Reaction 6.2 more and more rapidly.

Eventually, the rates of Reactions 6.1 and 6.2 must become equal and the concentrations of *A, B, C,* and *D* no longer change; *A* and *B* are formed by Reaction 6.2 exactly as fast as they are consumed by Reaction 6.1, and *C* and *D* are formed by Reaction 6.1 exactly as fast as they are consumed by Reaction 6.2. This condition of static concentrations is called the *equilibrium state of reaction.*

If the position of equilibrium is very far to the product side of the reaction equation, we say the reaction **goes essentially to completion.** If the equilibrium is very far to the reactant side, we say there is **no observable reaction.** It can happen that a reaction proceeds so slowly that it seems as if no changes are occurring, even though the system is far from its equilibrium position. Such a reaction is difficult to study unless some way can be found to speed it up. The equilibrium condition is best studied with moderately rapid reactions having equilibria close enough to the middle point that the concentrations of products and reactants are both large enough to be easily measured. A reaction that can reach a measurable state of equilibrium is indicated by using two-way arrows in the reaction equation:

$$aA + bB \rightleftharpoons cC + dD \qquad (6.3)$$

The double arrow indicates that the same equilibrium concentrations of **A, B, C,** and **D** can be approached from either side of the equation. The equilibrium condition is expressed quantitatively by the **equilibrium constant**, K_{eq}, which is written for Reaction 6.3 as:

$$K_{eq} = \frac{[C]^c [D]^d}{[A]^a [B]^b} \qquad (6.4)$$

where the exponents **a, b, c,** and **d** are the same values as the coefficients **a, b, c,** and **d** in Equation 6.3. K_{eq} is a characteristic constant for each reaction and varies only with temperature.

For a reaction to be in equilibrium means that the concentrations of all chemical species in the system do not change with time and that the temperature, volume, and pressure of the system also remain constant. On a molecular scale, equilibrium is not a static but a dynamic condition, where chemical changes still take place but are not detectable.

Both the forward and reverse reactions continue to occur when the system is in equilibrium. Their rates, however, have become exactly equal so that there is no *net* change in reagent concentrations. At equilibrium, the forward and reverse reactions release and absorb equal amounts of energy, so that there also is no *net* energy change and no tendency for the temperature to change .

Once equilibrium is reached, it can be maintained only if all relevant conditions are held constant. If reagent concentrations are changed or if the temperature, pressure, or volume are changed, the equilibrium will be disturbed and the system will undergo additional net changes until a new equilibrium is reached.

Le Chatelier's Principle

In 1888, Le Chatelier observed that a chemical system in equilibrium tends to oppose or counteract any imposed change.

Concentration Change

As an example of Le Chatelier's principle, assume Reaction 6.3 has reached equilibrium and then additional compound **D** is added to the system. The equilibrium is disturbed by the increase in **D** and the system must change until a new equilibrium is reached. Le Chatelier pointed out that the direction of the change will always tend to minimize or counteract the disturbance; thus, the addition of **D** must cause the reverse reaction to dominate so as to consume some of the extra **D** in the system. This is consistent with Equation 6.4. Increasing the concentration of **D** increases the value of the numerator. If K_{eq} is to remain constant, the new equilibrium that is reached must have reagent concentrations that give a larger denominator for Equation 6.4. This can occur if the reverse reaction of 6.3 dominates while the new equilibrium is attained, producing more **A** and **B**.

Adding additional reagent to a chemical system in equilibrium will cause the reaction which consumes that reagent to dominate, as the system seeks a new equilibrium. This minimizes the increase in concentration of the added reagent, in agreement with Le Chatelier's Principle.

If a reagent is added that does not react, there is no change in the equilibrium. The system cannot minimize the concentration change by seeking a new equilibrium.

Temperature Change

Most reactions are either **exothermic** (releasing heat energy to the surroundings) or **endothermic** (absorbing heat energy from the surroundings). A few reactions just chance to be **1lf** (neither releasing nor absorbing heat energy). What would be the effect of heating a chemical system after it has reached equilibrium? The system must move towards a new equilibrium in a way that absorbs thermal energy and minimizes the net temperature increase. Suppose that the forward direction of Reaction 6.3 is exothermic. Then, the reverse reaction must be endothermic. Heating the system will cause the endothermic reverse reaction to dominate, shifting the equilibrium to the left, so that part of the additional energy added to the system will be absorbed, minimizing the temperature rise.

- **Heating a chemical system in equilibrium will cause the endothermic reaction to dominate as a new equilibrium is attained, absorbing some of the excess energy and minimizing the increase in temperature.**

- **Cooling a chemical system in equilibrium will cause the exothermic reaction to dominate as a new equilibrium is attained, releasing additional energy and minimizing the decrease in temperature.**

- **If a reaction is thermoneutral, heating or cooling it has no effect on the position of equilibrium. The system cannot minimize the change in energy by seeking a new equilibrium.**

In this experiment, the influence of concentration and temperature changes on the equilibrium of Reaction 6.5 will be observed.

$$Cu(H_2O)_4{}^{2+} + 2\,Cl^- \;\rightleftharpoons\; Cu(H_2O)_2Cl_2 + 2\,H_2O \tag{6.5}$$
$$\text{blue} \qquad\qquad\qquad \text{green}$$

Cupric ion (Cu^{2+}) in water solution forms covalent bonds with 4 water molecules to form blue colored $Cu(H_2O)^{2+}$, called the **tetraaquacopper(II) complex ion**. Adding chloride ion to a cupric ion solution causes the forward reaction of 6.5 to occur, consuming part of the added chloride ions to form green colored $Cu(H_2O)_2(Cl)_2$, called the **diaquadichlorocopper(II) complex**, a soluble neutral complex. If only a little chloride ion is added, the equilibrium will be to the left side of Reaction 6.5, where blue $Cu(H_2O)_4{}^{2+}$ is in greatest concentration, and the solution will appear blue. As more and more chloride ion is added, the equilibrium shifts towards the right side and the increasing concentration of $Cu(H_2O)_2(Cl)_2$ causes the solution color to become increasingly greener. If enough Cl^- is added, the equilibrium of Reaction 6.5 moves far enough to the right to make the solution appear completely green.

Reaction 6.5 is not thermoneutral. By observing color changes that occur upon heating and cooling equilibrium solutions, you should be able to deduce which direction of Reaction 6.5 is exothermic and which endothermic.

PLAN OF EXPERIMENT

Influence of Concentration

The equilibrium of Reaction 6.5 is shifted from left to right by adding increasing amounts of Cl^-. The change in equilibrium position is evident because the solution color changes progressively from deep blue, through intermediate blue-green, to green. Then, the equilibrium is shifted back to the left by adding H_2O, and the color changes from green back to blue.

Influence of Temperature

Aqueous solutions of cupric and chloride ions having different equilibrium positions are heated and cooled. By observing the color changes that accompany the changes in thermal energy, you can determine which direction of Reaction 6.5 is exothermic and which endothermic.

Direct Observation of Energy Changes

You will have deduced the exothermic and endothermic directions of Reaction 6.5 by observing the color changes that accompany heating and cooling equilibrium systems and applying Le Chatelier's Principle. You can verify your conclusions by measuring the temperature change of two reactions:

1.   A measured amount of concentrated HCl is added to a measured volume of pure water:

$$HCl + H_2O \rightleftharpoons H^+(aq) + Cl^-(aq) \tag{6.6}$$

The temperature rise of the solution is measured. (Reaction 6.6 is exothermic in the forward direction.)

2.   Using the same amounts of reagent as in Step 1, concentrated HCl is added to an aqueous solution of cupric ion. This time, both Reactions 6.6 and 6.5 occur, because Reaction 6.6 produces chloride ion that reacts with the aquated cupric ion to form $Cu(H_2O)_2(Cl)_2$. The energy change due to Reaction 6.5 will be added to that from Reaction 6.6 and the temperature change will be different.

- **If the temperature change is larger, Reaction 6.5 must be exothermic, adding more thermal energy to the solution.**
- **If the temperature change is smaller, Reaction 6.5 must be endothermic, because it absorbed some of the energy released by Reaction 6.6.**

You can decide whether this experiment confirms or contradicts your conclusions based on the color changes with temperature about the exothermic and endothermic directions of Reaction 6.5.

SAFETY

1. Wear approved eye protection.

2. Concentrated hydrochloric acid (HCl) is very corrosive to skin and clothing. Its fumes can damage nose and lung tissues. Use it only in the hood and handle it with great care.

3. Dry ice can injure skin tissue. Do not touch it with your bare fingers.

PROCEDURE

A. Effect of Concentration

1.   Obtain 30 mL of 0.5 M $CuSO_4$ in a 50 mL Erlenmeyer flask and 5 g solid NaCl in a plastic weighing dish or a watch glass.

2.   Add the NaCl to the $CuSO_4$ solution, in approximately 1 g increments, using a spatula. Swirl the solution after each 1 g addition until the solid is dissolved. Observe the progressive color change of the solution. Record your observations on the data sheet.
Do not add more than 5 grams NaCl.

3.   Pour approximately 10 mL of the final solution into another 50 mL Erlenmeyer flask and set it aside for Part B., step 1.

4.   Using a 10 mL graduate cylinder, slowly add 10 mL H_2O, in increments of approximately 1 mL, to the remaining solution that was not set aside. Observe the progressive color change. Record your observations on the data sheet.

B. Effect of Temperature

1.   Place the flask containing 10 mL of solution set aside in Part A, step 3, on dry ice in an insulated bucket. Observe the color change that occurs after a few minutes. Remove it from the dry ice and allow it to warm to room temperature. Again, observe any color change. Record your observations on the data sheet.

2.   Obtain 30 mL 0.5 M $CuSO_4$ solution in a 50 mL Erlenmeyer flask and add 2 g NaCl. Swirl the solution until the solid is dissolved.

3.   Heat the solution on a hot plate at high setting, until a color change is observed. This might take about 5 min. Record your observations on the data sheet.
Do not boil the solution.

4.   Allow the solution to cool back to room temperature. Observe any color change and record your observations on the data sheet.

C. Direct Observation of Energy Changes

1.   Into a polystyrene coffee cup, carefully measure from a graduate cylinder 10 mL distilled water. Take the cup with its water and a thermometer to the fume hood, where burets are set up containing concentrated HCl.

2.   Place the cup under a buret and measure the water temperature. You may have to tip the cup so the thermometer bulb is adequately immersed in the solution. Record the temperature on the data sheet.

3.   Carefully measure 10 mL HCl from the buret into the cup. Stir the solution with the thermometer and immediately record the temperature. Take additional temperature readings at 5 second intervals, until they pass through a maximum value. Record the temperatures on the Data sheet.
Observe and record any color changes.

4.   Empty the acid solution into a sink with running water and thoroughly rinse out the cup with water. Drain the rinse water thoroughly, but it is not necessary to dry the cup.

5.   Carefully measure from a graduate cylinder 10 mL 0.5 M $CuSO_4$ solution into the cup. Take the cup and a thermometer to the fume hood, where burets are set up containing concentrated HCl.

6.   Place the cup under a buret and measure the temperature of the $CuSO_4$ solution. Record the temperature on the data sheet.

7.   Carefully measure 10 mL HCl from the buret into the cup. Stir the solution with the thermometer and immediately record the temperature. Take additional temperature readings at 5 second intervals, until they pass through a maximum value. Record the temperatures on the data sheet. Observe and record any color changes.

8.   Empty the acid-$CuSO_4$ solution into a sink with running water and rinse out the cup with water.

Name ______________________________________   Date ____________

EXPERIMENT 6
PRELABORATORY EXERCISE

1.   A solution in which Reaction 6.7 is occurring has reached equilibrium. The concentrations of species are such that the solution at equilibrium appears pink.

$$CoCl_4^{2-}(aq) + 6\,H_2O \rightleftharpoons 4\,Cl^-(aq) + Co(H_2O)_6^{2+}(aq) \qquad (6.7)$$
blue   pink

When the solution is cooled, no color change is observed; when it is heated, it turns blue.

   a. The forward direction of reaction is: **(a)** exothermic;   **(b)** endothermic;   **(c)** thermoneutral.

   b. If additional chloride ion is added to the equilibrium system, the equilibrium will:
        (a) shift in the forward direction;   **(b)** shift in the reverse direction;   **(c)** not shift at all.

2.   Write the equation for the equilibrium constant of Reaction 6.7. Note that the solution is aqueous, so that water is in great excess and the concentration of H_2O cannot change significantly. Under these conditions, $[H_2O]$ is essentially constant and is omitted from the equilibrium constant equation.

Name ____   Date _____________

EXPERIMENT 6

DATA
(Observe significant figures in all calculations.)

A. Effect of Concentration
Step 2 (addition of NaCl):

    a. Initial solution color: _____________________

    b. Final solution color: _____________________

    c. Manner of color transition (abrupt, continuous, any other observations):

Step 4 (addition of H_2O):

    a. Initial solution color: _____________________

    b. Final solution color: _____________________

    c. Manner of color transition (abrupt, continuous, any other observations):

    Answer question 1 on the Questions page.

B. Effect of Temperature
Step 1 (cooling first, then warming):

    a. Initial room temperature color of solution: _________________

    b. Color of cold solution: _________________

    c. Color of solution after warming back to room temperature: _________________

    d. Manner of color transitions (abrupt, continuous, any other observations):

Steps 3 and 4 (heating first, then cooling):

    a. Initial room temperature color of solution: _________________

    b. Color of hot solution: _________________

    c. Color of solution after cooling back to room temperature: _________________

    d. Manner of color transitions (abrupt, continuous, any other observations):

    Answer question 2 on the Questions page.

EXPERIMENT 6
DATA
(Observe significant figures in all calculations.)

C. Direct Observation of Energy Changes
Steps 2 and 3 (adding HCl to H_2O):
  Initial temperature of water: ___________________

Water temperature:
   immediately after adding HCl: ___________________

   5 seconds after adding HCl: ___________________

   10 seconds after adding HCl: ___________________

   15 seconds after adding HCl: ___________________

   20 seconds after adding HCl: ___________________

   25 seconds after adding HCl: ___________________

   30 seconds after adding HCl: ___________________

Maximum temperature reached: ___________________

Other observations:

Steps 6 and 7 (adding HCl to $CuSO_4$ solution):
  Initial temperature of $CuSO_4$ solution: ___________________

$CuSO_4$ solution temperature:
   immediately after adding HCl: ___________________

   5 seconds after adding HCl: ___________________

   10 seconds after adding HCl: ___________________

   15 seconds after adding HCl: ___________________

   20 seconds after adding HCl: ___________________

   25 seconds after adding HCl: ___________________

   30 seconds after adding HCl: ___________________

Maximum temperature reached: ___________________

Other observations:

Answer question 3 on the Questions page.

Name ____   Date _____________

EXPERIMENT 6
QUESTIONS

1. Part A: Effect of Concentration
Explain why the data of Part A either support or contradict Le Chatelier's Principle.

2. Part B: Effect of Temperature
On the basis of Le Chatelier's Principle and the data in Part B, write the reaction indicating the exothermic direction of Reaction 6.5.

Write the chemical equation that indicates the endothermic direction of Reaction 6.5.

3. Part C: Direct Observation of Energy Changes
Explain how the results of Part C. either verify or contradict your choice of the exothermic and endothermic directions of Reaction 6.5 in Part B.

EXPERIMENT 7
Ideality of Oxygen Gas at Room Temperature

PRELABORATORY PREPARATION
1. Do the Prelaboratory Exercise and turn it in at the beginning of your laboratory period.
2. Review the discussion of standard deviation in "Treatment of Experimental Data," page 2.
3. Review the technique for fitting glass tubes into rubber stoppers, Methods, Section VII, page 20.

INTRODUCTION
The **ideal-gas law,** Equation 7.1, quite accurately predicts the behavior of most gases, if the gas pressure is no higher than a few atmospheres and the gas temperature is well above its condensation temperature.

$$PV = nRT \tag{7.1}$$

P = pressure (atm); V = volume (L); n = moles of gas; T = temperature (K)
R = gas constant = 0.08206 L atm mol^{-1}K^{-1}

The units of P, V, and R are chosen here to be consistent and convenient for this experiment. Temperature must always be expressed in **degrees Kelvin.**

Two important assumptions are made in the derivation of the ideal-gas law:
1. Attractive and repulsive forces between gas molecules are insignificantly weak compared to their thermal energy.
2. The actual volume of the molecules themselves is negligibly small compared to the volume of their container, so that the space inside the container is essentially empty.

The meaning of assumption **1** is that pressure exerted by a gas upon the container walls is determined entirely by the temperature of the gas, i.e., the gas kinetic energy. If intermolecular forces were significant, they would influence the pressure by affecting the velocities of the gas molecules. Then, the correct gas pressure would not be predicted by Equation 7.1.

The meaning of assumption **2** is that compression of a gas is not limited by the size of the gas molecules. If molecular size were not negligible, molecular collisions would cause the gas to be less compressible and its correct volume would not be predicted by Equation 7.1.

If measured pressures and volumes do not fit Equation 7.1 at the experimental temperature, it indicates that intermolecular forces and molecular volumes cannot be disregarded and that the gas is not behaving ideally.

In this experiment, you will test whether these ideal gas assumptions apply to oxygen under your laboratory conditions by making a known amount of oxygen and measuring its pressure, volume, and temperature. You insert your measured data into the ideal-gas law to see if they give the correct value for the gas constant R. If they do, then your oxygen gas sample behaves as an ideal gas; if not, your gas sample is not ideal.

Testing for gas ideality
To determine whether or not a gas behaves ideally, it is necessary to have the following information, which allows a calculation of the gas constant R.
- a. the molar mass, M, of the gas,
- b. the mass, m, of the measured sample,
- c. values for P, V, and T, measured when the gas sample is at equilibrium.

These quantities are then plugged into Equation 7.1 and a value of R calculated. The more that the value for R, calculated from experimental data, differs from the handbook value of $R = 0.08206$ L atm mol^{-1}K^{-1} (which is taken to be the "correct" value), the less ideal the gas. Of course, in an actual experiment, you cannot expect to obtain the handbook value exactly, even if the gas is truly ideal. There always are measuring uncertainties, and an experiment can only indicate the ideality or nonideality of a gas within experimental limits of error.

The significance of an experimental result is unclear unless the experimental limits of error are estimated.

The following example shows how experimental limits of error influence the interpretation of data.

EXAMPLE:

Consider the following data, measured on a sample of helium gas. Measured values are given with their experimental uncertainties.

Sample data measured in an experiment on Helium:

molar mass He :	M_{He}	= 4.0026 (handbook value)
measured mass of sample :	m	= 0.5042 ± 0.0001 g
measured pressure of sample:	P	= 753.6 ± 0.1 torr
measured volume of sample:	V	= 3.052 ± 0.005 L
measured temperature of sample :	T	= 21°C ± 1°C = 294 ± 1 K

What do these values indicate concerning the ideality of helium gas, within the experimental limits of error? To use the experimental data for answering this question, rewrite Equation 7.1 as:

$$R = \frac{PV}{nT} \tag{7.2}$$

The number of moles, n, must be expressed in terms of the mass of the gas sample:

$$n = \frac{\text{mass of sample}}{\text{molar mass}} = \frac{n}{M}$$

Inserting this into Equation 7.2 gives:

$$R = \frac{MPV}{mT} \tag{7.3}$$

Now, use Equation 7.3 to calculate the value of **R** predicted by the experimental data.

$$R_{exp} = \frac{(4.003 \text{ g/mol})(753.6 \text{ torr})(3.052 \text{ L})}{(0.5042 \text{ g})(294 \text{ K})(760 \text{ torr/atm})} = 0.0817 \frac{\text{L atm}}{\text{mol K}}$$

Note that, of all the experimental values, the temperature measurement has the fewest significant figures (three) and this uncertainty in the temperature value limits the answer to 3 significant figures. We know that the handbook value is **R** = 0.0821 L atm mol^{-1} K^{-1}, to 3 significant figures, so that **R$_{exp}$** differs from **R$_{handbook}$** by

$$\frac{(0.0817 - 0.0821)}{(0.0821)} \times 100 = -0.487\%$$

Does this result indicate a small degree of nonideality for helium gas under our experimental conditions or is the discrepancy simply due to experimental error? We cannot tell until we determine the uncertainty in our calculated value for **R**, due to uncertainties in our measurements.

Determination of Uncertainty in Calculated Value of R

We must determine the largest and smallest values of **R** that are consistent with the error limits of the measurements.

1.   Get the largest possible value for **R** by making the numerator of Equation 7.3 as large as experimental uncertainty permits, and the denominator as small as experimental uncertainty permits. To do this, **add** the limits of error in the **numerator,** and **subtract** them in the **denominator.**

$$R_{max} = \frac{\{4.003 \text{ g/mol}\}\{(753.6 + 0.1)\text{torr}\}\{(3.052 + 0.005)\text{L}\}}{\{(0.5042 - 0.0001)\text{g}\}\{(294 - 1)\text{K}\}\{760 \text{ torr/atm}\}} = 0.0822 \text{ L atm mol}^{-1} \text{ K}^{-1}$$

2.   Get the smallest possible value for **R** by making the numerator of Equation 7.3 as small as experimental uncertainty permits, and the denominator as large as experimental uncertainty permits. To do this, **subtract** the limits of error in the **numerator**, and **add** them in the **denominator.**

$$R_{max} = \frac{\{4.003 \text{ g/mol}\}\{(753.6 - 0.1)\text{torr}\}\{(3.052 - 0.005)\text{L}\}}{\{(0.5042 + 0.0001)\text{g}\}\{(294 + 1)\text{K}\}\{760 \text{ torr/atm}\}} = 0.0813 \text{ L atm mol}^{-1} \text{ K}^{-1}$$

The measurement limits of error indicate that this experiment yielded a value for R that is somewhere between 0.0822 and 0.0813. With only a single determination, there is no reason to favor any particular value for R within this range.

The results of the experiment are correctly expressed by giving the average value for **R** between its uncertainty limits, to 3 significant figures, and including the plus-minus deviation from the average[1].

$$\text{average } R_{exp} = \overline{R}_{exp} = \frac{R_{max} + R_{min}}{2} = \frac{0.0822 + 0.0813}{2} = 0.08175$$

$$\overline{R}_{exp} = 0.0818 \pm 0.0005 \text{ atm mol}^{-1} K$$

The handbook value for R is 0.08206 L atm^{-1} K^{-1}, which lies between the maximum and minimum values measured in this experiment. Therefore, the results of the helium experiment are that no deviation from ideality was detected.

There are two ways to narrow the experimental uncertainty:

1.   Repeat the experiment with better equipment, so that uncertainties in the measured quantities are smaller. This approach requires good judgement. In the example above, the measurement that had the greatest effect on the experimental uncertainty was that of temperature, which had only 3 significant figures. This measurement must receive the first priority for improvement if the uncertainty limits are to be reduced by improving measurement precision.

2.   Repeat the measurement several times to calculate a more reliable average value. The limit of uncertainty for repeated measurements is the standard deviation, which becomes smaller as more measurements are made.

     Normally, both ways are practiced. An experimenter uses the best equipment available, and then makes enough repeat measurements to narrow uncertainty limits to an acceptable range.

PLAN OF EXPERIMENT

     You will measure the properties of oxygen gas. The oxygen is obtained by heating a solid compound, lead dioxide (PbO_2).[2] When reddish PbO_2 is heated above 290°C, it decomposes to orange-colored solid lead oxide (PbO) and gaseous oxygen (O_2):

$$2\,PbO_2(s) \xrightarrow{290°C} 2\,PbO(s) + O_2(g) \tag{7.4}$$

To test the ideality of O_2, you must measure its mass, volume, pressure, and temperature.

Mass of Oxygen

The only gas given off in the reaction is oxygen (*if the PbO$_2$ is dry*).

The mass of oxygen is equal to the mass lost by the solids in the reaction test tube.

Volume of Oxygen

The volume of oxygen is measured by causing O_2 to displace water from a bottle, so that the displaced water flows into a beaker. The volume of displaced water is measured in a graduate cylinder.

The volume of displaced water is equal to the volume of O_2 evolved.

Pressure of Oxygen

The pressure measurement is complicated slightly by the fact that the oxygen evolved is collected by trapping it in a bottle sealed with water. This causes the oxygen to be mixed with water vapor, which contributes to the total pressure.

---

[1] Actually, rounding to 3 significant figures gives:

$$R_{average} = 0.0818 \; (+0.0004, -0.0005) \text{ L atm}^{-1} K^{-1}.$$

By using the larger value for both deviations, we err in the safe and conservative direction, which always is best when reporting experimental results.

[2] Alternatively, potassium chlorate, $KClO_3$, may be used instead of PbO_2. It requires manganese dioxide, MnO_2, as a catalyst. The reaction is:

$$2\,KClO_3 \xrightarrow{MnO_2} 2\,KCl + 3\,O_2$$

Use about 3 g $KClO_3$, mixed very thoroughly with about 0.2 g MnO_2. Both reagents must be dry before using. Use the same apparatus and procedure as with PbO_2.

     SAFETY: Chlorates can react explosively with organic matter.

To find the partial pressure of oxygen, it is necessary to first measure the total pressure of the oxygen-water mixture and then subtract the partial pressure of water vapor.

The water vapor pressure depends only on the temperature of the water and can be read from a table in the chemical reference section in the front of this book.
The procedure for determining the *total pressure* of the oxygen-water vapor mixture, is to adjust the total pressure in the gas collecting bottle so that it is the same as the external atmospheric pressure. Then, the external atmospheric pressure is measured with a barometer. The total pressure in the collection bottle is due to oxygen gas and water vapor. After the total pressure is adjusted to atmospheric pressure, Dalton's law of partial pressures gives:

$$P_{atm} = P_{O_2} + P_{water\ vapor}$$

$P_{water\ vapor}$ depends only on the water temperature, and may be determined by measuring the water temperature and reading the water vapor pressure for that temperature from a table in the Reference Section in the front of this manual. Then, all the quantities are known for calculating the oxygen pressure:

$$P_{O_2} = P_{atm} - P_{water\ vapor}$$

Temperature of Oxygen
The temperature of your collected oxygen is the laboratory room temperature.
This assumes you allow your apparatus to cool until it comes to thermal equilibrium with the surroundings.

EVOLUTION OF OXYGEN
The decomposition reaction (7.4) goes slowly enough to allow close control over the amount of O_2 produced. Simply heat PbO_2 until the desired amount of O_2 has been collected. Then remove the heat to stop the reaction. The mass lost by the solid PbO_2 is equal to the mass of O_2 evolved.
If the PbO_2 contains absorbed water, water will evaporate during heating and add to the mass loss.
To prevent this error, the PbO_2 must be thoroughly oven-dried prior to use. Cool the PbO_2 in a dessicator so it is at room temperature before weighing.

SAFETY
1. Wear approved eye protection.
2. All lead compounds are toxic. Avoid inhaling PbO_2 dust.
3. Do not clamp the test tube tightly. If too much pressure should build up in the test tube, it is desirable that the stopper and test tube can separate and move in opposite directions, to avoid a glass shattering explosion.
4. Follow closely the technique for fitting glass tubing into rubber stoppers, Section XII, Methods. In particular, be sure to lubricate the tubing with water, protect your hands with a cloth or folded paper towel, and exert insertion force with extreme care. Be certain the hole through the stopper is large enough. If you make your own hole through the stopper, use a boring tool of the appropriate size.

PROCEDURE
1.   Obtain 15-20 g of dry, room-temperature PbO_2 in a clean, dry 6" test tube. (PbO_2 should have been dried for 3-5 days at 120°C and stored in a dessicator at room temperature.) Weigh the tube and contents accurately on an analytical balance. Record the mass on the Data sheet.

2.   Assemble the apparatus shown in Figure 7.1. Use a bottle or flask of 1/2 to 1 L capacity and a 400 mL beaker. The glass tip of an eyedropper with its constricted end will serve for glass tube **C**.

3.   Read Section XII, Methods, *before* fitting the glass tubing into the rubber stoppers. Lubricate the glass tubes and the stopper with water. Fill the bottle with tap water and assemble the apparatus, except **do not** attach flexible tube **A** to the test tube yet. Add enough tap water to the beaker to cover the constricted end of the glass delivery tube by about 1 cm.

4.   Blow into flexible tube **A** in order to force water into flexible tube **B**, filling it. Allow excess water to run into the beaker. Then raise and lower the beaker several times to move water back and forth through flexible tube **B**, in order to remove all air bubbles.
Be certain the constricted end of glass tube C remains below the beaker water surface during these manipulations.

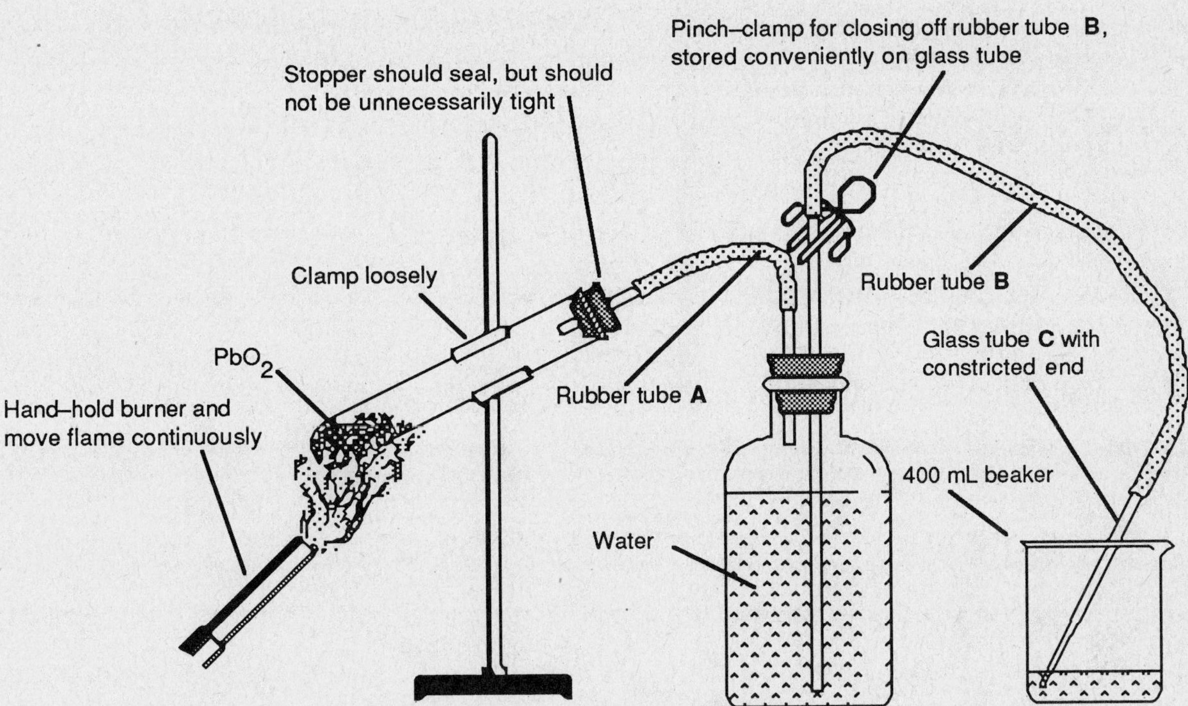

Figure 7.1. Apparatus for generating oxygen at known pressure, volume, and temperature.

5.   Raise the beaker until the water level in the bottle is a few mm below the short glass tube projecting into the bottle.
     Water must not enter this glass tube.

6.   Close flexible tube **B** with the pinch clamp and lower the beaker to the table top. The constriction in tube **C** prevents water below the pinch clamp from draining out. Do not worry if a few drops drain out.

7.   Attach the weighed test tube containing PbO$_2$ to the stopper on tube **A**. The stopper should fit just tightly enough to prevent gas leaks. Avoid making it too tight.
     Do not tighten the clamp around the test tube.
The clamp should support the tube but should not prevent the tube from sliding in the clamp. These precautions are intended to allow the test tube and stopper to separate before an explosion can occur, in case of a pressure buildup.

8.   Check for leaky connections by opening the pinch clamp. Some water will flow into the beaker, but if there are no leaks, there will be no *continuing* flow of water out of the bottle into the beaker, even with a large difference in water level heights.

9.   The pressure in the test tube and bottle must be the same as the external atmospheric pressure, so that you can measure it with a barometer. This condition is established by raising the beaker until the water levels in the beaker and bottle are the same.
     **Be certain the constricted end of glass tube C remains below the beaker water surface during these
     manipulations.**
Then close the pinch clamp on rubber tube **B**, to seal atmospheric pressure inside your apparatus, and remove constricted glass tube **C** from the beaker.

10.   Empty the beaker into a sink, draining it completely, **but do not dry it.** Leaving residual water on the beaker walls will improve your volume measurement, because approximately the same amount of water will remain on the walls when you pour the water, displaced into the beaker by evolved oxygen, into a graduated cylinder to measure its volume.

11.   Replace constricted glass tube **C** into the beaker.
Remove the pinch clamp. This is very important.
Do not worry if a small amount of water flows into the beaker before the flow stops. Leave this water in the beaker. This water will be measured as part of the displaced water and will not affect the accuracy of the experiment.
Obtain the instructor's approval of your apparatus before proceeding.

12.   Holding the gas burner in your hand, heat the PbO_2 gently at first, and then more strongly, until you can see that O_2 is being evolved fast enough to maintain a moderate flow rate of water from the bottle to the beaker.
Do not heat the PbO_2 in only one spot. Move the flame about to prevent PbO that is formed from melting and breaking the test tube.

13.   Continue heating the PbO_2 until you have collected about 300 mL of water in the beaker.

14.   Then remove the flame and allow the test tube to cool to room temperature. It will take about 10-15 minutes. As the tube cools, some water is drawn back into the bottle from the beaker, lowering the water level in the beaker.
Be careful that the tip of glass tube C remains below the water level in the beaker, during the cooling period.

15.   After the test tube has cooled to room temperature, repeat Step 9, equalizing the internal and atmospheric pressures by adjusting the beaker height until water levels in the bottle and beaker are equal.
Be certain the constricted end of glass tube C remains below the beaker water surface during these manipulations.

16.   When water levels in the beaker and bottle are equal, replace the pinch clamp on rubber tube **B**.

17.   Remove the test tube and its contents and weigh them carefully. Record the mass on the Data sheet.
The mass loss of the test tube contents is equal to the mass of oxygen evolved.

18.   Pour water collected in the beaker into a clean, dry 50 mL graduated cylinder and carefully measure the volume. Drain the beaker completely. Record the volume on the Data sheet.
The volume of water equals the volume of O_2 generated at room temperature and atmospheric pressure.

19.   Obtain a current barometric pressure reading to determine the atmospheric pressure. Your instructor will either provide the value or explain how to get it. Record the pressure on the Data sheet.
The barometric pressure reading is equal to the total pressure of the oxygen-water vapor mixture in the gas-collection bottle.

20.   Measure room temperature by allowing a thermometer to come to thermal equilibrium with the room air. Record the temperature on the Data sheet.
Assume that the thermometer reading is the temperature of your collected gas.

21.   Determine the vapor pressure of water at your room temperature from the table in the Reference Section of this manual. Record the value on the Data sheet.

22.   Do not attempt to clean your test tube. Your instructor will show you where to discard it.

23.   Calculate a value for R_{exp} from your data. Be sure to correct for the vapor pressure of water when determining the O_2 pressure. Record your value for R_{exp} on the Data sheet and also write it on the blackboard where your instructor indicates.

24.   Before leaving the laboratory, copy from the blackboard all the values for R_{exp} that were posted by your classmates. Enter these values, with your own value, on the Data sheet.

Name ______________________________________   Date ____________

EXPERIMENT 7
PRELABORATORY EXERCISE

1.   Oxygen gas is collected over water at 22.0°C. The pressure in the collection bottle is equilibrated with atmospheric pressure, which is measured to be 749.8 torr. What is the pressure of O_2 in the collection bottle?

____

2.   Calculate the value for R_{exp} (including the limits of uncertainty) which is indicated by the following measurements on a sample of carbon dioxide gas, CO_2:

$$M_{CO_2} = 44.01 \text{ (handbook value)}$$

$$m = 3.7446 \pm 0.0001 \text{ g}$$
$$P = 758 \pm 0.1 \text{ torr}$$
$$V = 1.898 \pm 0.005 \text{ L}$$
$$T = 0°C \pm 1°C$$

R_{exp} = ___________ ± ___________

What does this experiment indicate about the ideality of CO_2 gas under the experimental conditions?

Calculations:

Name _______________________________________   Date ____________

EXPERIMENT 7

DATA
(Observe significant figures in all calculations.)

1. Mass of test tube and PbO_2:   __________ ± __________

2. Mass of test tube and residue after heating: __________ ± __________

3. Mass of oxygen evolved: __________ ± __________

4. Volume of water displaced from bottle into beaker (equals O_2 volume):   __________ ± __________

5. Barometric pressure: __________ ± __________

6. Room temperature: __________ ± __________

7. Vapor pressure of water at room temperature: __________ ± __________

8. Pressure of O_2 (calculated from Dalton's law of partial pressures)

__________ ± __________

R_{exp} (calculated from above data) =   __________ ± __________

Calculations: (M_{O_2} = 32.00)

List the values of R_{exp} obtained by your classmates, plus your value.

1. __________ (your value)	11. __________	21. __________
2. __________	12. __________	22. __________
3. __________	13. __________	23. __________
4. __________	14. __________	24. __________
5. __________	15. __________	25. __________
6. __________	16. __________	26. __________
7. __________	17. __________	27. __________
8. __________	18. __________	28. __________
9. __________	19. __________	29. __________
10. __________	20. __________	30. __________

$R_{exp} \pm 1$ **std. dev.** (calculated from all measurements) =   __________ ± __________

Calculations:

EXPERIMENT 7
QUESTIONS

1.   What conclusion can you draw about the ideality of oxygen gas at your experimental conditions,

    a.   from your measurement alone?

    b.   from the average of all the measurements made in your class?

2.   Compare the limits of uncertainty for **R** calculated from your single measurement with uncertainty limits based on the measurements made by your entire class.

3.   What would have been the effect (+, -, or 0) on the calculated value of **R**, of each of the following situations?

    a. You neglected to take into account the vapor pressure of water.   _______

    b. The PbO_2 did not decompose completely.   _______

    c. Oxygen leaked out around the stopper of the test tube.   _______

    d. The PbO_2 contained an inert impurity (e.g. sand)   _______

    e. The PbO_2 contained an impurity that evolved oxygen (e.g., $KClO_3$)   _______

    f. The receiving beaker had not been drained thoroughly of water before the experiment began.   _______

    g. The PbO_2 was not completely dry before the experiment began.   _______

EXPERIMENT 8
Molar Mass of a Vapor by the Dumas Method

PRELABORATORY PREPARATION

Do the Prelaboratory Exercise and turn it in at the beginning of your laboratory period.

INTRODUCTION

The Dumas method is a simple and direct way to determine the molar mass of a volatile organic liquid. By itself, the Dumas method is not very accurate and is used only to obtain approximate results. **However, the Dumas method may be used in combination with an empirical formula, obtained by making an elemental analysis of the compound, to yield a molecular formula and an accurate molar mass for the vapor of the organic liquid.** If it is known that the compound neither dissociates nor associates (forming dimers) in its vapor form, then the molar mass and formula determined for the vapor is correct, also, for the liquid. The method is based on **Avogadro's law:**

Equal volumes of gases at the same temperature and pressure contain equal numbers of molecules.

Avogadro's law allows the proportionality constant **c** in Boyle's law, **PV = cT**, to be interpreted as the number of moles of gas (**n**), times a universal constant (**R**). This, of course, gives the ideal gas law:

$$PV = nRT \tag{8.1}$$

For this experiment, the quantity **n** = grams of sample/molar mass = m/M. When **m/M** is substituted for **n** in Equation 8.1, the equation can be rearranged to give Equation 8.2.

$$\text{molar mass} = M = m\,\frac{RT}{PV} \tag{8.2}$$

where **M** = molar mass, **m** = mass of sample (g), **T** = temperature (K), **P** = pressure (atm), **V** = volume (L), and **R** = 0.082 L atm K^{-1} mol^{-1}. With these units for **P** and **R**, Equation 8.2 allows the molar mass of the vapor to be calculated directly in terms of quantities which will be measured in this experiment.

EXAMPLE 1:

In a Dumas experiment, a volatile organic liquid was vaporized and measurements were made to determine the mass, pressure, volume, and temperature of the vapor. These measured quantities were found to be:

mass of vapor = 989 mg   pressure of vapor = 746.2 torr;
volume of vapor = 218 mL   temperature of vapor = 99.5°C = 372.7 K

Calculate the molar mass indicated by the experiment.

SOLUTION:

Substituting the measured quantities into Equation 8.2 gives:

$$M = \frac{(0.989\ \text{g})(0.0821\ \text{L atm}\ K^{-1}\ mol^{-1})(372.7\ \text{K})}{\left[\dfrac{746.2\ \text{torr}}{760\ \text{torr atm}^{-1}}\right]\left[\dfrac{218\ \text{mL}}{1000\ \text{mL L}^{-1}}\right]} = 141\ \text{g}\ mol^{-1}$$

The molar mass, as determined in the above example, must be considered only a rough approximation, because of the inherent poor precision of the Dumas experiment. One of the sources of error is the assumption that the vapor behaves as an ideal gas. This assumption alone can introduce an error of 5 to 10 percent, because organic vapors often are markedly nonideal at pressures near atmospheric and temperatures only a little higher than the normal boiling point. In addition, measurement uncertainties can introduce an additional margin of error. The Dumas experiment is, nevertheless, a useful technique when its approximate value for a molar mass is combined with additional information obtained from an elemental analysis of the organic compound. Then, the true molar mass and chemical formula can be determined quite accurately.

A Dumas measurement and an elemental analysis, in combination, can yield unambiguous values for the molar mass and chemical formula of an unknown compound.

The next example illustrates how to use an elemental analysis in combination with a Dumas determination, to improve the accuracy of both.

EXAMPLE 2:

The elemental analysis of the same organic compound used in the Dumas experiment of Example 1, shows it to be composed of: 17.8 mass% C, 2.1 mass% H, and 78.5 mass% Cl. (Don't be surprised that that this adds up to only 98.4 mass%, instead of 100 mass%. Experimental measurements can be "perfect" only by accident.)

(a)   Determine the empirical, or simplest, formula for the compound from the elemental analysis.
(b)   Use the results of (a), in combination with the answer to Example 1, to determine the correct molar mass and chemical formula.

SOLUTION:

(a)   Convert the mass percentages to the number of grams of each element that would be contained in 100 g of the compound. 100 g of compound will contain: 17.8 g C, 2.1 g H, and 78.5 g Cl. Change these masses to moles, to obtain the molar ratios of elements in the compound.

$$17.8\,g\ C = \frac{17.8\,g}{12.0\,g/mol} = \underline{1.48\ mol\ C};\ 2.1\,g\ H = \frac{2.1\,g}{1.01\,g/mol} = \underline{2.1\ mol\ H};\ 78.5\,g\ Cl = \frac{78.5\,g}{35.5\,g/mol} = \underline{2.21\ mol\ C}$$

This gives a molar ratio composition of: 1.48 C : 2.1 H : 2.2 1 Cl.

Note that the ratio contains non-integral numbers. Because the molecule cannot contain fractional atoms, we must find the smallest set of *integral* numbers that are in the same ratio as the molecular composition.

The easiest way to find such a set is:

1.   Divide each number in the non-integral set by the smallest number in the set.

$$\frac{1.48}{1.48} : \frac{2.1}{1.48} : \frac{2.21}{1.48} = 1.00 : 1.4 : 1.49$$

2.   Look for the smallest integral multiplier that brings all the numbers as close as possible to integers. **All** numbers in the set should be within about ±0.2 of being integral. Multiply the set 1.00 : 1.4 : 1.49 by 2, to obtain 2.00 : 2.8 : 2.98. Each number in the new set is within ±0.2 of an integer, so we may round the new set to the integer set:

2 : 3 : 3

Finding the Empirical Formula

A molecule of the compound must have C, H, and Cl in an atom ratio of 2 : 3 : 3, giving an **empirical formula** of: $C_2H_3Cl_3$.

Finding the Molecular Formula

The true **molecular formula** must be some integral multiple of the empirical formula. Therefore, possible molecular formulas are: $C_2H_3Cl_3$, $C_4H_6Cl_6$, $C_6H_9Cl_9$, etc.

We now compare molar masses of possible correct formulas with the molar mass that was determined in the Dumas experiment, to see which one matches most closely.

The Dumas experiment yielded a molar mass of 141 g mol^{-1}. The molar mass of $C_2H_3Cl_3$ is 133.5 g mol^{-1} and that of $C_4H_6Cl_6$ is 267 g mol^{-1}. The closest value to the Dumas result of 141 g mol^{-1} is 133.5 g mol^{-1}. The *only* molecular formula and molar mass that are compatible with both the Dumas experiment and the elemental analysis are:

$$C_2H_3Cl_3 \ \text{and} \ M = 133.5 \text{ g mol}^{-1}.$$

PLAN OF EXPERIMENT

As Example 1 demonstrated, the Dumas experiment consists of vaporizing a volatile organic liquid and measuring the mass, volume, pressure, and temperature of the vapor. An organic liquid, with a boiling point lower than that of water, is placed in a 250 mL Erlenmeyer flask that has only one, very small, opening in it. This liquid is boiled by placing the flask into a boiling water bath. As the sample vaporizes, excess vapor escapes through the small hole, preventing a pressure buildup. The motion of the vapor, moving from the liquid in the bottom of the flask out through the small hole, sweeps with it the air that originally was in the flask.

If you start with a large enough sample of liquid, all of the air eventually will be swept out and the flask will contain only the organic sample. Continued heating finally will evaporate all of the organic liquid and the flask will have only organic vapor in it.

The small size of the hole and the continued outward flow of vapor prevents any significant back-diffusion of air into the flask during the evaporation period. When the flask holds only the organic vapor, most of the needed measurements are easily found:

    a.   The **pressure** of the vapor is equal to the outside atmospheric pressure, having equilibrated through the small opening. The atmospheric pressure is found by reading it from a barometer.

    b.   The **volume** of the vapor is equal to the volume of the flask.

    c.   The **temperature** of the vapor is equal to the temperature of the water bath, because the vapor container is almost entirely immersed in the water bath.

    d.   The **mass** of the vapor is found by condensing the vapor back to a liquid and weighing it.

IMPORTANT EXPERIMENTAL PRECAUTIONS

Water Bath.

Do not allow the water–bath to boil strongly. Keep it just barely boiling

Otherwise, steam over the bath will obscure your observations of the escaping organic vapor. Also, you may lose so much water from the bath that a refill is needed to keep the sample flask covered with water. Only water already boiling can be added, so that the bath temperature does not drop. (See below, under **Pressure Determination, part (b).**

Pressure Determination.

Two precautions must be carefully observed.

1.   The internal and external pressures must be equilibrated through the small hole. As long as any liquid remains, the rapid production of vapor will cause the inside pressure to be higher than atmospheric, forcing vapor to flow out through the hole.

The inside and outside pressures are equal when all the liquid has evaporated and vapor has stopped flowing out through the hole.

One of the challenges of this experiment is to accurately observe the moment when vapor flow has ceased. When vapor no longer flows from the hole, continue heating for about two more minutes, to bring all the vapor into thermal equilibrium with the walls of the flask.

Only when no vapor has flowed out for about two minutes is the pressure of the vapor equal to the pressure of the outside atmosphere and the vapor temperature equal to that of the water bath.

2.   By Dalton's law of partial pressures, the total pressure in the flask is: $P(\text{total}) = P(\text{atm}) = P(\text{vapor}) + P(\text{air})$. It is essential that no air enter the flask during the heating period. If air is in the flask when the flow of vapor out through the hole ceases, $P(\text{air})$ will not be zero and you will find the wrong value for $P(\text{vapor})$.

You must insure that $P(\text{air}) = 0$ by paying close attention to the following:

-    **First,** you must be sure to start the experiment with enough liquid sample, to insure that all the air originally in the flask is swept out during the heating period.

-    **Second,** you must be careful that the temperature of the water bath *never* decreases once vaporization has begun.

Even a brief temperature drop will create a pressure drop in the flask, causing air to be drawn in from the outside.

If this were to happen when not enough liquid remained to make sufficient vapor for expelling this air, your vapor pressure measurement would be in error. On the other hand, it does not matter at all if the water bath temperature continues to rise during vaporization, toward the bath final temperature.

Volume Determination.
   The volume of the vapor is the same as the volume of its flask.
The flask volume is found by weighing it when dry and empty (full of air only), then filling it with tap water and weighing it again. It is important to carefully fill the flask completely with bubble-free water. Use a clean flask and fill it slowly to avoid air bubbles. The outside of the flask must be dry.
   The flask volume is the mass of the water filling the flask, divided by the density of water at the water temperature.
The density of water is tabulated in the Reference Section in the front of the manual. The mass of air filling the "empty" flask is negligible, compared with the mass of the water, and may be ignored.

Temperature Determination.
   You will use a volatile liquid that has a boiling temperature below that of water, so that it can be vaporized in a boiling water bath.
   The sample flask must be immersed in the water bath as completely as possible, to insure a uniform temperature throughout the vapor.
 While the organic sample is boiling, the temperatures of the liquid and the vapor just leaving the surface are somewhat cooler than the water bath. When the last liquid has evaporated, it will take a short time for the vapor temperature to equilibrate with the water bath temperature, through molecular collisions with the flask wall. This will take about 2 minutes. Then, the vapor temperature may be taken to be the same as the water bath temperature.

Mass Determination.
   If all goes well, you will reach a point in the experiment where you have a flask full of pure organic vapor at known pressure, volume, and temperature, with no liquid or air present. Vapor has ceased flowing from the flask opening, the pressure and temperature have equilibrated, and no air has reentered the container.
   At just this moment, you must stop heating the sample and quickly condense the vapor back to a liquid by immersing it in an ice bath.
We can find the mass of the liquid by weighing the sample flask containing the liquid and comparing its mass with that of the empty flask. When the sample is in liquid form, loss of sample by evaporation is greatly reduced during weighing. If the vent hole is small enough, losses from the flask after condensation will not cause serious error.
   Minimize evaporation loss after condensation by being careful to avoid agitation of the sample with rough movements.

Calculation of Sample Mass
   When the vapor is condensed back to a liquid, air will reenter the flask. The flask then contains air, organic liquid, and organic vapor. The **partial pressure** of the organic vapor is equal to the vapor pressure of your sample at its cooled temperature. The **mass** of the vapor that filled the flask before being condensed now is equal to the mass of the organic liquid, plus its vapor, in the cooled flask.
   The mass of the flask, plus its contents, includes the air that has reentered upon cooling. Because organic vapor is present after cooling, not as much air will reenter the flask as was originally present during the "empty" weighing of the flask. Therefore, the mass of vapor that filled the flask before cooling, is only **approximately** equal to the difference between the mass of the cooled flask with the sample in it and the mass of the empty flask; the flask containing the sample has less air in it, after cooling, than did the empty flask. Because the mass of the displaced air is a significant quantity, it is necessary to correct for the missing air[1].
   It is essential to correct for sample volatility by taking into account the mass of air displaced by sample vapor.

Correcting for Sample Volatility.
   The total pressure in the flask during weighing is: $P(atm) = P(vapor) + P(air)$, where $P(atm)$ = external atmospheric pressure, $P(vapor)$ = vapor pressure of the organic sample at room temperature, and $P(air)$ = partial pressure of air in the flask.
   - In the initial weighing of the "empty" flask: $P_i(vapor) = 0$, and $P_i(air)$ = atmospheric pressure.
   - In the final weighing of the flask plus sample:
        $P_f(vapor)$ = vapor pressure of sample liquid and $P_f(air) = P_i(air) - P_f(vapor)$.

---

[1] The volume of the condensed liquid sample also displaces some air. This volume, however, is very small compared to the total flask volume and it is not necessary to correct for it in this experiment.

Note that the partial pressure of air is *reduced* by an amount equal to P(vapor).
Proof of the last statement above is given by the following reasoning: The decrease in the partial pressure of air, $\Delta P(air)$, between weighings, is:

$\Delta P(air) = P(air, sample flask) - P(air, empty flask)$
$P(air, sample flask) = P(atm) - P(vapor),$ and $P(air, empty flask) = P(atm),$ so that:
$\Delta P(air) = [P(atm) - P(vapor)] - P(atm)$
$\Delta P(air) = -P(vapor)$ (The negative sign indicates that the partial pressure of air has decreased.)
The mass of this missing air is:

$$\Delta m(air) = \rho V[P(vapor)/P(atm)] \tag{8.3}$$

where: $\Delta m(air)$ = mass change of air in the flask, between the empty weighing and the weighing with sample;

ρ = density of air = 0.0012 g/mL, at room temperature near atmospheric pressure

V = volume of the flask, and P(atm) = atmospheric pressure determined from a barometer reading.

The mass of the missing air, $\Delta m(air)$, must be *added* to the mass gain of the sample flask in order to find the true mass of the condensed vapor, as in Equation 8.4.

$$m(vapor) = (mass of flask with sample) - (mass of empty flask) + \Delta m(air) \tag{8.4}$$

Here, m(vapor) = mass of organic vapor and is the sample mass used in Equation 8.2. The vapor pressure of your sample will be given to you by your instructor.

SAFETY
1. Wear approved eye protection.
2. Add one or two boiling chips to the water bath, to avoid bumping.
3. Many organic compounds and their vapors are toxic and skin irritating. Avoid inhalation and skin contact. Carry out the evaporation under a desk fume hood, if available. Otherwise, make one with a funnel and vacuum tubing attached to an aspirator.
4. Some unknowns may be halocarbons, which can react with blood alcohol to cause liver damage. Avoid consumption of alcohol for 24 hours before and after this experiment.
5. Some samples may be flammable. Pour your sample well away from any open flames and do not allow any flames near the venting vapor.
6. Place a split stopper around the thermometer to cushion it before clamping it in the water bath.

PROCEDURE
1.   Obtain a square of aluminum foil, about 6 x 6 cm square, a 15-20 cm length of soft copper wire, and a clean, dry 250 mL Erlenmeyer flask. The aluminum foil and wire will be used to close the flask opening, as in Figure 8.1.

2.   Place the aluminum foil on a piece of cardboard or pad of paper and pierce a tiny hole through the center with a needle or straight pin. Only the very tip of the pin should penetrate the foil.

3.   Weigh the flask, aluminum foil, and copper wire all together to the nearest milligram. Do not fold the foil or wire. Make this weighing twice, to get an average value. Record the masses on the Data sheet.

4.   Obtain about 8 mL of unknown sample from your instructor in a clean, dry, labeled, and rubber stoppered test tube. Mark the unknown identification number on your test tube and, also, at the top of the Data sheet. The sample vapor may be toxic; keep the sample stoppered until you use it.

5.   Obtain the elemental analysis and boiling temperature of your unknown from your instructor. Write this information at the top of your Data sheet.

6.   Introduce your sample into the Erlenmeyer flask.[2] Place the aluminum foil over the flask mouth, centering the tiny hole, and fold down the sides to cap the flask opening, as shown in Figure 8.1. Wrap at least two turns of copper wire tightly over the foil around the neck of the flask and twist the ends, to secure the foil and seal the flask.

---

[2] Your instructor may have you add a tiny bit of an azo dye, to color the sample and make it easier to tell when all the liquid has evaporated.

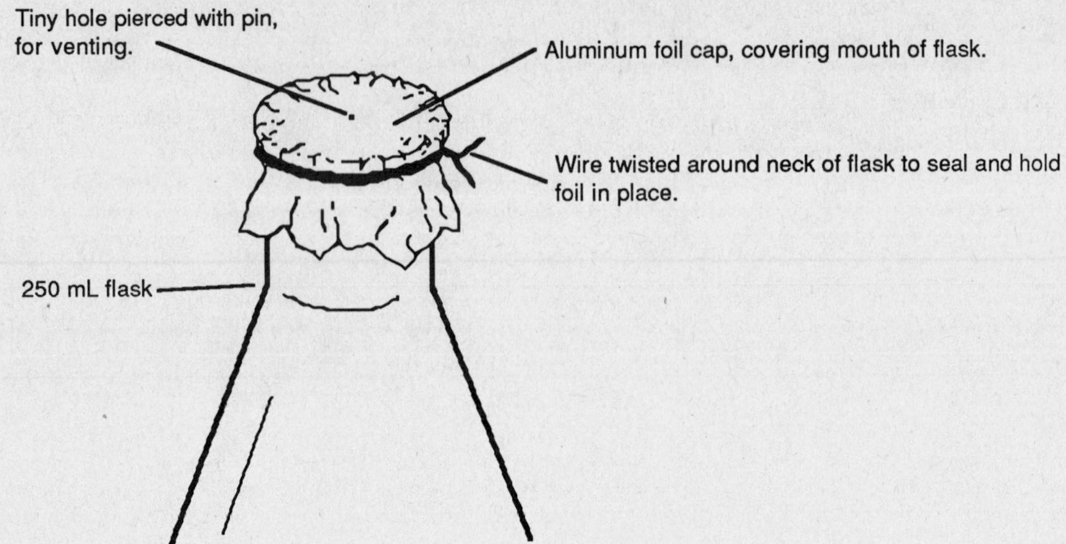

Figure 8.1.   Method for closing Dumas flask with a tiny vent hole. Place a 6 x 6 cm square of aluminum foil on a piece of cardboard and, with a needle or pin, make the smallest hole possible in its center. Cover an Erlenmeyer flask with the foil and hold it in place with a piece of copper wire.

7.   Set up a water heating bath, as in Figure 8.2; use a 1000 mL beaker for the bath and clamp the capped Dumas flask in its place in the beaker *before* adding any water. Position the flask so that it will be as completely immersed as possible.
   Clamp the flask in a slightly tilted position, so water vapor bubbles cannot collect under its bottom.

8.   Your instructor may have provided a plastic basin or sink full of ice water, for condensing the vapor, when heating is finished. If not, half fill another large beaker with cold water for this purpose.

9.   Fill the water bath beaker with water to within 1 cm of its top.
   Be careful not to get any water on the foil cap.
Add 3 or 4 boiling chips and 5 drops of concentrated hydrochloric acid to the bath water. The acid will prevent the formation of mineral deposits on the outside of the flask during heating. Turn on the aspirator, if you have improvised a funnel fume hood.

10.   Turn on the hot plate and heat the water bath slowly to boiling. At some point before the water boils, your sample in the flask should begin to boil. When this occurs, turn the hot plate lower to slow the heating rate.
   Do not let the water bath boil strongly. Keep it at a slow, gentle boil so water vapor will not obscure your view of the escaping organic vapor.
Observe the neck of the flask for organic condensate. If any sample condenses in the neck area, warm the neck carefully and slowly with an electric hot air blower (*not a flame !*), if available, to evaporate the droplets. Alternatively, you can make an aluminum foil collar around the flask, as in Figure 8.2.
   You may have to keep the bath water level high with additional water to prevent condensation in the flask neck. If you must add water to the bath, it is very important that you add hot water frequently in *small* quantities, so that boiling does not stop and the bath temperature does not decrease.
   Once heating begins, it is very important that the temperature of the bath and flask never be allowed to decrease, until you are ready to condense the sample vapor, so that air will not be drawn back into the flask.
Take care that boiling water does not splash up into crevices in your aluminum foil cap.

11.   Observe the escaping vapor by watching the flask vent hole against a bright background, such as a light or a window.
        A small test tube containing cold water makes a good vapor detector. Hold it over the vent and watch for vapor condensation.
Watch carefully, to determine when all the liquid has evaporated and vapor has stopped flowing from the vent.
        When vapor stops coming out, continue heating for 2 min more, to allow the vapor temperature to equilibrate with the bath temperature.

12.   During the 2 min equilibration time, read and record the temperature of the water bath on the Data sheet. Then remove the thermometer from the bath and swing the fume hood to one side.

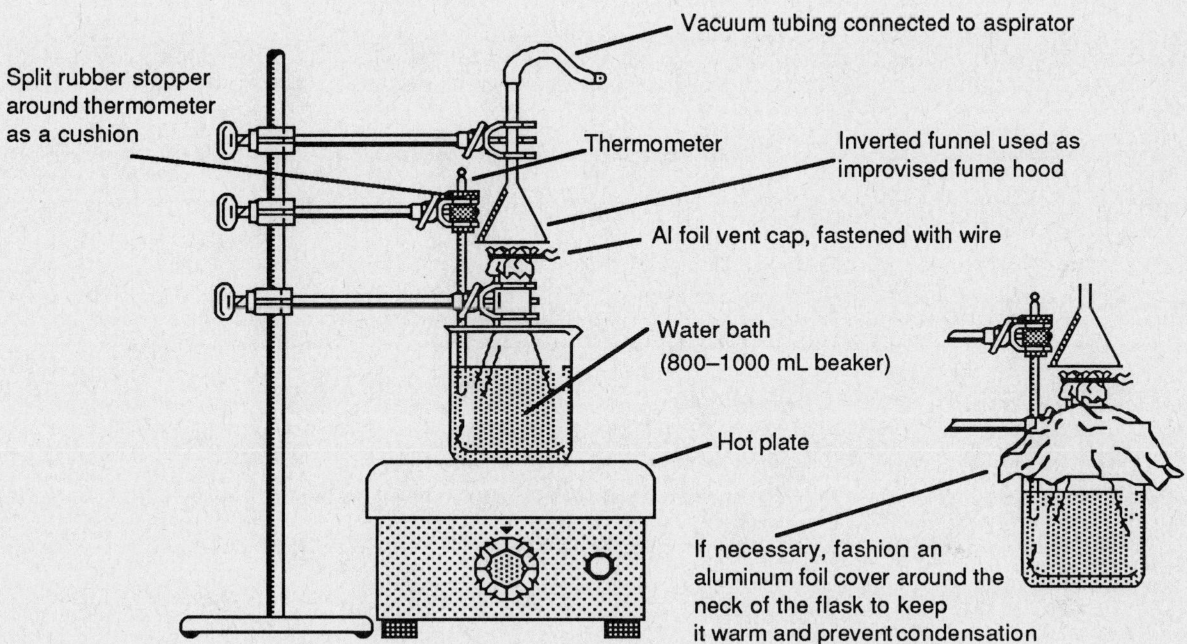

Figure 8.2.   Assembly for Dumas method of molecular weight determination.

13.   Two minutes after vapor flow has stopped, remove the flask from the bath by unfastening its clamp from the support rod and then using the clamp as a handle to carry the flask.

14.   Condense the vapor quickly by immersing just the bottom of the flask in cold water or an ice bath. Then, remove the flask from the quenching bath and let it cool to room temperature by standing for 10 minutes. During the cooling period, there will be a steady inflow of air, which prevents any significant diffusion of sample vapor out of the flask.

15.   While the flask is cooling, obtain the barometric pressure and record it on the Data sheet.

16.   After 10 min of cooling, gently dry the outside of the flask. Try not to agitate the sample. Allow it to stand another 5 min near the balances.

17.   Carefully avoiding agitation of the sample, weigh the flask with its condensed contents and the aluminum cap assembly, to the nearest milligram. Record this mass on the Data sheet.

18.   Carefully remove the copper wire and aluminum foil cap and weigh them to the nearest milligram. Record this mass on the Data sheet. If the foil cap and wire are not damaged, you may use them again for the duplicate measurement.

19.   Make a duplicate measurement, steps 3 to 18. Obtain another sample of the same unknown. If necessary, obtain new foil and wire for capping the flask. Weigh any new pieces to the nearest milligram and record the mass.
        Do not discard the condensate or clean the inside of the flask before the second measurement. New sample can be added to any residue without affecting the second measurement.
It is not necessary to reweigh the flask in its empty condition, because the mass measurements are, by far, the most precise of all the measurements in this experiment.

20.   After the second determination, discard the condensate in the container provided and rinse out the flask with tap water. Carefully dry the outside.

21.   Find the volume of the flask by filling it carefully with tap water from a wash bottle. Fill it right to the brim while it is standing on a triple beam balance. Be certain that no bubbles are trapped inside. Record the mass on the Data sheet. Empty the flask, refill it again, and repeat the weighing, to obtain an average value.

Name _______________________________________   Date _____________

EXPERIMENT 8
PRELABORATORY EXERCISE

1.   A Dumas experiment was performed on a volatile organic compound. The boiling point was found to be 42°C. The Dumas flask volume was 286 mL and the atmospheric pressure was 704.2 torr. The empty flask, with only air in it, weighed 239.5652 g. The flask with condensed sample in it weighed 240.3186 g. Using Equations 8.3 and 8.4, find the mass of condensed sample. (Density of air = 0.0012 g mL^{-1}.)

mass of condensed sample: _____________________

2.   In another Dumas experiment on a different compound, the water bath temperature was 98.7°C, the barometric pressure was 733.3 torr, the Dumas flask volume was 286 mL, and the mass of condensed liquid, already corrected for sample volatility, was 775.5 mg. Calculate the molar mass of the compound.

molar mass of the compound: _____________________

3.   Analysis of the same volatile organic compound used in Question 2 shows it to be 52.8 mass% carbon, 8.9 mass% hydrogen, and 36.4 mass% oxygen.
   a. What is its simplest integral empirical formula? (See Example 2, Introduction.)

empirical formula: _____________________

   b. Determine the true molar mass, consistent with both the elemental analysis and the Dumas experiment of Question 2.

true molar mass: _____________________

4.   Why must you wait 2 minutes after all the sample liquid has evaporated before condensing the sample vapor?

5.   Why should the water bath boil only very gently, instead of rapidly?

6.   Why will starting with too small a liquid sample cause a serious error?

7.   Why must the vent hole be as small as possible?

8.   Why is it important that the water bath temperature never decrease, once the sample has begun to vaporize?

9.   Why must the measured sample mass be corrected for sample volatility?

CALCULATIONS:

Name ____   Date ____________

EXPERIMENT 8

DATA
(Observe significant figures in all calculations.)

Unknown No.:   ____________

Boiling temperature of unknown: ____________

Elemental analysis (mass% composition):   ____

Measurement
1   **2**

1. Mass of flask assembly when empty: __________

2. Mass of room temperature flask
    (entire assembly + condensed sample):   __________   __________

3. Mass of vent cap + wire: __________   __________

4. Mass of condensed sample without correction
    for sample volatility: __________   __________

5. $\Delta m(air)$ (from Eq. 8.3):   __________   __________

6. Corrected mass of condensed sample: __________   __________

7. Temperature of boiling water bath: __________   __________

8. Barometric pressure: __________   __________

9. Mass of flask filled with water: __________   __________

10. Average mass of flask filled with water: __________

11. Water temperature: __________

    Density of water: __________

    Flask Volume: __________

12. Molar mass from Dumas experiment alone: __________   __________

    Average molar mass: __________

13. Integral empirical formula from elemental analysis: ________________________

14. True molar mass and formula: **molar mass:** ____________   **formula:** ________________

15. Percent error of Dumas measurement alone: __________

CALCULATIONS:

Name ____   Date ____________

EXPERIMENT 8
QUESTIONS

1.   State the error (+, -, 0) caused by each of the following on a molar mass determined by the Dumas method:

    a. Some water is trapped under the vent cap during weighing.   _____

    b. The vapor does not displace all of the air in the flask during evaporation.   _____

    c. During cooling, some vapor escapes from the flask.   _____

    d. The temperature of the vapor is a little lower than that of the water
        bath, when the flask is removed from the bath.   _____

2.   Is it possible for the molar mass of the vapor to be different than the molar mass of the liquid form of the same substance? Explain.

3.   Is the quantity of air in the flask at the end of the experiment the same as it was at the beginning? How does this affect the determined molar mass?

EXPERIMENT 9
Colligative Properties:
Determination of Molar Mass by
Freezing Point Depression

PRELABORATORY PREPARATION
Do the Prelaboratory Exercise and turn it in at the beginning of your laboratory period.

INTRODUCTION
Whenever a substance is dissolved in solvent, the vapor pressure of the solvent is lowered. As a result, the boiling temperature and freezing temperature are changed. The freezing point of the solution is lowered and, if a nonvolatile substance is the solute, the boiling point is raised, relative to the pure solvent. The magnitude of these changes depends only on the number of solute particles (molecules or ions) present in a given mass of solvent.
Properties of a solvent that depend only on the number of solute particles dissolved, and are independent of the nature of the particles, are called colligative properties.

Two practical applications of colligative properties are the sprinkling of household salt on icy streets and sidewalks to melt the ice by lowering its freezing temperature, and the addition of antifreeze to automobile cooling water to both lower its freezing temperature and raise its boiling temperature. For scientific purposes, colligative properties are very useful for determining molar masses of unknown compounds and the degree of association of known compounds in solution.
The relation between the molar mass of an unknown non-dissociating[1] solute and the freezing point depression of a solvent is given by:

$$\Delta T_f = K_f\, m \tag{9.1}$$

where ΔT_f = (freezing point of solution) - (freezing point of pure solvent), K_f = freezing point constant for the solvent, and m = **molality** of the solute in the solvent. Molality of solute is defined as:

$$m = \frac{\text{moles of solute}}{\text{kg of solvent}} = \frac{\text{grams of solvent/molar mass of solute}}{\text{kg of solvent}}$$

To determine molar masses by freezing point depression of a solvent, the K_f value of the solute must have been previously determined. This is done by dissolving a measured amount of a *known* compound into the solvent and measuring the new freezing point. Since both the solute molality and freezing point depression are known, K_f can be found from Equation 9.1 for the solvent being calibrated. It is important that the solute used for calibration does not dissociate, so that the number of solute particles is a well defined quantity, being equal to the number of molecules added.
The freezing point constant, K_f, is numerically equal to the number of degrees the freezing point of the solvent is lowered, when one mole of solute particles is present in 1000 g of solvent.

Freezing point constants of a few common solvents, determined by the procedure just described, are given in Table 9.1.

---

[1]The freezing point depression depends upon the number of solute *particles* added to the solvent. If the solute dissociates in the solvent, like an electrolyte in water, the number of particles is greater than the number of undissociated molecules originally added and the true molar mass will not be obtained.

Table 9.1: Freezing Point Constants for Some Common Solvents.

Solvent	Freezing point (°C)	K_f (kg solvent/mol solute)
Water	0.0	1.86
Naphthalene	80.2	6.9
Benzene	5.5	4.90
Camphor	178.4	37.7
para- dichlorobenzene	53.1	7.10
tert- butanol	25.5	9.1

EXAMPLE 1:
Determining the Amount of Freezing Point Depression from Kf, Molality, and Degree of Dissociation.

a. From the K_f values given in Table 9.1, you can determine that a solution made of one mole of methyl alcohol (undissociated) in 1000 g of water will have a freezing point of -1.86°C.

b. If one mole of the salt $BaCl_2$, which dissociates completely in water, were dissolved in 1000 g of water, there would be 3 moles of ion particles present. Thus, the freezing point would be depressed by 3 times 1.86 degrees, giving a freezing temperature for the solution of -5.58°C.

EXAMPLE 2:
Finding Molar Mass from the Freezing Point Depression.
A solution containing 15.00 g glucose in 100 g water is found to freeze at -1.53°C. What is the molar mass of glucose?

ANSWER:
From Table 9.1, K_f (water) = 1.86. Therefore the molality of the solution must be

$$m = \frac{\Delta T_f}{K_f} = \frac{1.53}{1.86} = 0.820 \text{ molal}$$

which means that there are 0.820 moles of glucose per 1000 g water in the solution. The solution, as prepared, contains 15.00 g glucose per 100 g water. Multiply the amounts by 10, to get the mass of glucose per 1000 g water. There would be 150.0 g glucose per 1000 g water, and our calculation of molality shows that this 150 g glucose must be equal to 0.820 moles of glucose. Therefore, the molar mass of glucose is:

$$M(glucose) = \frac{150 \text{ g}}{0.820 \text{ mol}} = 183 \text{ g mol}^{-1}$$

The true molar mass of glucose, determined from its formula $C_6H_{12}O_6$, is 180 g mol^{-1}. The percent error is:

$$\frac{(183 - 180)}{180} \times 100\% = 1.7\%$$

PLAN OF EXPERIMENT

The solvent into which you will dissolve an unknown compound, is *t*-butyl alcohol[2]. You will have to calibrate your thermometer; although it reads temperature *changes* quite accurately, it is not necessarily calibrated to give an accurate value for a particular temperature.

Because you cannot assume that your thermometer is sufficiently accurate to give the handbook value for the freezing temperature of your solvent, you must calibrate it by measuring your solvent freezing point.

To do this, you have to measure the cooling curve for your solvent, found by observing the rate at which the liquid and solid phases cool. The cooling curve is a time vs. temperature graph of the cooling behavior of your sample. This procedure is necessary because the freezing point often is difficult to determine by simple visual observation, due to the phenomenon of undercooling and the fact that solidification can occur over a broad temperature range.

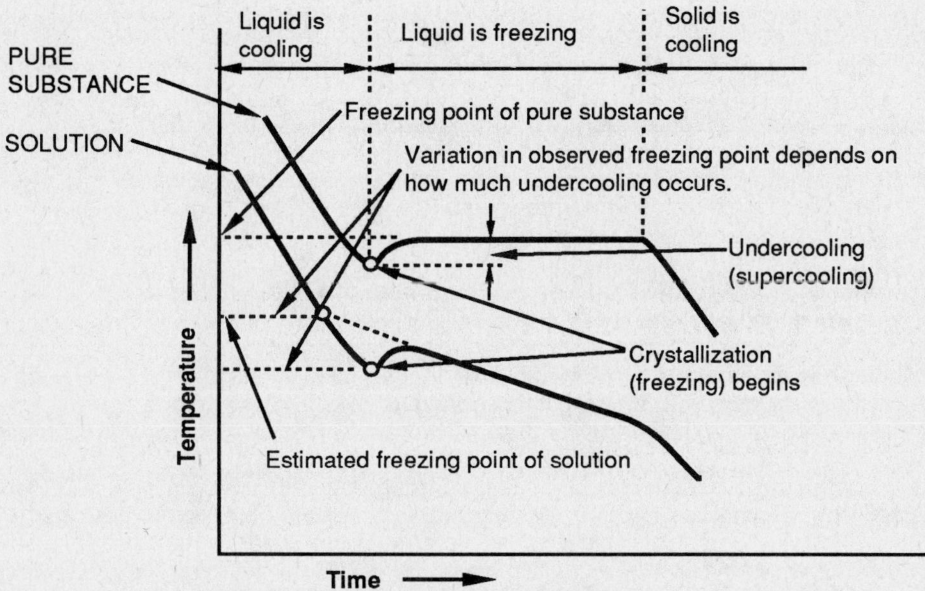

Figure 9.1. Typical cooling curves for pure and impure substances.

Figure 9.1 illustrates how to use the cooling curve to find the freezing point, by extrapolating the slope of the freezing part of the curve back to its intersection with the liquid cooling part. Notice, in Figure 9.1, that the cooling curves for a pure solvent and a solution are different. The freezing portion of the pure solvent is horizontal and straight, while that for the solution slopes downward. The reason for this is that, when a solution freezes, the solvent frequently freezes out first. The solid that forms is usually pure, or nearly pure, solvent, so that the solid material has a different composition than the liquid.

As solidification progresses and solvent leaves the liquid phase, the remaining liquid becomes increasingly more concentrated in solute.

The increase in solute concentration continues to lower the freezing point during solidification, and produces the sloping cooling curve in the crystallization region. If the slope is large, there can be an ambiguity in determining the exact onset of freezing.

Another cause for ambiguity in the freezing point is the phenomenon of undercooling, illustrated in the cooling curves of Figure 9.1. Sometimes undercooling can be minimized by "seeding" the solution with a small crystal of the solute, dropped into the solution at the expected freezing point. Stirring the solution also helps to reduce the amount of undercooling. Figure 9.1 shows the graphical method for circumventing these problems.

---

[2] Napthalene or *p*-dichlorobenzene may be used as solvents instead. Their freezing points and K_f values are compatible with the procedure described here. Exposure to *t*-butyl alcohol vapor may be somewhat less hazardous than for the other two solvents, and, therefore, we have recommended its use.

SAFETY

1. Wear approved eye protection.

2. Vapors of organic compounds always are potentially hazardous. Avoid breathing the vapors. If possible, heat the solvent in a fume hood.

PROCEDURE

A. Calibration of thermometer with pure *t*-butyl alcohol.

1. Weigh a large test tube to 0.01 g. Record the mass on the Data sheet.

2. Add about 25 g *t*-butyl alcohol and weigh the test tube and its contents. Record this mass on the Data sheet.

3. Fill a 600 mL beaker nearly full of tap water to serve as a water bath for melting the solvent. Clamp the test tube in the water bath, as shown in Figure 9.2, and heat the bath to melt the solvent. (*t*-butyl alcohol will melt at about 25°C, *p*-dichlorobenzene around 75°C, and naphthalene around 85°C.)

4. When most of the *t*-butyl alcohol has melted, insert your thermometer and stirrer into the test tube. The loop of the stirrer should encircle the thermometer bulb. Be sure the thermometer bulb is well covered with liquid *t*-butyl alcohol.

5. When all the *t*-butyl alcohol has melted, remove heat from the bath and dry the outside of the test tube with a towel. You will use the bath again.

6. Place the test tube in an empty 125 or 250 mL Erlenmeyer flask for measuring the cooling rate. The flask shields the tube from air currents, slowing the cooling rate and helping to keep the temperature uniform in the liquid.

7. Stir the liquid continuously as it cools, to minimize undercooling and help maintain a uniform temperature. Read and record the temperature every 30 seconds. If solidification is not complete when cooling has reached near room temperature, add small increments of cold tap water to the flask in which the test tube stands to continue the cooling curve.
    When the freezing point of *t*-butyl alcohol is reached, crystals will start to form and the temperature will remain constant. Shortly after this, the solvent will have solidified to a degree where you no longer can stir it.

8. Your cooling curve should resemble the "pure substance" curve in Figure 9.1. Repeat this cooling curve, by remelting and then cooling the same sample. Use the average melting point from your two measurements.

B. Determining the molar mass of an unknown compound.

1. Obtain an unknown sample from your instructor and record its number on the Data sheet.

2. Weigh the unknown, in its container, to 0.01 g. Record the mass.

3. Add about 3 g of unknown (use about 2 g if the solvent is naphthalene or *p*-dichlorobenzene) to the test tube containing the *t*-butyl alcohol for which you have just measured a cooling curve. Be certain that none of the unknown sticks to the side of the test tube.
    If some unknown does stick to the side of the test tube, immerse the test tube in the hot water bath to melt its contents. Then mix the unknown into the solvent by tilting and rotating the test tube.

4. Reweigh the unknown sample container and record the mass. Calculate the mass of solute added to the solvent.

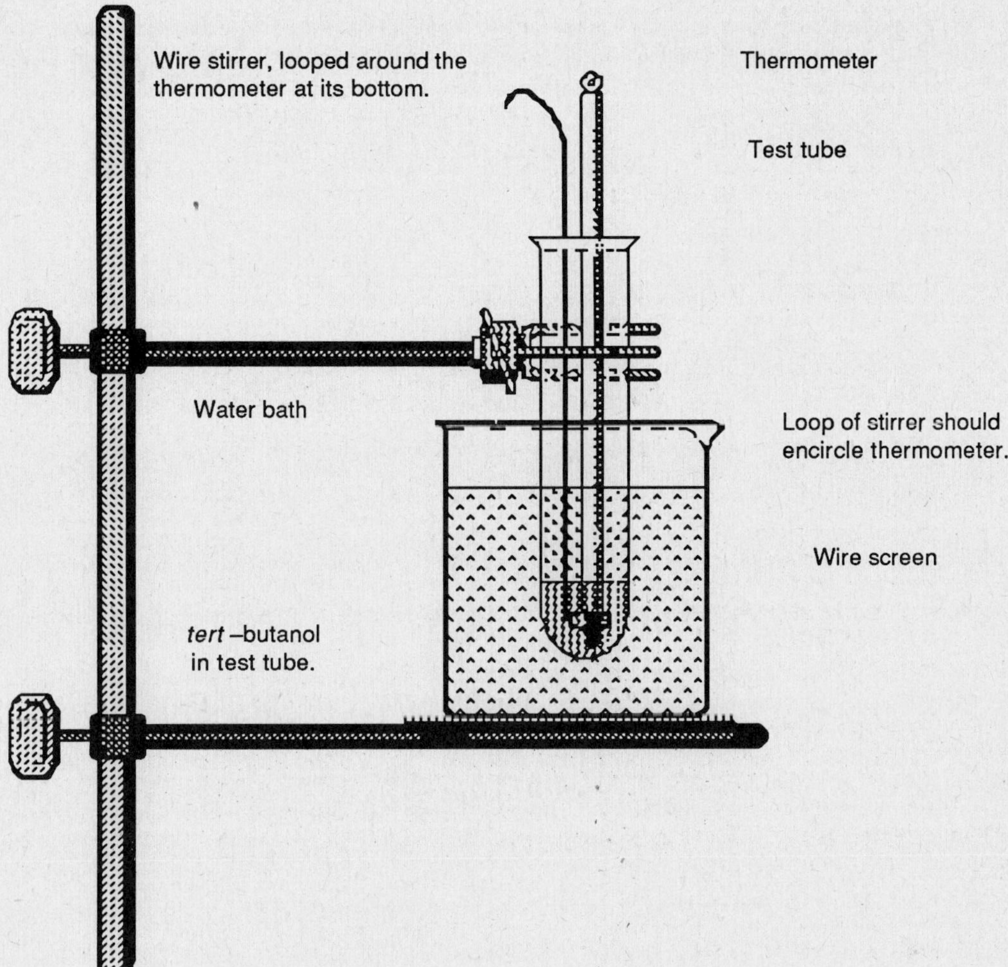

Figure 9.2. Apparatus for measuring solution cooling curves.

5. Heat the test tube in the water bath, as before, until all solid material is melted.

6. Remove the test tube from the bath and dry the outside. Allow the solution to air dry in an empty flask, while stirring the solution and recording the temperature every 30 seconds, as before. The cooling curve should resemble the "solution" curve in Figure 9.1.

7. Repeat the solution cooling curve measurement by remelting and cooling the same solution, so you have two freezing point values to average.

8. Calculate the molar mass of your unknown.

9. Dissolve your *t*-butyl alcohol solution in warm water and dispose of it into the container provided. (Naphthalene and *p*-dichlorobenzene can be remelted in the hot water bath for disposal. Remaining solid material can be scraped out of the test tube with a dry brush.)

Name ____   Date ____________

EXPERIMENT 9
PRELABORATORY EXERCISE

1.   Use Table 9.1 to calculate the molar mass of an unknown substance, if the freezing point of a solution containing 1.00 g of the unknown in 100 g of camphor is 14.8°C.   ____________________

2.   What is the expected freezing temperature of a solution containing 2.00 g CCl_4 in 100 g benzene? Use Table 9.1.   ____________________

3.   Explain briefly why the *solution* cooling curve, in Figure 9.1, slopes downward in the region following the beginning of crystallization.

Calculations:

Name _______________________________________   Date ____________

EXPERIMENT 9

DATA
(Observe significant figures in all calculations.)

Part A.

1. Mass of empty test tube: _______________

2. Mass of test tube with solvent: _______________

3. Mass of solvent: _______________

4. Temperature-time data:

RUN 1		RUN 2			**Calculations:**
t(s)	temp(°C)	t(s)	temp(°C)		
_____	_____	_____	_____		
_____	_____	_____	_____		
_____	_____	_____	_____		
_____	_____	_____	_____		
_____	_____	_____	_____		
_____	_____	_____	_____		
_____	_____	_____	_____		
_____	_____	_____	_____		
_____	_____	_____	_____		
_____	_____	_____	_____		
_____	_____	_____	_____		
_____	_____	_____	_____		
_____	_____	_____	_____		
_____	_____	_____	_____		
_____	_____	_____	_____		

5. Freezing point of solvent,   Run 1: ___________   Run 2: ___________

6. Average freezing point of solvent: ___________

DATA

Part B.

1.   Unknown No.:   ___________________

2.   Mass of unknown, in its container :   ___________________

3.   Mass of unknown and container, after transferring
some unknown to the solvent: ___________________

4.   Mass of unknown transferred to solvent: ___________________

5.   Temperature-time data:

RUN 1		RUN 2		Calculations:
t(s)	temp(°C)	t(s)	temp(°C)	
___	___	___	___	
___	___	___	___	
___	___	___	___	
___	___	___	___	
___	___	___	___	
___	___	___	___	
___	___	___	___	
___	___	___	___	
___	___	___	___	
___	___	___	___	
___	___	___	___	
___	___	___	___	
___	___	___	___	
___	___	___	___	
___	___	___	___	

6.   Freezing point of solvent,   Run 1: ___________   Run 2: ___________

7.   Average freezing point of solvent: ___________

8.   Freezing point depression: ___________

9.   Molar mass of unknown: ___________

Name ____   Date ____________

EXPERIMENT 9
QUESTIONS

1.   What error in your calculated molar mass (+, -, 0) would be caused by the following:

    a.   Some solvent is lost during heating by volatilization.   ____________

    b.   Some unknown is lost during transfer to the solvent.   ____________

    c.   The thermometer used for calibration of the solvent
    and measurement of the solution reads 1°C too low.   ____________

2.   Assuming that NaCl is completely ionized, what would be the freezing point of a 1.00 molal solution of NaCl in water?   ____________

3.   If you have only a small amount of an unknown (perhaps about 0.1 g), that is soluble in all of the solvents listed in Table 9.1, which solvent should you choose for the most accurate molar mass determination by the freezing point depression method? Explain your choice.

Calculations:

Name ____ Date ____________

EXPERIMENT 9
(Additional graph paper is at the back of this book.)

EXPERIMENT 10
Water of Hydration

PRELABORATORY PREPARATION
Do the Prelaboratory Exercise and turn it in at the beginning of your laboratory period.

INTRODUCTION
Water has a strong attraction for many compounds because of its polar character and electronic structure. Because water is a normal component of our atmosphere, most compounds will contain some dissolved or adsorbed water. In some cases, water molecules become chemically bound in a compound, usually an ionic salt, so that the water molecules become part of the crystal lattice. **In such compounds, called hydrates, the water molecules are in a definite proportion relative to the other atoms and must be included as part of the chemical formula.**

Copper(II) sulfate pentahydrate[1] is such a compound. It may be written either as $CuSO_4(H_2O)_5$ or $CuSO_4 \cdot 5H_2O$.

The water in hydrates often is loosely bound and may be driven off by heating the solid. If $CuSO_4(H_2O)_5$ is heated until all the water is driven off, the remaining $CuSO_4$ crystals are called anhydrous copper(II) sulfate. If the anhydrous crystals are allowed to stand in air, they will absorb water from the air continuously, until the pentahydrate is formed. If an aqueous solution of Cu^{2+} and SO_4^{2-} ions is evaporated, $CuSO_4(H_2O)_5$ will crystallize directly. The behavior described for $CuSO_4(H_2O)_5$ is typical of many other hydrates. Some examples are $NiSO_4(H_2O)_7$, $NaCO_3(H_2O)_{10}$, $CaSO_4(H_2O)_2$, $NaBr(H_2O)_2$, and $CoCl_2(H_2O)_6$. At any given pressure, the temperature at which a particular hydrate will lose its water completely is different for each salt.

Some hydrates lose their water at room temperature and atmospheric pressure; these are called **efflorescent.** Hydrates which are stable at room temperature and one atmosphere, but are not yet saturated with water to the stoichiometric limit, will absorb water from the atmosphere; these are called **hygroscopic.** Hygroscopic salts often are used as drying agents, called **dessicants.**

When a hydrate is heated, it may lose its water in several stages, forming a series of hydrates with regular crystalline structures that contain progressively smaller proportions of water. Color changes often accompany a change in degree of hydration. For example, $CoCl_2(H_2O)_6$ is red, $CoCl_2(H_2O)_2$ is violet and anhydrous $CoCl_2$ is blue. Intermediate species, such as $CoCl_2(H_2O)_3$, are unstable and do not form regular crystal structures. **At any particular temperature, the degree of stable hydration is well defined, and the stable form of the hydrate has a definite formula.**

PLAN OF EXPERIMENT
First you examine the behavior of a group of compounds when they are heated, to determine which are hydrates and which are not. Then you drive all the water from an unknown hydrate, to determine the mole ratio of water in its stable structure.

---

[1] IUPAC nomenclature: Copper(II)sulfate-water (1/5)

SAFETY
  Wear approved eye protection.

PROCEDURE
A. Testing for Hydrates
1.   Obtain several different samples from the following group of compounds. Your instructor will tell you how many.

sodium chloride	strontium chloride	copper(II) sulfate
potassium chloride	cobalt(II) chloride	magnesium sulfate
magnesium chloride	sodium tetraborate	sodium sulfate
chromium(III) chloride	sodium acetate	sucrose

Write the names of your compounds into the table on the Data sheet. The degree of hydration for these compounds ranges from zero to twelve water molecules.

2.   Place about 0.2 g of each compound into separate small dry test tubes. Mark the test tubes for identification.

3.   Gently heat each sample, in turn, over a burner flame and observe the results. Write your observations in the table on the data page. Try to infer which of the compounds are hydrates, using the criteria below.

Some Characteristics of Hydrate Behavior when Heated:
1.   **Evolution of water**, which may condense on the cooler upper
            wall of the test tube.
2.   **Color changes** may occur in the solid as water is lost.
3.   **Decomposition** may accompany loss of water. The decomposition products often form
            acids or bases upon reaction with the evolved water of hydration.[2] Test the evolved
            moisture with litmus or pH paper.
4.   After heating, the **solid residue often dissolves** in water, frequently with a color change.

5.   If after heating, the solid residue crystals are allowed to cool and stand in humid air for an
            hour or two, the **crystals may reabsorb enough water from the air to return to their hydrated
            color**. Fine crystals will rehydrate more rapidly than coarse crystals. If room humidity is low,
            breathing with moist breath on the crystals will speed rehydration. Spreading residue
            crystals out on a shallow dish also helps.

**Some compounds that are not hydrates will decompose on heating to form water as a product.
Unlike water of hydration, this water does not exist in the compound as bound water molecules, but
is formed by atom rearrangements during decomposition. The residue from such compounds is
generally *water insoluble*, even on heating.**

B. Determining the Formula of a Hydrate
1.   Obtain a sample of $BaCl_2(H_2O)_x$. You will try to find the value of **x** and determine the mole ratio of water in this compound.

2.   Strongly heat a clean, dry crucible and cover, supported in a clay triangle on a ring stand, for 1-2 min, to drive off any adsorbed moisture.
            From this point on, handle the crucible and cover with tongs only.

---

[2] As an example, strong heating of calcium oxalate monohydrate, $CaC_2O_4(H_2O)$, can yield products that form a basic solution with the water of hydration. The water of hydration is lost around 226°C. Continued heating to 500°C decomposes the anhydrous $CaC_2O_4(H_2O)$ according to Reaction 10.1:

$$CaC_2O_4(H_2O) \longrightarrow CaCO_3 + CO \qquad (10.1)$$

At 800°C, $CaCO_3$ decomposes by:

$$CaCO_3 \longrightarrow CaO + CO_2 \qquad (10.2)$$

CaO dissolves in the water of hydration to make a basic solution:

$$CaO + H_2O \rightleftharpoons Ca^{2+} + 2\,OH^- \qquad (10.3)$$

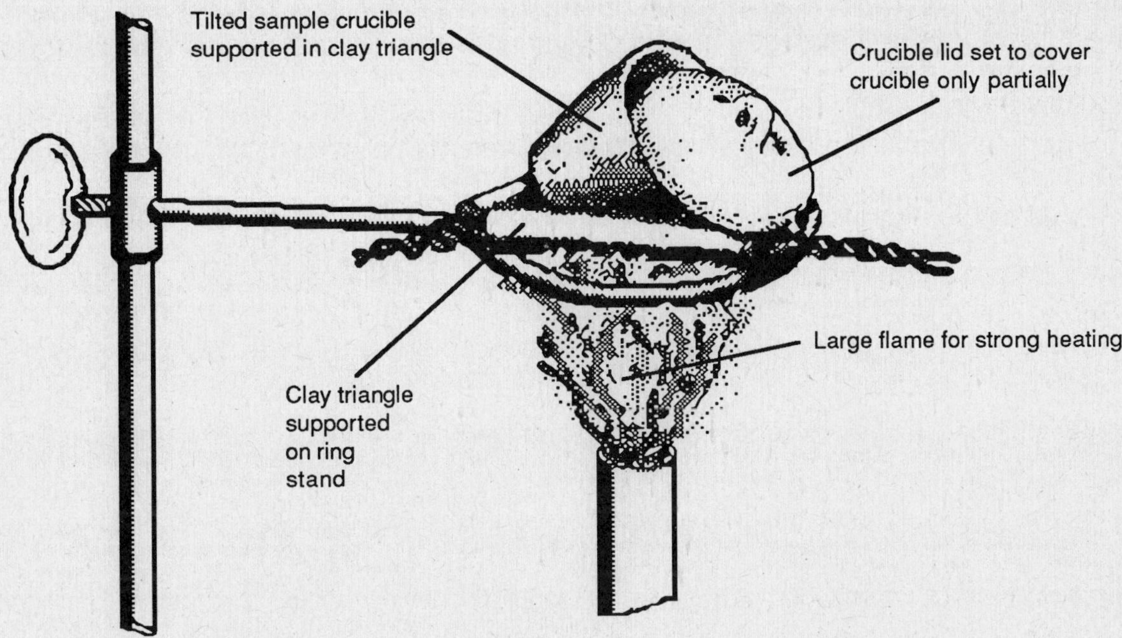

Figure 10.1. Heating a partially covered crucible, which limits access to air but allows water vapor to escape.

3. Allow the crucible and cover to cool. Then, weigh them together on an analytical balance, to 0.1 mg. Record the mass on the data sheet.

4. Place 3 or 4 g of your $BaCl_2(H_2O)_x$ sample in the crucible, replace the cover, and reweigh accurately. Record the mass on the data sheet.

5. Place the covered crucible and its contents on the triangle support. Partially uncover the crucible (handle the cover with tongs only) to allow water vapor to escape, as illustrated in Figure 10.1.

6. Heat the sample very gently, for about the first five minutes, to prevent spattering.

7. When the sample is dry enough that spattering is unlikely, heat strongly for about 15 min more.

8. Then cover the crucible completely with its lid and allow it to cool for 10 min.

9. When the crucible is at room temperature, weigh the crucible, lid, and contents. Record the mass on the data sheet.

10. Reheat strongly with cover partially off, as before.

11. Cover completely, cool, and reweigh accurately, as before. Record the mass on the data sheet.

12. Continue reheating, cooling, and reweighing, until two successive weighings are within ±0.003 g of one another, indicating complete decomposition. Record this final mass value on the data sheet.

13. If instructed to do so make a second determination with a new sample of $BaCl_2(H_2O)_x$.

14. From the mass loss data, calculate the number of moles of water per mole of $BaCl_2$, in your sample.

15. Store the dried and weighed sample of anhydrous $BaCl_2$ until the next laboratory period. Then reweigh it and calculate the percent of lost water which has been regained. The exact value obtained will depend on the environmental humidity, as well as the hygroscopic characteristics of $BaCl_2$.
Record the percent rehydration value from step 15 at the top of the data sheet for the next experiment that you turn in. Label this data as *Percent Rehydration from Experiment 10.*

Name _________________________________   Date ____________

EXPERIMENT 10
PRELABORATORY EXERCISE

1.   The formula for a hydrate of sodium phosphate is $Na_3PO_4(H_2O)_{12}$.

    a.   If all the water were driven off by heating, how much mass would be lost from a sample which weighed 2.5433 g before heating?   _________________

    b.   What is the mass% water in the hydrate?   _________________

2.   A hydrate of cobalt chloride, $CoCl_2(H_2O)_x$, is heated until it reaches a constant mass. The original sample weighed 1.6884 g, before heating. After heating, its mass was 1.0856 g. What is the formula of the hydrate?

Calculations:

Name _______________________________________　　Date _____________

EXPERIMENT 10

DATA
(Observe significant figures in all calculations.)

A. Testing for Hydrates
Enter the names of the compounds you are testing in the top row of open spaces.

Compound name						
Is moisture evolved?						
Is there a color change on heating?						
Results of pH test of moisture						
Is there water soluble residue?						
Is there a color change on dissolving?						
Is there a color change on standing?						
Is the compound a hydrate?						

B. Determining the Formula of an Unknown Hydrate

Determination

1　　　　　　　　　　　2

1.　Mass of empty crucible + cover: _____________　_____________
2.　Mass of crucible, cover and sample: _____________　_____________
3.　Mass of hydrate sample: _____________　_____________
4.　Mass of sample after each heating: _____________　_____________

_____________　_____________

_____________　_____________

_____________　_____________

_____________　_____________

5.　Final mass of anhydrous residue: _____________　_____________
6.　Mass of water originally present: _____________　_____________
7.　Moles of water originally present: _____________　_____________
8.　Moles of $BaCl_2$ originally present: _____________　_____________
9.　Mole ratio of water to $BaCl_2$: _____________　_____________
10.　Average mole ratio of water to $BaCl_2$: _____________
11.　Formula of hydrate: _______________________________

Use other side of page for calculations.

Name ____   Date ____________

EXPERIMENT 10
QUESTIONS

1.   **a.** If the hydrate is not completely dehydrated when your final weighing is made, how will this affect the calculated molar ratio of water to $BaCl_2$? (+, 0 , -) :   _________

   b. Explain your reasoning used in part a.

2.   How many grams of water are needed to completely rehydrate 15 g of anhydrous $BaCl_2$ to the hydrate of barium chloride that you identified as your sample?   ______________

Calculations:

EXPERIMENT 11
Preparation and Reactions of Sodium Carbonate and Sodium Hydrogen Carbonate

PRELABORATORY PREPARATION
1. Do the Prelaboratory Exercise and turn it in at the beginning of your laboratory period.
2. Review vacuum filtering technique and quantitative transfer of filtrate, Sections VIII and IX, Methods, pages 22-23.

INTRODUCTION

All the alkali metals form carbonates that have the general formula M_2CO_3, and hydrogen carbonates that have the general formula $MHCO_3$, where **M** stands for any alkali metal. The two types of alkali metal carbonates have very different properties.

All alkali metal hydrogen carbonates decompose upon heating to give the normal (non-hydrogen) carbonate, carbon dioxide and water:

$$2\,MHCO_3(s) \rightleftharpoons M_2CO_3(s) + CO_2(g) + H_2O(l) \tag{11.1}$$

The temperature required for Reaction 11.1 depends on the alkali metal ion, in the order:

$$Cs^+ > Rb^+ > Na^+ > Li^+$$

where Cs^+ requires the highest and Li^+ the lowest temperature. Sodium hydrogen carbonate, a subject of this experiment, starts to decompose thermally around 100°C, at one atmosphere of pressure.

All normal carbonates are more resistant to decomposition than the hydrogen carbonates and must be heated to much higher temperatures before decomposing. All normal carbonates, except for those with NH_4^+ and alkali metal cations, are insoluble in water near room temperature.

Sodium carbonate and sodium hydrogen carbonate are the most important commercially of the alkali metal carbonates. Sodium carbonate forms a series of hydrates (see Experiment 10). It crystallizes from an aqueous solution as the decahydrate, $Na_2CO_3(H_2O)_{10}$, which is known as washing soda or sal soda, and is used as a cleansing agent. The anhydrous form, Na_2CO_3, is known as soda ash and has many industrial applications, including the manufacture of glass, soaps, paper, and textiles.

Sodium hydrogen carbonate ($NaHCO_3$), also known as sodium bicarbonate or ordinary baking soda, also has many uses in industry and the home. It dissolves in water as a base and will neutralize acid solutions, evolving CO_2 gas in the process:

$$NaHCO_3 \xrightarrow{\text{H}_2\text{O}} Na^+ + HCO_3^- \tag{11.2}$$
$$HCO_3^- + H_3O^+ \longrightarrow 2\,H_2O + CO_2(g) \tag{11.3}$$

Reactions 11.2 and 11.3 are the basis for using sodium hydrogen carbonate in the manufacture of effervescent beverages, as a stomach antacid medicament, and as a leavening agent in baking.

Sodium carbonate and sodium hydrogen carbonate may be produced by the Solvay process, developed in 1863 by Ernest Solvay, in which carbon dioxide is bubbled through a solution of ammonia and sodium chloride:

$$NH_3(aq) + CO_2(aq) + H_2O \longrightarrow NH_4^+(aq) + HCO_3^-(aq) \tag{11.4}$$

$$NaCl(aq) + NH_4^+(aq) + HCO_3^-(aq) \xrightarrow{<15°C} NaHCO_3(s) + NH_4Cl(aq) \tag{11.5}$$

At temperatures of 15°C, or lower, sodium hydrogen carbonate precipitates and can be filtered off. Heating the sodium hydrogen carbonate to around 100°C decomposes it to anhydrous sodium carbonate:

$$2\,NaHCO_3 \rightleftharpoons Na_2CO_3 + H_2O + CO_2(g) \tag{11.6}$$

PLAN OF EXPERIMENT

In Part A, you prepare sodium hydrogen carbonate by a procedure which is equivalent to the Solvay process, and calculate the theoretical and percent yields. Then, in part B, you convert $NaHCO_3$ to Na_2CO_3 by heating it, and determine the percent yield. Finally in part C, you investigate some reactions of the two carbonate compounds.

EXAMPLE:

Na_2CO_3 was prepared according to Reactions 11.4-11.6, using 12.77 g NaCl. All other reagents were in excess.

    a.   Calculate the theoretical yield of Na_2CO_3 .

    b.   If the measured yield of Na_2CO_3 was 8.98 g, calculate the percent yield.

SOLUTION:

The molar mass of NaCl is 58.45 g mol^{-1}, so

$$12.77 \text{ g} = 12.77g/58.45 \text{ g mol}^{-1} = 0.2185 \text{ moles.}$$

In Reaction 11.5, moles of $NaHCO_3$ produced equals moles of NaCl that react.

The theoretical yield of any product is the maximum amount that can be produced, based on the amounts of reactants that are used.

Therefore, 0.2185 moles represents the theoretical yield of $NaHCO_3$, because NaCl is the limiting reactant, all others being in excess.

In Reaction 11.6, the moles of Na_2CO_3 produced equals 1/2 the moles of $NaHCO_3$ reacting. Therefore,

    a. The theoretical yield of Na_2CO_3 is 0.2185/2 = 0.1092 moles. The molar mass of $NaHCO_3$ is 84.02 g mol^{-1}, so that:

$$\text{theoretical yield of } Na_2CO_3 = (0.1092 \text{ mol}) (84.02 \text{ g mol}^{-1})$$
$$\text{theoretical yield of } Na_2CO_3 = \mathbf{9.178 \ g}$$

    b. The measured mass of Na_2CO_3 recovered was 8.98 g. Therefore:

$$\text{percent yield of } Na_2CO_3 = \frac{8.98}{9.18} \times 100\% = \mathbf{97.8\%}$$

SAFETY

1.   Wear approved eye protection.

2.   Dry ice can cause severe "freeze burns." Do not touch it with your bare hands.

3.   The fumes from concentrated ammonium hydroxide are very irritating. Always handle this reagent in a fume hood and avoid inhalation of its vapor.

PROCEDURE
A.   Preparation of Sodium Hydrogen Carbonate, $NaHCO_3$

1.   Using weighing paper, weigh on a triple-beam balance about 15 g sodium chloride, NaCl, to the nearest 0.01 g. Record the mass on the Data sheet and transfer the sample to a 250 mL beaker.

2.   Prepare about 25 mL of ice-cold distilled water by immersing a beaker containing distilled water in an ice-water bath.

3.   Prepare a warm-water bath by heating a 500 mL beaker, 3/4 full of water, to a luke-warm temperature of 25° to 30°C.

4.   **In a hood,** add to the beaker containing the NaCl sample, 50 mL of concentrated ammonium hydroxide, NH_4OH, in 5 mL portions with continuous stirring. Then return the sample to your desk.

5.   Add several small pieces of dry ice (solid CO_2) with continuous stirring. **Do not touch the dry ice with your bare hands.**
From time to time, warm the reaction beaker to around 10°C or 15°C by immersing its bottom in the warm-water bath. NaHCO$_3$ should precipitate.

6.   After precipitation is complete, add 1 or 2 more small pieces of dry ice and allow the solution to become cold. Stir continuously until the bubbling has stopped.

7.   Weigh a circle of filter paper before placing it in a Buchner funnel and record its mass. Vacuum filter the cold solution, transferring the filtrate to the filter paper quantitatively.

8.   After filtering, remove the vacuum hose from the filter flask, but leave the aspirator turned on.

9.   Wash the precipitate by pouring 3 to 5 mL of ice-cold distilled water uniformly over the filtrate with the suction off. Let the water soak in, then attach the vacuum hose to apply suction, drawing the wash water through the sample and filter. Repeat for a total of three washings.

10.   Carefully remove the filter paper from the filter funnel. Open out the filter paper and dry the paper with its precipitate in an electric oven at 110°C.

11.   Weigh the dried precipitate on its filter paper to 0.01 g on a triple-beam balance. Record the mass and calculate the percent yield.

B.   **Preparation of Na$_2$CO$_3$ from NaHCO$_3$**
1.   Heat an empty evaporating dish on a clay triangle to a dull red color for 3 or 4 min to "burn off" any surface residue and moisture.

2.   Cool the dish to room temperature and, handling it with tongs, weigh it on a triple-beam balance to 0.01 g. Record the mass on the Data sheet.

3.   Set aside for later use in Part C a small labeled test tube containing about 1 g of the sodium hydrogen carbonate prepared in Part A.

4.   Transfer the rest of your sodium hydrogen carbonate to the evaporating dish. Weigh the dish and sample on a triple-beam balance to 0.01 g and record the mass on the Data sheet.

5.   Place the dish and sample on the clay triangle and begin to heat it gently with a low flame, to prevent spattering.

6.   Increase the flame height gradually during the next 5 min, until you are heating the sample strongly. Stir continuously with a spatula while heating. Continue heating strongly for 20 min.

7.   Remove the flame and allow the dish and sample to cool to room temperature.

8.   When cool, weigh the dish and contents to 0.01 g and record the mass on the Data sheet.

9.   Repeat the strong heating process for 10 min. Cool and reweigh, as before.

10.   Repeat step 9 until two successive weighings are the same, within ±0.03 g. Then, record the mass of the dish and contents on the Data sheet.

C.   **Some Reactions of Carbonates and Hydrogen Carbonates**
1.   Transfer the small portion of NaHCO$_3$, set aside in part B, to a watch glass. Add 1 or 2 drops of dilute (3 M) HCl. Answer question 1 on the Question sheet.

2.   Dissolve the NaCO$_3$, made in part B, in about 5 mL distilled water. Warm the water, if necessary, to aid dissolution. Transfer the solution to a small test tube. Add about 5 mL of 1 M CaCl$_2$ to the test tube. Answer question 2 on the Question sheet.

Name ____   Date ____________

EXPERIMENT 11
PRELABORATORY EXERCISE

1. $NaHCO_3$ was prepared according to Reactions 11.4 and 11.5, using 30 g NaCl. Assume that NaCl is the limiting reagent.

        a. What is the theoretical mass yield of $NaHCO_3$?   __________

        b. How many grams of CO_2 are consumed if all of the NaCl reacts?   __________

        c. If the actual yield was 32.4 g $NaHCO_3$, what was the percent yield?   __________

2.   The $NaHCO_3$ from question 1 was heated, forming Na_2CO_3 by Reaction 11.6. Based on the actual yield of $NaHCO_3$ in problem 1, what is the maximum mass of Na_2CO_3 that can be made?

____

Calculations:

Name _______________________________________   Date ____________

EXPERIMENT 11

DATA
(Observe significant figures in all calculations.)

A. Preparation of NaHCO$_3$

1.   Mass of NaCl   _______________

2.   Mass of filter paper: _______________

3.   Mass of filter paper + dried NaHCO$_3$:   _______________

4.   Mass of dried NaHCO$_3$:   _______________

5.   Theoretical mass yield, based on
      mass of NaCl used: _______________

7.   Percent yield: _______________

B. Preparation of Na$_2$CO$_3$

1.   Mass of evaporating dish: _______________

2.   Mass of dish + NaHCO$_3$ sample: _______________

3.   Mass of NaHCO$_3$:   _______________

4.   Mass of dish + Na$_2$CO$_3$ after successive heatings

      to constant mass: 1. _______   2. _______   3. _______   4. _______

5.   Final mass of dish + Na$_2$CO$_3$:   _______________

6.   Mass of Na$_2$CO$_3$ produced: _______________

7.   Percent yield: _______________

Calculations:

Name ____   Date ____________

EXPERIMENT 11
QUESTIONS

1.   **a.** What do you observe in part C, step 1?

     b. What is the gaseous product?

     c. The other products are water and sodium chloride. Write the equation for the reaction between $NaHCO_3$ and HCl.

2.   **a.** What do you observe in part C, step 2?

     c. What is the precipitate?

     d. Write the equation for the reaction between Na_2CO_3 and $CaCl_2$.

3.   In part A, step 9, why was it necessary that the water used to wash the precipitate be ice-cold?

EXPERIMENT 12
Conversion of a Carbonate to a Chloride

PRELABORATORY PREPARATION
1. Do the Prelaboratory Exercise and turn it in at the beginning of your laboratory period.
2. Review vacuum filtering technique and quantitative transfer of filtrate, Sections VIII and IX, Methods, pages 22-23.

INTRODUCTION
The laws of conservation of matter and definite proportions originally were established by careful quantitative laboratory experiments. Now that they are uniformly accepted as basic principles, these laws can be used to determine reaction yields and to help analyze what is happening in new chemical experiments. This experiment is intended to confirm the law of definite proportions and to illustrate its application in predicting the yield of a reaction product.

PLAN OF EXPERIMENT
Sodium chloride, $NaCl$, can be made from sodium carbonate, Na_2CO_3, and hydrochloric acid, HCl, via Reaction 12.1:

$$Na_2CO_3 + 2\,HCl \longrightarrow 2\,NaCl + H_2O + CO_2(g) \tag{12.1}$$

For Reaction 12.1, it is easy to perform the experiment in a manner that allows the amount of Na_2CO_3 that reacts and the amount of $NaCl$ that is formed to be readily measured.

1. By using an excess of HCl, all the starting amount of Na_2CO_3 is forced to react.
2. After reaction is complete, the only product that remains in the reaction vessel is solid $NaCl$, because:
   a. The CO_2 product leaves the reaction crucible as a gas.
   b. H_2O and excess HCl are evaporated off by strong heating of the reaction crucible.

By knowing the starting mass of Na_2CO_3 and using the law of conservation of matter, you can predict the maximum amount of $NaCl$ that theoretically can be made, the *theoretical yield*. By comparing the theoretical yield with your experimental yield, you can calculate your percent yield.

EXAMPLE:
Na_2CO_3 reacts with an excess of HCl according to Reaction 12.1.
a. If the starting amount of Na_2CO_3 was 0.5127 g, calculate the theoretical yield of $NaCl$.
b. If the actual yield was 0.4926 g, what was the percent yield?

SOLUTION:
The reaction is: $Na_2CO_3 + 2\,HCl \longrightarrow 2\,NaCl + H_2O + CO_2(g)$

From the reaction equation, the theoretical moles of $NaCl$ produced is 2 times moles of Na_2CO_3 that react. To find how many moles and grams of $NaCl$ can be made by complete reaction of the starting 0.5127 g of Na_2CO_3, you first must determine the molar masses of Na_2CO_3 and $NaCl$:

$$\text{molar mass } Na_2CO_3 = 105.99 \text{ g}$$
$$\text{molar mass } NaCl = 58.44 \text{ g}$$

$$\text{Starting moles of } Na_2CO_3 = \frac{0.5127}{105.99} = 0.004837$$

Theoretical moles NaCl produced = 2 times moles Na_2CO_3 reacting = 2(0.004837)
$$= \mathbf{9.674 \times 10^{-3} \text{ mol}}$$

Theoretical mass NaCl produced = $(9.674 \times 10^{-3} \text{ mol})(58.44 \text{ g mol}^{-1})$ = **0.5654 g**
The actual yield was 0.4926 g. Therefore,

$$\textbf{Percent yield} = \frac{0.4926}{0.5654} \times 100\% = \mathbf{87.13\%}$$

Since different students will start with different amounts of Na_2CO_3, a comparison of the data from several students can serve to demonstrate the law of definite proportions. Equation 12.1 shows that the mole ratio of NaCl to Na_2CO_3 should be a constant value of 2, if the law of definite proportions is obeyed. Therefore the mass ratio should be a constant value of $2(58.44)/105.99 = 1.10$, regardless of what starting mass of Na_2CO_3 was used, as long as it all reacts. This requires that HCl be in excess. By comparing the mass ratios obtained by several other students in your laboratory, you either can confirm the law of definite proportions, or conclude that the average student in your laboratory has not yet developed good laboratory technique.

Your instructor will prepare a data table on the blackboard. When you have completed your experiment, enter your results into this table. Before you leave the laboratory, copy the data from 6 other students onto your data sheet table. You can illustrate the law of constant proportions with this data, if the other students have performed the experiment properly.

SAFETY

Wear approved eye protection.

PROCEDURE

1.   Support a clean, dry crucible and cover in a clay triangle on a ring stand. Strongly heat the crucible and cover to red heat for 2 min, to drive off any moisture and residue.

After cleaning, handle the crucible and cover only with tongs, to avoid adding moisture and oils from your hands.

Remove heat and allow the crucible and cover to cool to room temperature.

2.   When cool, accurately weigh the crucible and cover on an analytical balance. Record the mass on the data sheet.

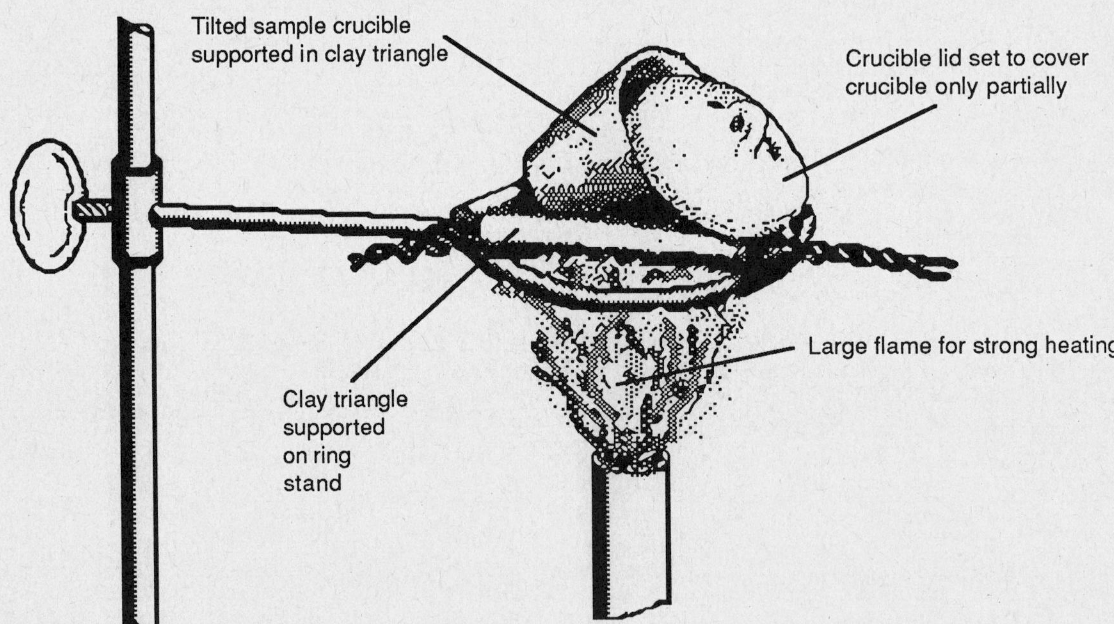

Figure 12.1. Heating a partially covered crucible, which limits access to air but allows liquids to evaporate.

3.   Add about 0.3 g of dry Na_2CO_3 to the crucible, replace the cover and reweigh accurately. Record the mass on the data sheet.

4.   Obtain about 3 mL concentrated HCl in a stoppered small test tube. With an eyedropper, add HCl slowly, dropwise, to the Na_2CO_3 in the crucible. The fizzing reaction is due to the formation of CO_2 gas.
Be careful not to add the HCl too rapidly or the sudden evolution of CO_2 can cause loss of sample by spattering.
Continue to add HCl until no further production of CO_2 is detected. *Then, add 5 drops more.* Using tongs to handle it, swirl or rock the crucible gently to help mix the reagents during this process.
Do not stir the reagents with a rod, which would retain adhered sample.

5.   Arrange the crucible with contents, partially covered to allow evaporation, on a clay triangle, as in Figure 12.1. Warm the crucible gently to evaporate the liquid.
Do not allow the sample to spatter.

6.   When the sample is dry, cover the crucible completely and heat it strongly for another 5 min. Then, allow the crucible to cool to room temperature.

7.   Accurately weigh the crucible, cover, and product. Record the mass on the data sheet.

8.   Do a duplicate experiment if there is time.

9.   Enter the masses of your reagent Na_2CO_3 and product NaCl on the blackboard table, as well as on your Data sheet.

Name ____   Date _____________

EXPERIMENT 12
PRELABORATORY EXERCISE

1. In the conversion of sodium carbonate to sodium chloride by Equation 12.1, 0.7668 g Na_2CO_3 reacted with an excess of HCl.

    a. What is the theoretical mass yield of NaCl?   _________________

    b. The actual yield was 0.8206 g NaCl. What was the percent yield?   _________________

2. Based on the actual yield of problem 1, what volume of CO_2 gas is produced, at 25°C and one atmosphere pressure?

    ____

Calculations:

Name ____   Date ____________

EXPERIMENT 12

DATA
(Indicate the correct units for measured quantities.)

Determination

	1	**2**

1. Mass of empty crucible and cover:

2. Mass of crucible, cover, and Na_2CO_3 sample:

3. Mass of Na_2CO_3 sample:

4. Theoretical yield of NaCl,
based on Na_2CO_3 mass:

5. Mass of crucible, cover and NaCl product:

6. Mass of NaCl product obtained:

7. Percent yield:

In the table below, copy data for 6 students from the blackboard. Add your own data. Use only one determination from each student.

Student	Starting mass of Na_2CO_3	Mass of NaCl product	$\dfrac{\text{mass NaCl}}{\text{mass } Na_2CO_3}$
1			
2			
3			
4			
5			
6			
self			

Calculations:

EXPERIMENT 13
Classes of Chemical Reactions: Oxidation-Reduction, Acid-Base Neutralization, Precipitation, Metathesis, and Decomposition Reactions in a Cyclic Sequence

PRELABORATORY PREPARATION
1. Do the Prelaboratory Exercise and turn it in at the beginning of your laboratory period.
2. Review suction filtering, quantitative transfer of filtrate, and decanting, Section VIII, Methods, page 20.

INTRODUCTION
A chemical reaction is a process in which the reacting molecules rearrange their atoms and electrons into a new pattern, creating different molecules which are the reaction products.

All chemical reactions have one characteristic in common: they alwaysoccur so as to increase the stability of the chemical system, under the conditions of pressure and temperature that prevail during the process.

Simply stated, the reaction products always are more stable than the reagents. The differences between one reaction and another lie in the details of the atom and electron rearrangements.

A group of different reactions that have similar "reactant-to-product" rearrangement details, are conveniently collected together into a **class of reaction.** This experiment is a series of related reactions that represent the five different reaction classes of the title.

PLAN OF EXPERIMENT
In this experiment, five different classes of reactions are illustrated by a sequence of chemical changes that converts copper metal into several different chemical forms, and then returns the copper to its metal form again. As you perform the experiment, you can observe how processes that occur at the molecular level often can be detected and identified by large scale effects, such as color changes and precipitate formation. At the end of the experiment, a comparison of the starting and ending amounts of copper metal will allow a calculation of the percent yield from the entire cycle.

Because the reaction cycle involves reactions of copper with several different compounds, the overall plan of the experiment is best seen by looking first at the chemical changes which occur to the copper alone. The transformations of only the copper species are given by Equation 13.1, which is simplified by omitting all other species involved in the cycle. The separate steps are discussed in the following section.

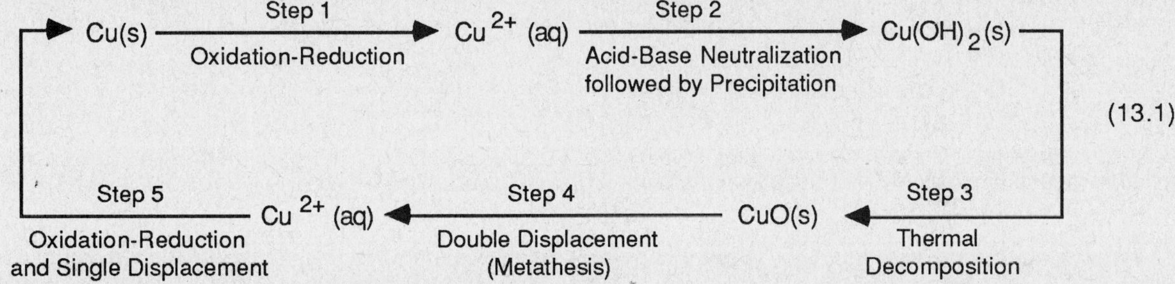

$$(13.1)$$

REACTION STEP 1: OXIDATION-REDUCTION
The essential process occurring in **oxidation-reduction** reactions (also called **redox** reactions) is a transfer of electron charge from one atom to another. Only the valence shell electrons are involved. The valence electron charge distribution around an atom is described formally by the oxidation number of the atom. The oxidation number is always a small integer, which may be positive, zero, or negative.

The electron charge distribution of any atom in its pure elemental form is arbitrarily assigned an oxidation number of zero. Every atom in a non-elemental molecule has an oxidation number assignment that is related to the *change* in electron charge around it, relative to its elemental state.

If an atom *loses* electron charge during a reaction, its oxidation number changes to a more positive value, and we say the atom has been *oxidized*.

If an atom *gains* electron charge during a reaction, its oxidation number changes to a more negative value, and we say the atom has been *reduced*.

See your textbook for additional details concerning the assignment of oxidation numbers to atoms in molecules, as used in the following examples.

EXAMPLE 1:

In Step 1 of Equation 13.1, $Cu(s)$ loses electron charge to become $Cu^{2+}(aq)$. Step 1 involves dissolving solid copper in nitric acid (HNO_3). The ionic reaction of Step 1 is[1]:

$$Cu(s) + 4\,H_3O^+ + 4\,NO_3^- \longrightarrow Cu^{2+} + 2\,NO_2(g) + 6\,H_2O + 2\,NO_3^- \qquad (13.2)$$

oxidation numbers: 0   +1,-2   +5,-2   +2   +4,-2   +1,-2   +5,-2

The oxidation number of the Cu atoms changes from 0 in elemental $Cu(s)$, to +2 in Cu^{2+}, an increase of two units to a more positive value; therefore, $Cu(s)$ is oxidized. Nitrogen atoms change from +5 in NO_3^- to +4 in NO_2, a decrease of one unit to a more negative value; therefore, N atoms are reduced. No other atoms change their oxidation number. Notice that only half of the nitrogen atoms, those that go into NO_2, are reduced. The other half do not change their oxidation number and remain in the form of NO_3^-.

The total net change in oxidation numbers must balance out to zero.
In Reaction 13.2, each copper atom that changes by +2 units must transfer the electron charge to 2 nitrogen atoms, which change by -1 unit each.

It is a general principle that whenever an atom loses electron charge and is oxidized, as does Cu in Reaction 13.2, other atoms must gain an equal amount of electron charge. These other atoms, therefore, must change their oxidation numbers to more negative values and be reduced.

An oxidation process must always be coupled with a reduction process.
The element oxidized and the element reduced are referred to as the
oxidation-reduction couple.

Because oxidation-reduction processes involve changes in the valence shell electron distribution, they generally affect the bond-forming capabilities of the reacting atoms. In Reaction 13.2, reduction of the nitrogen in NO_3^- causes it to change the nature of its bonding to oxygen, and the molecule rearranges its structure to become nitrogen dioxide, NO_2, an orange-brown toxic gas.[2] When you perform Step 1 in this experiment, the evolution of NO_2 gas is readily visible. At the same time, the reaction solution will turn blue because of the formation of $Cu^{2+}(aq)$.

---

[1] Reaction 13.2 would remain balanced if 2 molecules of NO_3^- were subtracted from each side of the ionic reaction, giving: $Cu(s) + 4\,H_3O^+ + 2\,NO_3^- \longrightarrow Cu^{2+} + 2\,NO_2(g) + 6\,H_2O$. The reason for including the extra NO_3^- in 13.2 is made clear from the complete reaction: $Cu(s) + 4\,HNO_3 \longrightarrow Cu(NO_3)_2 + 2\,NO_2(g) + 2\,H_2O$. It is seen that 4 moles of HNO_3 are needed for complete reaction with 1 mole of $Cu(s)$ and that 1 mole of the salt $Cu(NO_3)_2$ is formed. In the aqueous solution, the cupric salt is dissociated into its ions, as in 13.2.

[2] Nitrogen dioxide is one of the major components of urban air pollution, where it is formed by oxidation-reduction reactions in the atmosphere. The high combustion temperatures in automobile engine cylinders allow a direct oxidation-reduction reaction of atmospheric N_2 and O_2, injected along with the fuel.

$$N_2 + O_2 \longrightarrow 2\,NO$$

oxid. nos.:   0   0   +2,-2

The NO product is emitted into the atmosphere with the automobile exhaust, where it reacts further to produce NO_2 by a complex series of reactions. Overall, the net atmospheric reaction also is an oxidation-reduction process:

$$2\,NO + O_2 \longrightarrow 2\,NO_2$$

oxid. nos.:   +2,-2   0   +4,-2

Methods for controlling NO_2 pollution, such as catalytic converters and special engine designs that give better fuel mixture control and lower combustion temperatures, are based on an understanding of these oxidation-reduction processes.

REACTION STEP 2 (FIRST PART): ACID-BASE NEUTRALIZATION

By the Brønsted-Lowry model, acid-base reactions are those where a proton, H^+, is transferred from one reactant to another. They are, in a sense, complimentary to oxidation-reduction reactions, in which electron charge is transferred from one reactant to another. The molecule that contributes the proton is the **acid**, and the molecule that accepts the proton is the **base**.

Water can fill both roles: it can either donate or accept a proton, acting either as an acid or a base.

This behavior is illustrated by Reactions 13.3 and 13.4.

$$HCl + H_2O \longrightarrow Cl^- + H_3O^+ \tag{13.3}$$

A proton is transferred from HCl to H_2O: HCl is the acid and H_2O is the base.

$$NH_3 + H_2O \longrightarrow NH_4^+ + OH^- \tag{13.4}$$

A proton is transferred from H_2O to NH_3: H_2O is the acid and NH_3 is the base.

Substances that can act either as an acid or a base are called **amphoteric.** Its amphoteric character is one reason why water is a particularly important solvent for acid-base reactions. One water molecule can even transfer a proton to another water molecule, so that water can be an acid and a base in the same reaction:

$$H_2O + H_2O \longrightarrow H_3O^+ + OH^- \tag{13.5}$$

Reaction 13.5 always occurs to some extent, and even pure water always contains some H_3O^+ and OH^-, in equal concentrations. When an acid is added to water, as in Equation 13.3, additional H_3O^+ is formed. The solution resulting from Equation 13.3 is called an acid solution, because it contains more H_3O^+ (called **hydronium ion**) than OH^- (called **hydroxyl ion**). The addition of a base to water, as in Equation 13.4, causes an excess of OH^-, resulting in a basic solution. The acidic or basic strength of an aqueous solution depends on the relative concentrations of H_3O^+ and OH^-.

$[H_3O^+] > [OH^-]$ **solution is acidic**

$[H_3O^+] < [OH^-]$ **solution is basic**

$[H_3O^+] = [OH^-]$ **solution is neutral**

An acidic or basic solution can be neutralized by adding either acid or base, whichever is needed, until $[H_3O^+] = [OH^-]$.

EXAMPLE 2:

In Step 2 of Reaction 13.1, an acidic solution containing Cu^{2+} is neutralized by adding a solution of the base sodium hydroxide (NaOH). Solid NaOH is an ionic crystal with Na^+ and OH^- ions arranged in rigid lattice positions. When dissolved in water, the ions separate, contributing $OH^-(aq)$ for the neutralization reaction:

$$NaOH(s) \xrightarrow{\;H_2O\;} Na^+(aq) + OH^-(aq) \tag{13.6}$$

The point of neutralization is reached when enough OH^- has been added so that $[OH^-] = [H_3O^+]$. The point of neutralization is identified by the color change of an indicator dye, like litmus, which has different colors in acid and base solutions.

REACTION STEP 2 (SECOND PART): PRECIPITATION

After the acid solution containing Cu^{2+} has been neutralized, the addition of still more OH^- allows a precipitation reaction to occur. Any hydroxyl ion that is added after the neutralization point is reached will react with Cu^{2+} to form insoluble cupric hydroxide, $Cu(OH)_2$:

$$Cu^{2+}(aq) + 2\,OH^- \longrightarrow Cu(OH)_2(s) \tag{13.7}$$

Reaction 13.7 does not involve transfer of electron charge (there are no changes in oxidation number) or transfer of protons. Thus, it is neither an oxidation-reduction nor acid-base reaction. The process occurring is the combination of $Cu^{2+}(aq)$ with OH^- to form an insoluble precipitate of $Cu(OH)_2$. When all the Cu^{2+} has precipitated, excess OH^- accumulates and turns the solution basic.

The completion of Reaction 13.7 is signaled with a dye indicator that changes color when the solution turns basic.

REACTION STEP 3: THERMAL DECOMPOSITION

Thermal decomposition occurs when a compound is heated to a temperature high enough to make it unstable and cause it to break apart into two or more smaller molecules or fragments.

As the temperature increases, molecules move with higher velocities and collide with greater force. Eventually some temperature will be high enough that chemical bonds are broken by the force of molecular collisions, and a rearrangement of atoms can take place. If the atoms rearrange into two or more smaller fragments, we say the original molecule has thermally decomposed.

In Reaction Step 3, heating cupric hydroxide causes it to decompose to cupric oxide and water.

$$Cu(OH)_2(s) \longrightarrow CuO(s) + H_2O \tag{13.8}$$

To form these products, bonds from the copper atom to an OH group, and from an oxygen to a hydrogen atom must be broken, leaving H and OH fragments. The water molecule product is formed as these fragments rearrange during the decomposition process. When rearrangements occur to form very stable products like water, internal energy is released and becomes available for furthering the reaction. In such cases, the thermal energy requirements for the reaction may be lowered considerably. Reaction 13.8, for example, takes place readily when $Cu(OH)_2$ is merely warmed in boiling water.

REACTION STEP 4: DOUBLE DISPLACEMENT

Double displacement (also called *metathesis*) occurs when two different reactants exchange atoms.

In Reaction Step 4, CuO(s) is dissolved in sulfuric acid solution:

$$CuO(s) + H_2SO_4 \longrightarrow CuSO_4 + H_2O \tag{13.9}$$

Reaction 13.10 is an example of a **double displacement** reaction, because the oxygen atom in CuO is replaced by a sulfate (SO_4) group, and the sulfate group in H_2SO_4 is replaced by the oxygen atom. Because $CuSO_4$ is a soluble salt, the effect of Step 4 is to return copper into solution as the cupric ion, as indicated by the ionic equation for Step 4:

$$CuO(s) + 2 H_3O^+ \longrightarrow Cu^{2+}(aq) + 3 H_2O \tag{13.10}$$

You should compare Reaction 13.10 with 13.2, where pure Cu(s) is dissolved in nitric acid solution. Cupric ion is a product in both cases. The important difference is that Reaction 13.10 involves no oxidation number changes; the oxidation number of copper in both CuO and Cu^{2+} is +2.

REACTION STEP 5: OXIDATION-REDUCTION

Only oxidation-reduction reactions can cause the oxidation numbers of atoms in the reacting compounds to change.

Once the oxidation number of Cu(s) was changed from 0 to +2 in Reaction Step 1, it obviously is necessary to include another oxidation-reduction reaction at the end of the cycle, in order to change it back to 0 again, so as to complete the cycle and end with elemental copper. In order to reduce Cu^{2+} and change its oxidation number from +2 back to 0, electron charge must be donated to the cupric ion from another reactant. In step 5, magnesium metal is added to the sulfuric acid solution containing Cu^{2+}. This initiates two different oxidation-reduction reactions that occur in Step 5.

Oxidation-Reduction Reaction No. 1:

Electron charge is transferred from Mg atoms to Cu^{2+}:

$$Cu^{2+}(aq) + Mg(s) \longrightarrow Cu(s) + Mg^{2+}(aq) \tag{13.11}$$
$$\phantom{Cu^{2}} +2 0 0 +2$$

The oxidation-reduction couple is Cu^{2+} and Mg(s). Mg reduces Cu^{2+} to Cu(s), and, in turn, is oxidized by Cu^{2+} from Mg(s) to Mg^{2+}.

The completion of Reaction 13.11 is indicated by the disappearance of the blue solution color, caused by the presence of $Cu^{2+}(aq)$.

OXIDATION-REDUCTION REACTION NO. 2:

Another oxidation-reduction reaction occurs simultaneously with Reaction 13.12, in which Mg(s) reduces hydrogen atoms in the hydronium ions to hydrogen gas.

$$Mg(s) + 2 H_3O^+ \longrightarrow Mg^{2+}(aq) + H_2(g) + 2 H_2O \tag{13.12}$$
$$ 0 +1 +2 0$$

Reactions 13.11 and 13.12 convert all the added solid magnesium into the solvated ion Mg^{2+}, so that Cu(s) is the only solid compound remaining in the reaction vessel. The solid copper may be collected by filtering, in order to determine the percent yield of copper recovered from the entire cycle.

EXAMPLE 3:

If the mass of copper metal dissolved in Step 1 was 0.507 g and the mass of copper collected in Step 5 was 0.498 g. What was the percent yield?

SOLUTION:

$$\text{Percent yield} = \frac{0.498}{0.507} \times 100\% = \mathbf{98.2\%}$$

A percent yield smaller than 100% usually means some combination of the following:
   a. Some of the sample was lost during the experiment.
   b. The reaction does not form only one product.
   c. The reaction does not go to completion.
It also is perfectly possible to end with a percent yield that appears to be greater than 100%. This can happen if extraneous material is collected along with the copper metal, or if weighing errors are made. This experiment is a test of your experimental skills for, with good experimental technique, yields between 97 and 103 percent can be obtained.

SAFETY

1. Wear approved eye protection.

2. Concentrated nitric acid is very damaging to skin and clothing, and its fumes are toxic. All procedure
   steps involving nitric acid should be performed in a fume hood. Avoid breathing the vapors and handle
   the acid with great care.

3. Toxic NO2 gas is evolved in Reaction Step 1. Avoid breathing the vapors, even though the procedure
   is performed in a fume hood.

4. Sodium hydroxide and sulfuric acid are corrosive reagents. Handle them with care, avoiding skin and
   clothing contact.

5. Flammable hydrogen gas is evolved in Reaction Step 5. Perform this step in a fume hood or under your
   improvised funnel hood. Keep flames away from your apparatus during this step.

PROCEDURE
(Work with a partner. A triple-beam balance is adequate for all weighings, which are made to 1 mg accuracy.)

Reaction Step 1
1a.   Weigh a clean watch glass and record its mass on the Data sheet.

1b.   Obtain a piece of pure copper wire weighing approximately 0.5 g. Fine wire will dissolve more rapidly. If your wire is larger than #18, cut it into small pieces after any necessary cleaning.
      The wire must be clean.
If necessary, clean it with steel wool, rinse it with water and dry it with a towel.

1c.   Place the wire on the watch glass and weigh them together. Record the total mass on the Data sheet and then place the copper sample into a clean 250 mL beaker. If it has not been cut up, coil it to lie flat in the beaker bottom.

1d. In a fume hood or under a funnel hood as shown in Figure 13.1, *slowly and carefully* add to the copper sample about 4 mL concentrated nitric acid. Pour the acid from a 10 mL graduated cylinder.
      If the copper dissolves readily, go to step 1f; if not, go to step 1e.

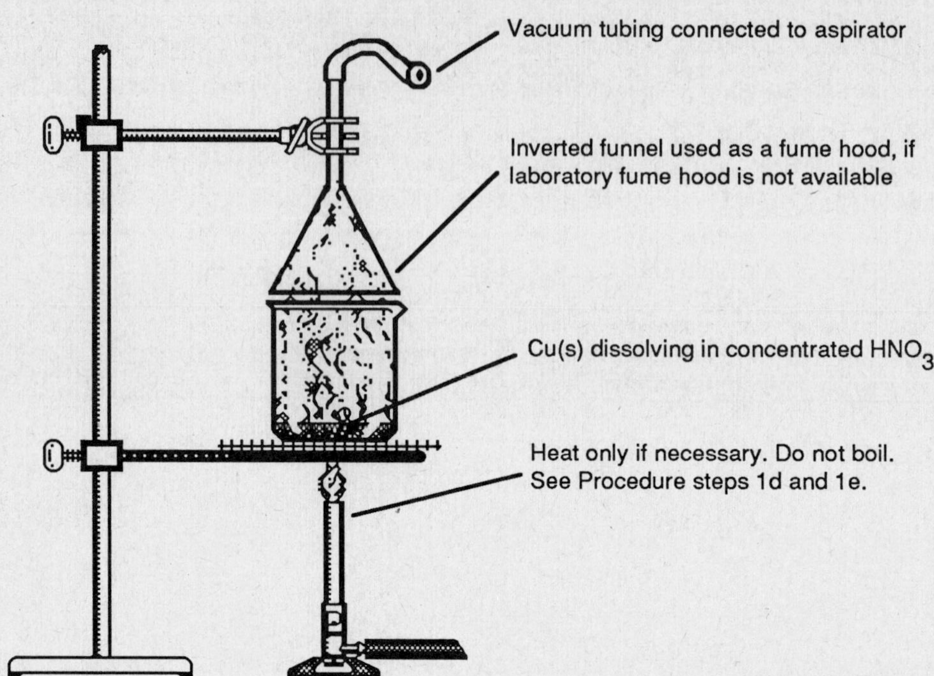

Figure 13.1. Improvised funnel hood for dissolving copper wire in nitric acid. Be sure aspirator is turned on.

1e.   If the copper does not dissolve readily in step 1d, add another 5 mL of concentrated HNO_3 to the beaker. Heat the beaker and contents on a hot plate or over a gas burner **in a fume hood or under an improvised funnel hood, as in Figure 13.1.** Heat the sample gently until the copper is dissolved.

Do not boil the acid solution.

1f.   Notice the color of the gas that is evolved. Record your observations on the Data sheet. **Avoid inhaling any of the gas.**

1g.   After the copper has dissolved, add 15 mL distilled water to dilute the sample for Reaction Step 2.

1h.   Record the color of the solution.

Reaction Step 2
2a.   Add 6 M NaOH dropwise from a buret or dropper, with continuous stirring. As you add NaOH, test the solution repeatedly for alkalinity by dipping the tip of a clean glass stirring rod into the solution and then touching the rod to a piece of red litmus paper that is lying on a paper towel.
When a drop of solution turns the litmus paper blue, The acid solution has been neutralized and turned alkaline. Additional OH^- added after neutralization will cause $Cu(OH)_2$ to precipitate.

A precipitate probably will begin to form *before* the neutralization point. Think about why this happens. You are asked to explain it on the Question page.
Stop adding NaOH when the litmus test indicates an alkaline solution. Record any color changes that occur in the solution.

2b.   Dilute the solution to about 100 mL with distilled water, in preparation for Reaction Step 3.

Reaction Step 3
3a.   Add 2 or 3 boiling chips to the beaker and boil it gently, while stirring, for 6-7 min on a hot plate or over a gas burner. Record any color or physical changes that occur.
It is important to stir continually while heating, to prevent bumping.

3b. After 6-7 min of gentle boiling, remove the beaker from the heat and place it in a cold water bath to speed cooling. Leave the stirring rod in the beaker. In another clean 250 mL beaker, begin heating about 150 mL of distilled water to use later for decanting and rinsing a filter residue.

3c. Insert a funnel through a one-hole rubber stopper that fits snugly into a suction flask. Assemble a suction filtering apparatus and connect it to an aspirator. Rinse a clean filter paper with distilled water from a wash bottle, pressing the filter paper against the funnel wall with your finger to squeeze out any air pockets between the glass and the paper.

3d. After 5-10 min of cooling, the solid sample should have settled out, leaving a clear supernatant liquid. Apply suction to the filter and carefully decant about 2/3 of the liquid through the filter, pouring the liquid down the stirring rod as shown in Figure 0.13, Section VIII, Methods. Using the water you have been heating, add about 100 mL of very hot distilled water to the beaker, swirl the contents, let the solid settle again, and decant through the filter once more. Keep the remaining 50 mL of heated water hot for later use.

3e. Filter the sample, transferring the precipitate and the boiling chips to the filter paper. Discard the filtrate, which should be a clear yellow liquid, without cloudiness.

3f. Quantitatively transfer the last traces of solid material from the beaker into the filter, using a stream of distilled water from a wash bottle. With the water stream, rinse as much solid as possible from the upper parts of the filter down into the bottom of the filter cone. Rinse off the stirring rod into the filter.

3g. Use the rest of the distilled water you have been heating to wash the solid collected on the filter paper. Pour about 5 mL of the hot water into the filter so that it just covers the solid residue and allow it to drain through.
   Pour the wash water slowly and carefully, trying to keep the solid undisturbed in the bottom of the filter cone.
Repeat this washing procedure 3 or 4 times.

3h. Support the filter and funnel, with the solid residue still on the filter paper, on a ring stand over a 250 mL beaker in preparation for Reaction Step 4. Describe the solid residue on the Data sheet.

Reaction Step 4
4a. Dissolve the solid residue of CuO by carefully pouring about 10 mL of 3 M H_2SO_4 directly over the residue on the filter paper, allowing it to drain through the filter into a collection beaker. (The boiling chips on the filter paper will not dissolve.) Record on the Data sheet any changes that occur. The solid probably will not be completely dissolved after all the acid has passed through. To dissolve the CuO completely, use a clean Pasteur dropping pipet (a fine-tipped eye-dropper) to recycle portions of the filtrate from the collection beaker back through the residue on the filter paper.
   Fill and handle the dropper very carefully so as not to lose any of the liquid.
Repeat this procedure as often as necessary to dissolve the solid.

4b. When all the solid is dissolved, use the dropper to rinse the filter paper with cold distilled water, until you cannot see any of the faintly blue Cu^{2+}(aq) on the paper. This may require 15-20 mL of distilled water. All of the copper should now be in the collection beaker as Cu^{2+}(aq).

4c. Save the acid solution of Cu^{2+} (the filtrate) in the collection beaker for Reaction Step 5.

Reaction Step 5
5a. Obtain about 0.2 - 0.3 g magnesium metal (40 mesh or finer, or magnesium ribbon cut up into small pieces).

5b. In a hood or under the funnel hood, slowly add magnesium pieces to the acid solution containing Cu^{2+} and stir rapidly with a glass rod. You may have to raise the funnel an inch or two to do this.
   Hydrogen gas is evolved copiously in this step. Be certain that no open flames are near your work area.
The magnesium will dissolve and a fine precipitate of Cu(s) will appear.

5c.   The reaction is complete when the blue color (due to the presence of Cu^{2+}(aq) disappears.
If any blue color remains after all the magnesium has dissolved, Cu^{2+} still remains in solution. Add more magnesium until the solution is colorless.

5d.   Much of the Cu(s) precipitate is too finely divided to settle out. Coagulate it by gentle heating for about one minute, while stirring gently. All the solid material should settle to the bottom of the beaker.
Do not boil the solution during coagulation.
After all the solid has settled, allow the solution to cool before continuing.

5e.   Test for the presence of Cu^{2+}(aq) in the supernatant liquid over the precipitate by transferring a few drops from the supernatant to a small test tube containing 1-2 mL of concentrated ammonium hydroxide. Use your dropping pipet for this.
A blue color indicates Cu^{2+}(aq).
Record your observations on the Data sheet. If Cu^{2+}(aq) is indicated, add more Mg to the supernatant, to complete the precipitation.
If additional precipitation is necessary, you must recover the copper that was transferred to the ammonium hydroxide test solution.
Rinse the test tube and pipet with distilled water from a wash bottle, collecting the rinse water and ammonium hydroxide into the supernatant acid solution. Repeat the Cu^{2+}(aq) test, with new ammonium hydroxide test solution, after the added Mg has reacted.
Dissolve any excess solid magnesium that might remain by adding a small amount of concentrated HCl.

5f.   Be certain all the copper metal precipitate has settled. Decant the supernatant liquid carefully, being careful not to lose any solid.
Do not try to pour off *all* the liquid. It is better to leave a little liquid with the solid than to lose any solid.

5g.   Wash the copper metal precipitate by decantation at least 3 times, following the procedure in Section XIII, Methods. Instead of pouring the decanted liquid through a filter, simply pour it into a 500 mL beaker.
Do not pour the supernatant directly into a sink. By collecting it in a beaker, you still can recover any solid that might accidently be poured off with the liquid.

5h.   Weigh a clean dry evaporating dish and record the mass.

5i.   Transfer all of the copper precipitate from the beaker to the evaporating dish. Use a rubber policeman on a glass stirring rod to sweep out the solid. Rinse out the beaker with distilled water from a wash bottle. See Figure 0.13d, Section VIII, Methods, for a similar procedure.
Be sure to rinse the policeman into the evaporating dish.

5j.   Let the solid settle in the evaporating dish and then carefully decant about 2/3 of the water. Use a dropping pipet to remove as much of the remaining water as possible, without touching any of the solid.

5k.   If a drying oven is available, place the dish in it at 120°C for 20 min to dry the copper.
Place the dish in the oven on a small piece of paper towel with your name written on it.

When the dish and its contents are dry, remove it from the oven and set it aside to cool to room temperature. Handle the hot dish with tongs or a folded paper towel used as a hot pad.

If a drying oven is not available, dry the copper by the following procedure:
  i.   Prepare a heating bath by placing a 250 mL beaker, 3/4 full of tap water, on a hot plate or on a wire gauze on a ring stand over a gas burner.
  ii.   Place the evaporating dish with the wet precipitate on top of the beaker and heat the water to boiling. Steam will warm the dish and dry the copper.
  iii.   Then set the dish aside to cool. Handle the hot dish with tongs or a folded paper towel. Wipe off any moisture on the outside of the dish with a paper towel.

5l.   When cool, weigh the dish and its contents and record the mass.

5m.   Repeat steps 5k and 5l until two successive weighings are within 2 mg of each other.

Name ______________________________________   Date ____________

EXPERIMENT 13
PRELABORATORY EXERCISE

1.   Heating solid barium carbonate to around 1200°C produces the following reaction:

$$BaCO_3(s) \longrightarrow BaO(s) + CO_2(g)$$

What class of reaction is this?   ______________________________________

2.   In a footnote to Reaction Step 1, oxidation-reduction reactions are described that produce NO_2 as an atmospheric pollutant. In the reaction

$$2\,NO + O_2 \longrightarrow 2\,NO_2$$

identify which compounds are oxidized and which are reduced; identify which atoms in the compounds change their oxidation state and give the magnitude and direction (+ or -) of the change for each atom.

    Compound oxidized: ________

    Atom that changes oxidation number is: ________

    Magnitude and sign of oxidation number change: ________

    Compound reduced: ________

    Atom that changes oxidation number is: ________

    Magnitude and sign of oxidation number change: ________

3.   Suppose your copper wire sample weighed 0.531 g at the beginning of this experiment and the recovered copper at the end weighed 0.522 g. What was your percent yield?

                                                              ____

Calculations:

Name ____   Date ____________

EXPERIMENT 13

DATA
(Observe significant figures in all calculations.)

A. Reaction Step 1

1c.   Mass of watch glass with copper metal: ________________

1a.   Mass of watch glass alone: ________________

       Mass of copper metal: ________________

1f.   Color of gas evolved: ________________

1h.   Color of solution: ________________

B. Reaction Step 2

2a.   Color change in solution: ________________

C. Reaction Step 3

3a.   Observations:

3h.   Observations:

D. Reaction Step 4

4a.   Observations:

E. Reaction Step 5

5e.   Observations:

5m.   Mass of evaporating dish and precipitate: ______________

5h.   Mass of evaporating dish alone: ______________

       Mass of copper precipitate: ______________

       Percent of copper recovered: ______________

Calculations:

Name ____   Date _____________

EXPERIMENT 13
QUESTIONS

Reaction Step 1
1.   In Procedure step 1f, what is the gas evolved?   __________

2.   In Procedure step 1h, what species causes the color of the solution?   __________

Reaction Step 2
3.   In Procedure step 2a, a precipitate is supposed to form after the acid solution has been neutralized by added base. Frequently, however, the precipitate starts to form before the point of neutralization is reached. Explain how this can happen.

Reaction Step 3
4.   In Procedure step 3e, why should the filtrate be clear, without cloudiness?

5.   In Procedure step 3g, what is being washed from the solid precipitate?

Reaction Step 4
6.   In Procedure step 4b, what species is indicated by the color of the solution, after the solid has been dissolved?

Reaction Step 5
7.   In Procedure step 5a, what reason might cause the solution to remain slightly blue in color, after all the magnesium has dissolved?

8.   In Procedure step 5d, what is being washed from the copper metal precipitate?

EXPERIMENT 14
Solubility of Salts:
Temperature Dependence

PRELABORATORY PREPARATION
Do the Prelaboratory Exercise and turn it in at the beginning of your laboratory period.

INTRODUCTION
The solubility of salts[1] in water is a topic of great practical and theoretical importance, and it continues to be a subject of active research. Because measuring the solubility of compounds can be a simple procedure, solubility experiments appeared early in the history of chemistry. At the beginning of the nineteenth century, the concept of molecule formation from atoms bonding together was just emerging. At that time, many experiments were performed to investigate how the properties of liquids, especially water, were changed when various substances were dissolved into them. It was found that freezing and boiling temperatures, electrical conductivity, vapor pressure, and osmotic pressure of aqueous solutions depended strongly on the nature and amount of the solute. In 1887, Arrhenius proposed that these experiments were best explained if salts, acids, and bases dissolve by dissociating into ions.

During this same period, careful measurements were being made on how salt solubility is affected by experimental variables such as temperature, pressure, solute concentration, and the presence of common and foreign ions.[2] All of these experiments were important for the development of modern theories of chemical equilibrium. In this experiment, you will determine the temperature dependence of solubility for an unknown solute.

Heat Changes Accompanying the Solution of a Salt
You probably have experienced the spontaneous heating of a solution obtained by diluting concentrated acids, such as sulfuric, hydrochloric, or nitric, with water. Dissolving solid sodium hydroxide in water also releases considerable heat. Less noticeable, perhaps, but easily observed, is the fact that dissolving some substances in water **cools** the solution. Potassium nitrate, KNO_3, is such a compound.

In general, whenever a compound dissolves in a solvent, heat energy is either released (heating the solution) or absorbed (cooling the solution).

---

[1] Salts are ionic solids that dissolve in water by dissociating into positive and negative ions, *other than H^+ and OH^-*. Substances that yield H^+ and OH^- are called acids and bases, respectively.
An example of a salt is NaCl, which dissolves by
$$NaCl \xrightarrow{H_2O} Na^+(aq) + Cl^-(aq)$$
An example of an acid is HCl, which dissolves by
$$HCl \xrightarrow{H_2O} H^+(aq) + Cl^-(aq)$$
An example of a base is NaOH, which dissolves by
$$NaOH \xrightarrow{H_2O} Na^+(aq) + OH^-(aq)$$
Salts, acids, and bases also are called electrolytes, because they produce aqueous solutions that conduct electricity. Such solutions are called **electrolytic solutions.**

[2] Foreign and common ions refer to ions contributed by substances different from the salts being measured. For example, if the measured salt is AgCl, which ionizes to Ag^+ and Cl^-, adding some KCl to the solution introduces the foreign ion K^+ and the common ion Cl^-.

The heat change reflects energy changes in the system, due to any dissociation that occurs and to the new attractive forces established between solute and solvent.

The discussion above suggests that dissolving a solid salt in a solvent can be regarded as a reversible chemical reaction:

$$\text{solid salt solute + solvent} \rightleftharpoons \text{solution} \tag{14.1}$$

The forward reaction is the dissolving of solid solute into solution, and the reverse reaction is the precipitation of dissolved species back into solid form. The heat change accompanying Reaction 14.1 is called the **enthalpy of solution**, written ΔH_{sol}.

Reactions that release heat are said to be *exothermic.* **For exothermic reactions, ΔH_{sol} is a negative quantity, because the reactants** *lose* **energy to the surroundings, heating the solution.**

Reactions that absorb heat are said to be *endothermic.* **For endothermic reactions, ΔH_{sol} is a positive quantity, because the reactants** *gain* **energy from the surroundings, cooling the solution.**

For any chemical change, the enthalpy changes for the forward and reverse reactions will have opposite signs, i.e., $\Delta H(\textbf{forward}) = -\Delta H(\textbf{reverse})$.

This means that the enthalpies of the forward and reverse reactions of 14.1 must have opposite signs. If the forward reaction is endothermic, then the reverse reaction must be exothermic, i.e.:

As more and more solid solute is added to a solvent, a point is always reached when no more solute will dissolve and if any more solute is added, it will remain in solid form. A solution into which no more solute will dissolve is called **saturated.** The solute concentration when saturation is reached depends on the temperature. The equilibrium condition of Reaction 14.1 exists only for saturated solutions.

The only way to be certain that a solution is saturated, is to have some undissolved solid present that has come to equilibrium with the solution.

Influence of Temperature on Solubility

Everyone is familiar with the fact that water-soluble substances dissolve more quickly in hot water than in cold water. It frequently is assumed that **more** of a solute will dissolve in hot water than in cold water and it may come as a surprise that this is not always the case. Many compounds become **less** soluble as their solutions are heated; the sulfate salts shown in Figure 14.1 are just two of many examples.

The change in solubility with temperature is related to the *sign* **of ΔH_{sol}, by Le Chatelier's principle:**

When a chemical system in equilibrium has its equilibrium disturbed by some imposed change (such as a temperature, pressure, or concentration change), the system will seek a new equilibrium by reacting in the direction that tends to counteract and minimize the imposed change.

To illustrate, consider the case where KNO_3 crystals are dissolved in water:

$$KNO_3 \xrightarrow{\text{H}_2\text{O}} K^+(aq) + NO_3^-(aq)$$

ΔH_{sol} is positive for this process, indicating that heat is absorbed in the endothermic forward reaction where solid dissolves into the solvent. When this system is in equilibrium, **lowering** the temperature causes the system to seek a new equilibrium by reacting in the direction that releases heat, to counteract the disturbance and raise the lowered temperature back towards its original value. Therefore, the reverse exothermic reaction is promoted and solute will precipitate from solution. If the temperature is **raised,** the forward endothermic reaction is promoted, because it will absorb some of the added heat energy and minimize the temperature rise. Thus, raising the temperature causes more solute to dissolve.

It should be clear from the above discussion that if you know how the solubility of a compound changes with temperature, you can predict the sign of its enthalpy of solution, using Le Chatelier's principle.

- **With reversible reactions, the forward and reverse reactions always have opposite signs.**

- **If heating a saturated solution causes more solute to dissolve, the forward dissolution reaction must be endothermic and the reverse precipitation reaction must be exothermic. Therefore, the sign of ΔH_{sol} is positive.**

- **If heating a saturated solution causes dissolved solute to precipitate, the forward dissolution reaction must be exothermic and the reverse precipitation reaction must be endothermic. Therefore, the sign of ΔH_{sol} is negative.**

The following example illustrates how to apply this principle.

EXAMPLE:

In a solubility experiment, it is found that the maximum amount of lithium carbonate, Li_2CO_3, that can be dissolved into 100 g of water at 10°C is 1.43 g.

Thus, the solubility of Li_2CO_3 at 10°C is 1.43 g/100 g H_2O.

When the saturated solution of Li_2CO_3 is heated from 10°C to 100°C, solid Li_2CO_3 precipitates, indicating a lower solubility at the higher temperature. This experiment shows that the forward dissolution reaction is exothermic and that the sign of ΔH_{sol} is negative.[3] The reverse precipitation reaction absorbs heat and tends to counteract the temperature increase.

PLAN OF EXPERIMENT

You will measure the solubility of an unknown substance at several temperatures. From a graph of solubility vs. temperature, you can predict the sign of ΔH_{sol} for the unknown. The measurement involves dissolving a known amount of solute into hot water, observing whether precipitation occurs on cooling or heating, and measuring the temperature at which saturation occurs and precipitation begins.

SAFETY

Wear approved eye protection.

PROCEDURE

Make all weighings to 1 mg on a triple beam balance.

1.   Obtain an unknown solid sample in a clean test tube and record its number on the Data sheet.

2.   Weigh a clean, empty weighing bottle or 10 mL beaker. Add the correct amount of sample to the container.
For unknown no. 1, weigh about 7 mg of sample into the container.
For unknown no. 2, weigh about 5 mg of sample into the container.
Record the mass of sample and container.

3.   Transfer the solid to a 125 mL Erlenmeyer flask and reweigh the empty weighing container and any solid that remains in it. Record the empty mass.

4.   **For unknown no. 1, pipet 5 mL of distilled water into the flask containing the unknown. For unknown no. 2, pipet 10 mL of distilled water into the flask containing the unknown.**

5.   Clamp a thermometer vertically to a support rod, with its bulb hanging free. Use a split rubber stopper or split rubber tubing between the thermometer and clamp. Clamp it high enough to slide your flask underneath without tilting. Adjust the assembly so it is easy to read the thermometer. Later, you will immerse the thermometer bulb in you sample solution, as it cools.

6.   Gently warm the flask and solution on a hot plate or over a burner to dissolve the solid.
This is done with the flask *away* from the thermometer on the support rod.

7.   When all the solid has dissolved, remove the flask from heat and bring it up around the thermometer, holding it tilted so the thermometer bulb is fully immersed in the solution. Allow the solution to cool, keeping the solution in motion by gentle shaking and swirling the flask.
The thermometer bulb must remain immersed at all times.

8.   Watch carefully for the first signs of precipitation. Note the onset of precipitation and record the temperature at which crystals first appear.
Go back and forth through the crystallization temperature several times, by reheating and recooling the solution, to get a good measurement.

9.   Repeat steps 4-8 four more times, each time adding additional water to the flask before heating.
For unknown no. 1, add 5 mL each time.
For unknown no. 2, add 10 mL each time.

---

[3]For most salts, ΔH_{sol} is positive. However, there are enough exceptions, such as this example where ΔH_{sol} is negative, that you cannot safely *assume* a positive sign.

10.   From the known mass of sample and the known volume of water for each run, calculate the concentrations of each solution at the onset of precipitation. Record the concentrations as **g of solute per 100 g water.** Assume that the density of water is 1.000 g/mL.

>   **The calculated values are the saturation concentrations at the temperature where precipitation just begins.**

Notice that the mass of solute remains constant, but the mass of water changes in each of the 5 measurements.

11.   Plot solubility of your unknown, using **g solute per 100 g water** on the y-axis and **temperature** on the x-axis.

12.   On the data sheet, report the solubility of your unknown at 30°C and at 50°C, reading the values from your graph.

13.   Predict the sign of ΔH_{sol} for your unknown, on the Data sheet.

Name ____   Date ____________

EXPERIMENT 14
PRELABORATORY EXERCISE

Figure 14.1 shows the temperature dependence of solubility for 4 different salts in water. In the table below the figure, give the sign (+ or -) of ΔH_{sol} for each salt and state whether heat is evolved or absorbed by the solution when the salt is dissolved.

Salt	Sign of ΔH_{sol} (+ or -)	Heat change on dissolving in water (make a checkmark in the correct column)	
		heat absorbed	heat released
KNO_3			
Na_2SO_4			
$NaCl$			
$Ce_2(SO_4)_3$			

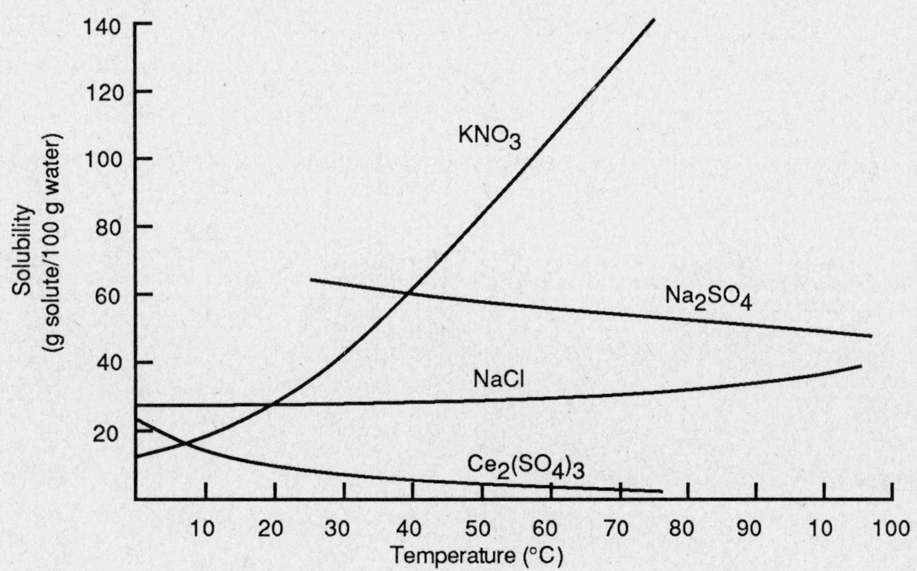

Figure 14.1.   Temperature dependence of solubility for several salts.

Name ____   Date _____________

EXPERIMENT 14

DATA
(Observe significant figures in all calculations.)

1. Unknown number: _________________

2. Mass of weighing bottle: _________________

3. Mass of weighing bottle + solid: _________________

4. Mass of weighing bottle + remaining solid after sample transfer: _________________

5. Mass of solid sample: _________________

Run No.	Volume of water added in each run, (mL)	Total volume of water (mL)	Total mass of water[a] (g)	Solubility[b]	Temperature of initial crystallization
1					
2					
3					
4					
5					

[a] Assume that total mass of water = volume in mL, which amounts to taking the density as 1.000 g/mL.

[b] Use: solubility = mass of solute dissolved per 100 g water.

Solubility at 30°C: _____________________

Solubility at 50°C: _____________________

(Read solubilities from graph on next page.)

Calculations:

DATA

Graph of solubility vs. temperature for unknown No. _________

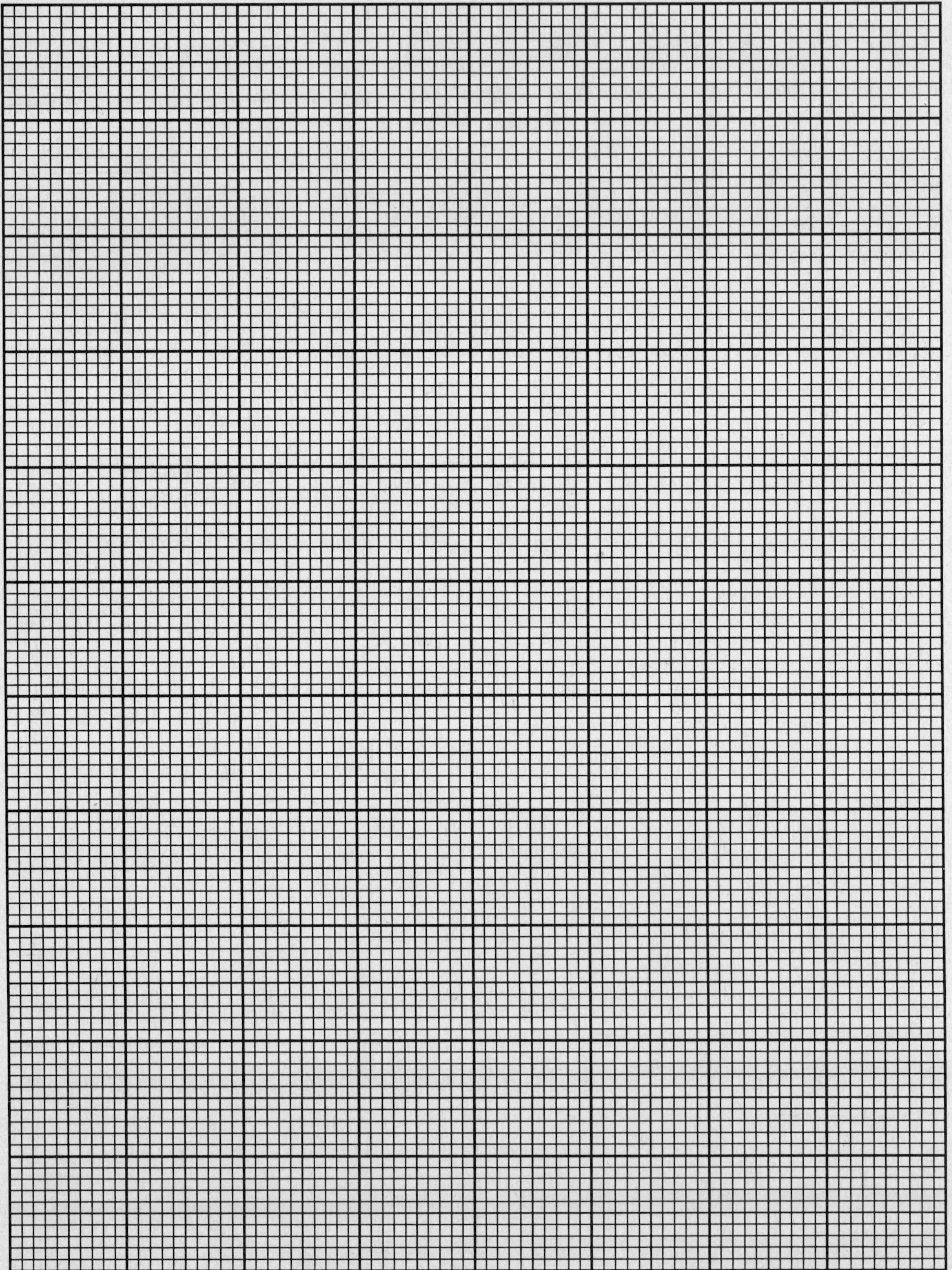

Name ____   Date ____________

EXPERIMENT 14
QUESTIONS

1.   If your thermometer was inaccurate and gave too high a temperature for every reading, would your error in determining the solubility at 30°C have been +, 0, or -?   _______

Explain:

2.   If you had failed to notice the very beginning of precipitation and didn't measure the temperature until a significant amount of the solute had precipitated, would your error in determining the solubility at 30°C have been +, 0, or -?   _______

Explain:

EXPERIMENT 15
Determination of a Solubility Product: Effects of Common and Foreign Ions on Solubility

PRELABORATORY PREPARATION
1. Do the Prelaboratory Exercise and turn it in at the beginning of your laboratory period.
2. Read the Introduction to Experiment 14. Know the meaning of the terms **solubility** and **saturated solution.**
3. Review use of the buret and pipet, Section IV, Methods, pages 16-18; the techniques of decanting and transferring a precipitate, Section VIII, Methods, pages 20-23; and the use of the spectrometer, Section XIII, Methods, pages 28-30; and .
4. Review the discussion of standard deviation, Chemical Calculations, page 1.

INTRODUCTION
When a solid salt is dissolved in water, the attractive forces that held the solid together before dissolution continue to act, working to reverse the dissolving process and bring the solute molecules back together again into crystalline form. However, new forces between solute ions and water molecules act to keep the solute dissolved. Because of this "tug-of-war," salt dissolution reactions always are reversible.

Dissolving a salt to make a solution is a reversible chemical reaction:

$$\text{solid salt solute + solvent} \rightleftharpoons \text{solution} \tag{15.1}$$

Removal of solute ions from the crystalline solid into solution is the forward reaction, and precipitation of solid solute out of solution is the reverse reaction.

Reaction 15.1 attains equilibrium when the rates of the forward and reverse reactions are exactly equal.

A necessary condition for Reaction 15.1 to be in equilibrium is that some solid solute be present in the system. Only then is the forward reaction possible.

When solid solute is present and Reaction 15.1 has attained equilibrium, the solution is said to be **saturated** with the solute.

Solubility Product
The solubility product is a form of equilibrium constant that expresses the concentration relationships among the dissolved ions in a saturated solution. Suppose we made a saturated aqueous solution of the salt CaF_2. The equilibrium reaction is:

$$CaF_2(s) \xrightleftharpoons{H_2O} Ca^{2+}(aq) + 2\,F^-(aq) \tag{15.2}$$

The solubility product expression for Reaction 15.2 is: $K_{sp} = [Ca^{2+}][F^-]^2$.

The solubility product applies *only* to saturated solutions.

Reaction Quotient
The behavior of a salt solution, such as that made by Reaction 15.2, can be predicted from a mathematical quantity called the **reaction quotient.** Consider the dissolution reaction of the generalized salt A_aB_b,

$$A_aB_b(s) \xrightleftharpoons{H_2O} a\,A^{m+}(aq) + b\,B^{n-}(aq) \tag{15.3}$$

The reaction quotient, **Q**, is defined as:

$$Q = [A^{m+}]^a[B^{n-}]^b \tag{15.4}$$

The reaction quotient has exactly the same mathematical form as the solubility product. The difference is that **Q** applies to *any* ionic solution, whether it is saturated or not. **Q** describes the ion product even while dissolution or precipitation reactions are taking place, in which case the value of **Q** will change with time. **Q** is more general than K_{sp}. It describes the ion product at any moment and for any combination of ion concentrations and does not have any particular value that is a characteristic of the solute.

K_{sp} is a special case of Q. It is the particular value of Q that represents the ion concentrations at saturated equilibrium.

The usefulness of Q comes from comparing its value with the tabulated value of K_{sp} for the same solute.

- **If $Q = K_{sp}$, solid solute and solution are in *equilibrium* and the solution is saturated.**

- **If $Q < K_{sp}$, the solution is *unsaturated*. This means that aquated ion concentrations are too low for equilibrium between solid and solution and, as long as any solid is present, solid solute will dissolve until $Q = K_{sp}$.**
- **If $Q > K_{sp}$, the solution is *supersaturated*. This means that aquated ion concentrations are too high for equilibrium and solid solute must precipitate until $Q = K_{sp}$.**

While a salt is being dissolved in water, the aquated ion concentrations increase until saturation is reached or until all the solid has dissolved, whichever occurs first. Equation 15.4 indicates that the value of Q will increase correspondingly, until saturation is reached and $Q = K_{sp}$. Values for K_{sp} of different compounds are tabulated and collected in handbooks and textbooks. The units for Q and K_{sp} depend on the **a** and **b** exponents in Equation 15.4, and, thus, on the stoichiometry of the equilibrium reaction.

When concentration is given in mol L^{-1}, the units of Q are $(mol\ L^{-1})^{a+b}$, where *a* and *b* are found from Equations 15.3 and 15.4.

Example 1 illustrates the use of Equation 15.4.

EXAMPLE 1:

Calculate the saturated ion concentrations in a solution of lead chloride, at 20°C, given that $K_{sp}(PbCl_2) = 1.6 \times 10^{-5}$ at 20°C.

SOLUTION:

The equilibrium reaction is:

$$PbCl_2(s) \underset{}{\overset{H_2O}{\rightleftharpoons}} Pb^{2+}(aq) + 2\ Cl^-(aq) \qquad (15.5)$$

Equation 15.5 corresponds to 15.1, with $PbCl_2(s)$, lead chloride, as the solute and H_2O the solvent. The solution contains $Pb^{2+}(aq)$ and $Cl^-(aq)$. When equilibrium exists and the solution is saturated, $Q = K_{sp}$. This condition allows the concentrations of Pb^{2+} and Cl^- to be calculated if the value of the solubility product is known at the equilibrium temperature.

Applying Equation 15.4 to Reaction 15.5, the expression for the reaction quotient is

$$Q = K_{sp} = [Pb^{2+}][Cl^-]^2 \qquad (15.6)$$

In a handbook, we find that $K_{sp}(PbCl_2) = 1.6 \times 10^{-5}$, at 20°C.

at 20°C. Assume that pure $PbCl_2$ was dissolved in pure water, so there are no other sources of Pb^{2+} and Cl. When equilibrium is attained, the dissolved concentration of Cl^- must be twice that of Pb^{2+}, by Reaction 15.5. Let **$[Pb^{2+}] = x$** and **$[Cl^-] = 2x$**. At equilibrium, when the solution is saturated, Equation 15.6 becomes:

$$K_{sp}(PbCl_2) = [x][2x]^2 = 1.6 \times 10^{-5}$$
$$4x^3 = 1.6 \times 10^{-5}$$
$$x = 1.6 \times 10^{-2}\ mol\ L^{-1}$$

$[Pb^{2+}] = x = 1.6 \times 10^{-2}\ mol\ L^{-1}$

$[Cl^-] = 2x = 3.2 \times 10^{-2}\ mol\ L^{-1}$

Values for solubility products are determined experimentally by measuring solute concentrations under saturation conditions. Once they are known, they are tabulated and used to find equilibrium concentrations, as in Example 1.

Solubility

Whereas *solubility product* is related to a measurement of the concentrations of dissolved species, *solubility* is related to a measurement on the solid solute.

Solubility of a substance is the maximum amount that will dissolve in a given amount of solvent, to make a saturated solution.

Solubility can be calculated from the solubility product, and, like K_{sp}, is dependent on the temperature.

For a salt dissolved in water:

- ***solubility product* is related to the concentrations of dissolved ions in a saturated solution.**
- ***solubility* refers to the maximum amount of solid salt that dissolves to make a saturated solution.**

It is important to note that, although solubility and solubility product are closely related, they have different numerical values and units.

Examples 2 and 3 illustrate the difference between solubility product and solubility, for the two chloride salts, silver chloride (AgCl), and lead chloride ($PbCl_2$). The simpler case of AgCl is considered first.

EXAMPLE 2:
Solubility of Silver Chloride
Find the solubility of solid AgCl in water at 20°C.

SOLUTION:
At 20°C, $K_{sp}(AgCl) = 1.8 \times 10^{-10}$. The equilibrium reaction is

$$AgCl(s) + H_2O \rightleftharpoons Ag^+(aq) + Cl^-(aq) \tag{15.7}$$

Therefore, the solubility product expression is

$$K_{sp}(AgCl) = [Ag^+][Cl\text{-}] = 1.8 \times 10^{-10} \ (mol \ L^{-1})^2 \tag{15.8}$$

Notice that the units of $K_{sp}(AgCl)$ are $(mol \ L^{-1})^2$ because the **a** and **b** exponents in 15.8 both are unity, so that a + b = 2. (See discussion preceding Example 1.)

The units for solubility are different from those for solubility product. Solubility units usually are chosen to be either **moles of solute per liter of solution** (the most convenient units for calculations), or **grams of solute per 100 g of solvent** (the most convenient units for making up saturated solutions in the laboratory).

If AgCl is the only solute in the solution, the moles of $Ag^+(aq)$ must be equal to the moles of $Cl^-(aq)$, by Reaction 15.7, *and this quantity also is equal to the moles of AgCl that have dissolved..* Let S_{AgCl} stand for the solubility of AgCl:

$$S_{AgCl} \ (mol \ L^{-1}) = [AgCl(dissolved)] = [Ag^+] = [Cl^-]$$

For substituting into K_{sp}, write the ion concentrations in terms of S_{AgCl}. Substituting S_{AgCl} into K_{sp}, Equation 15.8 gives:

$$K_{sp}(AgCl) = [Ag^+][Cl^-] = (S_{AgCl})(S_{AgCl}) = S^2_{AgCl} = 1.8 \times 10^{-10}(mol \ L^{-1})^2$$

$$\mathbf{S_{AgCl}} = \{1.8 \times 10^{-10}(mol \ L^{-1})^2\}^{1/2} = \mathbf{1.3 \times 10^{-5} \ mol \ L^{-1}}$$

The units of solubility may be changed from **mol L^{-1}** to **g per 100 g H_2O** by using the molar mass of AgCl = 143.34 .

$$\mathbf{S_{AgCl}} = (1.3 \times 10^{-5} \ mol \ L^{-1})(143 \ g \ mol^{-1}) = \mathbf{1.9 \times 10^{-3} \ g \ L^{-1}}$$

We may take the density of water as 1 g mL^{-1}, with sufficient accuracy, so that 100g H_2O = 100 mL H_2O. One hundred mL is 1/10 of a liter, so the amount of AgCl that dissolves in 100 mL of water is 1/10 of the amount that will dissolve in 1 liter.

The solubility in g per 100 g H_2O is just 1/10 the solubility in g per liter:

$$\mathbf{S_{AgCl}} = \frac{\mathbf{1.9 \times 10^{-3}}}{\mathbf{10}} = \mathbf{1.9 \times 10^{-4} \ g \ per \ 100 \ g \ H_2O}$$

EXAMPLE 3:
Solubility of Lead Chloride
Find the solubility of solid $PbCl_2$ in water at 20°C.

SOLUTION:
At 20°C, $K_{sp}(PbCl_2) = 1.6 \times 10^{-5}$. The equilibrium reaction is the same as Reaction 15.5 in Example 1:

$$PbCl_2(s) + H_2O \rightleftharpoons Pb^{2+}(aq) + 2 \, Cl^-(aq) \tag{15.5}$$

This case is slightly more complicated than that of AgCl, because of the stoichiometry of the equilibrium reaction. The solubility product expression is:

$$K_{sp}(PbCl_2) = [Pb^{2+}][Cl^-]^2 = 1.6 \times 10^{-5} \ (mol \ L^{-1})^3 \tag{15.9}$$

Notice that the units of $K_{sp}(PbCl_2)$ are different from those of $K_{sp}(AgCl)$.

If $PbCl_2$ is the only solute in the solution, the moles of $Cl^-(aq)$ must be equal to twice the moles of $Pb^{2+}(aq)$, from Reaction 15.5. The moles of $PbCl_2(s)$ that dissolve must be equal to the moles of $Pb^{2+}(aq)$ and, also, to 1/2 the moles of $Cl^-(aq)$.

Let S_{PbCl_2} stand for the solubility of $PbCl_2$:

$$S_{PbCl_2} \text{ (mol L}^{-1}) = [PbCl_2(\text{dissolved})] = [Pb^{2+}] = \frac{1}{2} [Cl^-]$$

For substituting into K_{sp}, write the ion concentrations in terms of S_{PbCl_2}:

$$[Pb^{2+}] = S_{PbCl_2} \text{ and } [Cl^-] = 2\, S_{PbCl_2}$$

Then, substituting into K_{sp}, Equation 15.9, gives:

$$K_{sp}(PbCl_2) = [Pb^{2+}][Cl^-]^2 = (S_{PbCl_2})(2\, S_{PbCl_2})^2 = 4\, S_{PbCl_2}^3 = 1.6 \times 10^{-5}(\text{mol L}^{-1})^3$$

$$S_{PbCl_2} = \left\{ \frac{1}{4} \times 1.6 \times 10^{-5}(\text{mol L}^{-1})^3 \right\}^{1/3} = \mathbf{1.6 \times 10^{-2} \ mol \ L^{-1}}$$

The molar mass of $PbCl_2$ is 278.12, so that:

$$S_{PbCl_2} = (1.6 \times 10^{-2} \text{mol L}^{-1})(278 \text{ g mol}^{-1}) = 4.4 \text{ g L}^{-1}$$

As in Example 2, 1/10 of this amount will dissolve in 100 g H_2O (taking the density of water as 1 g mL^{-1}).

$$S_{PbCl_2} = \mathbf{0.44 \ g \ per \ 100 \ g \ H_2O}$$

Notice that the relation between solubility and solubility product depends upon the number of ions into which a salt dissociates when it dissolves. The two particular cases of Examples 2 and 3 are generalized for other cases in Table 15.1.

Table 15.1.   Relation between solubility and solubility product, for several generalized salts of composition A_aB_b in pure water.

a	b	salt type	Equations for K_{sp} and S in pure water
1 1 Examples: AgCl, NaCl, CaS		AB	$K_{sp} = [A^{m+}][B^{n-}] = (S)(S) = S^2$ $S = K_{sp}^{1/2}$
1 2 Examples: CaCl$_2$, PbI$_2$		AB$_2$	$K_{sp} = [A^{m+}][B^{n-}]^2 = (S)(2\, S)^2 = 4\, S^3$ $S = \{K_{sp}/4\}^{1/3}$
1 3 Examples: AlCl$_3$, CrBr$_3$		AB$_3$	$K_{sp} = [A^{m+}][B^{n-}]^3 = (S)(3\, S)^3 = 27\, S^4$ $S = \{K_{sp}/27\}^{1/4}$
2 3 Examples: Al$_2$S$_3$, Mn$_2$(C$_2$O$_4$)$_3$		A$_2$B$_3$	$K_{sp} = [A^{m+}]^2[B^{n-}]^3 = (2\, S)^2(3\, S)^3 = 108\, S^5$ $S = \{K_{sp}/108\}^{1/5}$
2 1			

The missing entries at the bottom of the table are to be filled in as part of the Prelaboratory Exercise.

Common Ion Effect

A **common ion** in a solution is an ion that is the same as one formed from the solute, but is supplied from a different source. It might already be in the solvent or could be added later.

If the solution contains a common ion, the solubility of the solute is decreased.[1]

For example, salts containing Cl^-, such as $NaCl$, $AgCl$, $CuCl_2$, etc., will be less soluble in a weakly acidic aqueous solution of HCl than in pure water, because the acid solution already contains the common ion Cl^-. One important use of the common ion effect is to lower toxic metal ion concentrations in water supplies and waste effluents. The decrease in solubility can be calculated by using the solubility product, see Example 4.

EXAMPLE 4:

Calculate the solubility of AgCl in a 1.0×10^{-3} M HCl solution, and compare with its solubility in pure water.

SOLUTION:

The common ion is Cl^-, supplied by both HCl and AgCl. The reactions contributing ions to the solution are:

$$HCl + H_2O \underset{H_2O}{\rightleftharpoons} H_3O^+ + Cl^-(aq) \tag{15.9}$$

$$AgCl(s) \rightleftharpoons Ag^+(aq) + Cl^-(aq) \tag{15.10}$$

To use the solubility product, we must find the total concentration of Cl^- supplied by Reactions 15.9 and 15.10: $[Cl^-_{total}] = [Cl^-_{HCl}] + [Cl^-_{AgCl}]$. $[Cl^-_{HCl}]$ is determined by the molarity of the acid solution. Therefore,

$$[Cl^-_{HCl}] = 0.0010 \text{ mol L}^{-1}$$

$[Cl^-_{AgCl}]$ is determined by how much AgCl has dissolved in the solution. This quantity is unknown, but it can be calculated. We know that:

$[Cl^-_{AgCl}] = [Ag^+] = S'_{AgCl}$, where S'_{AgCl} designates the unknown solubility of AgCl in the 0.0010 M HCl solution.

Substitute these quantities into the expression for K_{sp}, which always depends on *total* ion concentrations:

$$K_{sp} = [Ag^+_{total}][Cl^-_{total}] = (S'_{AgCl})(0.0010 + S'_{AgCl}) = 1.8 \times 10^{-10} \tag{15.11}$$

Note that $[Ag^+_{total}] = [Ag^+_{AgCl}]$ because there is no other source of Ag^+. All that remains is to solve 15.11, which is a quadratic equation, for S'. Multiplying 15.11 out in full gives:

$$S'^2 + 0.0010 \, S' - 1.8 \times 10^{-10} = 0$$

This is solved using the quadratic formula, $x = \{-b \pm (b^2 - 4ac)^{1/2}\}/2a$, to give:

$$S' = \text{solubility of AgCl in the 0.0010 M HCl solution} = \{-0.001 \pm [10^{-6} - 4(1)(-1.8 \times 10^{-10})]^{1/2}\}/2a$$

$$\mathbf{S' = 1.8 \times 10^{-7} \text{ mol L}^{-1}}$$

(Only the plus value of the plus/minus term in the quadratic equation can be used because it is physically necessary that S' be a positive quantity.)

---

[1] This statement always is true when the common ion is in low concentration. However, it sometimes happens that a common ion in high concentration will form complexes with the solute, actually increasing its solubility. For example, the solubility of AgCl in water decreases, as expected, if additional Cl^- is present from another source, as long as $[Cl^-]$ is less than about 3×10^{-3} mol L^{-1}. If $[Cl^-]$ is increased above this value, the solubility of AgCl increases. In a concentrated solution of $[Cl^-] = 1.0$ mol L^{-1}, the solubility of AgCl is nearly the same as in pure water. This is attributed to reactions forming additional soluble species, such as

$$AgCl(s) + Cl^- \rightleftharpoons AgCl_2^-(aq)$$

$$AgCl(s) + 2\,Cl^- \rightleftharpoons AgCl_3^{2-}(aq)$$

$$AgCl(s) + 3\,Cl^- \rightleftharpoons AgCl_4^{3-}(aq)$$

The products of these reactions are called chloride complexes of silver chloride. As usual, working with dilute solutions spares us from having to consider such complications.

Foreign Ion Effect

A foreign ion is one that is different from the solute ions, and does not form a salt of low solubility with any of the solute ions.

For example, adding potassium nitrate, KNO_3, to a solution of AgCl would contribute the foreign ions K^+ and NO_3^-. Both of the new ion-pairs that now become possible, K^+ with Cl^- and Ag^+ with NO_3^-, form very soluble salts. The solids KCl and $AgNO_3$ will not precipitate. Furthermore, adding KNO_3 contributed no common ions. So, at first glance, it would appear that the addition of KNO_3 should have no effect whatsoever on the solubility of AgCl. Actually, however, the solubility of AgCl will increase slightly.

The presence of foreign ions in aqueous solutions generally increases the solubility of solute salts of low solubility.

This effect is rather complicated and difficult to calculate accurately. We will make no effort to do so. It is not difficult, however, to propose a qualitative explanation. Salts dissolve best in polar solvents. Foreign ions increase the polar character of the water solvent, effectively creating a different solvent that has stronger forces of attraction for the solute ions. Le Chatelier's principle does not predict this behavior. Figure 15.1 shows how the solubilities of two salts of low solubility, AgCl and $BaSO_4$, increase as the foreign salt KNO_3 is added to their aqueous solutions.

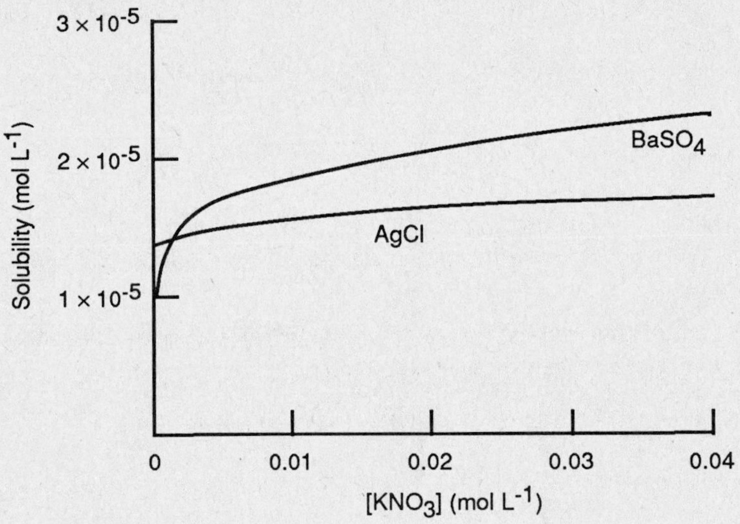

Figure 15.1. Solubility of AgCl and $BaSO_4$ in aqueous KNO_3 solutions.

PLAN OF EXPERIMENT

You make a saturated solution of the slightly soluble salt, PbI_2. By measuring the ion concentrations in that saturated solution, you can calculate the solubility product for the salt. Two methods for determining the ion concentrations are described: Part **A** uses the **titration method** and Part **B** uses the **spectrophotometric** method. Your laboratory instructor will tell you which method to use. You might use both methods, making one determination by each. After obtaining the solubility product, you measure the solubility of the salt in solutions containing common and foreign ions.

Titration Method

In the titration method, a measured volume of KI with a known concentration is titrated with a solution of $Pb(NO_3)_2$ until precipitation of PbI_2 just begins. At this point, the solution is saturated in Pb^{2+} and I^- and the concentrations of these ions are known accurately, as is the volume of the solution. From these data, the solubility product of PbI_2 can be calculated.

Spectrophotometric Method

In the spectrophotometric method, the response of the spectrophotometer is calibrated to allow a direct measurement of $[I^-]$. A saturated solution is prepared by dissolving solid PbI_2, which you have crystallized and purified into water until equilibrium is attained. From the stoichiometry of the dissolution reaction, you know the ratio of Pb^{2+} to I^- in solution. A sample of the solution is analyzed in the spectrophotometer to measure $[I^-]$, furnishing the data needed for calculating the solubility product.

SAFETY

1. Wear approved eye protection.

2. All lead compounds are toxic. Avoid inhalation of lead salt dusts and wash your hands after any contact
    with lead salts.

PROCEDURE
A. Titration Method
Your instructor may want you to make a second determination and average your results.
(Steps 1-4 are necessary for both Parts A and B.)

1.   From a buret, add 40.00 mL of 1.00×10^{-2} M KI solution to a 250 mL Erlenmeyer flask. Record the
exact molarity and volume of KI used on the data sheet.

2.   Titrate the KI solution with 2.00×10^{-2} M $Pb(NO_3)_2$. Stir constantly while titrating.
        **Watch carefully for the end point, which is the first appearance of either a faint yellow color or small
        sparkling crystals.**
The exact nature of the precipitate depends on the solution temperature. Crystals occur at around 28°C or
higher. Below 28°C, a faintly yellow colloidal precipitate is more likely.

3.   Record the exact amount of $Pb(NO_3)_2$ added and its exact molarity.
        The solution is now saturated in Pb^{2+} and I^-.

4.   Measure and record the solution temperature. Save the solution for use in Part B, if you are to do
Part B.

5.   Calculate K_{sp} for PbI_2. Assume that a negligible amount of Pb^{2+} and I^- was removed from the
solution by the small amount of PbI_2 that precipitated.

6.   Calculate the solubility (**S**) of PbI_2 in **grams per 100 mL of water.**

B. Spectrophotometric Method
(If you did not do Part A, begin with Steps 1–4 of Part A. Then continue below.)

1.   To the solution saved from **Part A,** add 20 mL 0.02 M $Pb(NO_3)_2$ solution.

2.   Stopper the flask and shake the mixture vigorously for about 2 min. Solid PbI_2 should precipitate.
Decant and discard most of the supernatant liquid.

3.   Transfer the solid PbI_2 and the remaining liquid to a test tube suitable for centrifuging. Centrifuge
the tube and contents for 2 min.

4.   Discard the liquid from the centrifuged tube, add 3-5 mL distilled water, stopper the test tube, and
shake the contents vigorously to wash excess Pb^{2+} and I^- from the solid.

5.   Centrifuge again and discard most of the liquid.
        You now have prepared pure, solid PbI_2.

6.   Add distilled water until the test tube is about 3/4 full. Stopper and shake the tube and its contents
vigorously for about 15 min. (Shake the tube for 1-2 min and then let it, and your arm, rest for an equal time,
over a total period of 15 min.)
        This lengthy and vigorous shaking is necessary to saturate the solution with dissolved PbI_2.

7.   Centrifuge the tube and its contents for 3 min. Then, decant most of the liquid into another clean,
dry centrifuge tube. Save the test tube containing the solid for use in Parts C and D.

8.   Centrifuge the tube containing the decanted liquid once again.
You are trying to obtain a clear solution with no color or solid material remaining.
Solid particles floating on the solution surface may be removed by dipping a small piece of clean, dry paper towel into the top few millimeters of solution and skimming off any solids.
If color remains, centrifuge for a longer time.
If solids settle out, decant the liquid into another test tube and centrifuge again.
Repeat this process until you have a clear, colorless solution with no solids. This is a saturated solution of PbI_2.

9.   Obtain a clean, dry spectrophotometer tube. Wipe the outside clean of fingerprints and smudges with cleaning tissue.
Do not use towels or handkerchiefs, which might leave lint on the tube.
Do not handle the lower half of the tube, because any dirt or fingerprint smudges on this part of the tube will scatter the light beam and cause an error in your light absorbance measurement.

10.   Rinse the spectrophotometer tube twice with a **small** amount (about 0.5 mL) of your centrifuged solution. Discard the rinse solution. Shake any remaining drops out of the tube and wipe the outside dry.

11.   Pipet 3.0 mL 0.02 M of KNO_2 (potassium *nitrite*) into the rinsed spectrophotometer tube and then add 2 drops of 6 M HCl.

12.   Measure and record the temperature of your clear, saturated PbI_2 solution.

13.   Pipet enough of your PbI_2 solution into the spectrophotometer tube to fill it to the level marked on the tube (or the level indicated by your instructor).

14.   Holding the tube at the top, rock it gently to mix the reagents. Insert it into the spectrophotometer and measure the absorbance of your solution at the absorbance maximum in the vicinity of 455 nm. Carefully follow the procedure of Section XVII, Methods. Record the absorbance value.

15.   If required by your instructor, repeat steps 1-14 for a second measurement.

16.   Using a calibration curve provided by your instructor, determine the concentration of I^- in your saturated PbI_2 solution. Record this value.

17.   Calculate the **solubility product** and **solubility** of PbI_2.

18.   For a more reliable result, obtain the values for K_{sp} determined by 5 of your labmates and determine the average value and standard deviation of six (counting your own) independent measurements.

C. Effect of Common Ions
1.   Use the solid PbI_2 saved in Step 7 of Part B. With a spatula, place approximately 1/2 the total solid into another clean, dry, centrifuge tube. Save the remainder of your solid sample for Part D and use the portion just transferred in the following procedure.

2.   To the centrifuge tube containing PbI_2 solid, add 10^{-3} M $Pb(C_2H_3O)_2$ solution (lead acetate solution; either hydrated or anhydrous salt may be used) until the tube is 3/4 full. Stopper and shake the tube vigorously as in Part B, Step 6. Repeat Steps 7-14 of Part B.

3.   Calculate the solubility of PbI_2 in 10^{-3} M $Pb(C_2H_3O)_2$ solution.

D. Effect of Foreign Ions
1.   Use the solid sample saved from Step 1, Part C. Add to this sample 10^{-3} M $NaNO_3$ (sodium nitrate) solution until the tube is 3/4 full. Stopper and shake the tube vigorously as in Part B, Step 6. Repeat Steps 7-14 of Part B.

2.   Calculate the solubility of PbI_2 in 10^{-3} M $NaNO_3$ solution.

Name ____ Date ____________

EXPERIMENT 15
PRELABORATORY EXERCISE

1.　　For PbF_2 at 18°C, $K_{sp} = 3.2 \times 10^{-8}$.
Calculate the solubility of PbF_2 at 18°C, in **mol L^{-1}** and in **g per 100 mL H$_2$O**.

$$S_{PbF_2} = \text{\underline{\hspace{3cm}}} \text{ mol L}^{-1}$$

$$S_{PbF_2} = \text{\underline{\hspace{3cm}}} \text{ g per 100 mL H}_2\text{O}$$

2.　　Table 15.1 is not completed for a salt of type **A$_2$B**, such as $Li_2(SO_4)$. Fill in the blank spaces of the table for an A$_2$B type salt. Enter your answers here also. Be sure to indicate the correct units.

For type A$_2$B salt:　　K_{sp}　=　_________________________________

　　　　　　　　　　　　　　　S　=　_________________________________

Calculations:

Name ____   Date ____________

EXPERIMENT 15

DATA

(Observe significant figures in all calculations.)

Exact molarity of $Pb(NO_3)_2$ solution: ____________________

Exact molarity of KI solution: ____________________

A. Titration Method

	Measurement 1	Measurement 2
1. Volume of KI solution:	___________	___________
2. Volume of $Pb(NO_3)_2$ solution:	___________	___________
3. Total volume of saturated solution:	___________	___________
4. Moles of I^-:	___________	___________
5. Concentration of I^- in saturated solution:	___________	___________
6. Average concentration of I^-:	___________	
7. Moles of Pb^{2+}:	___________	___________
8. Concentration of Pb^{2+} in saturated solution:	___________	___________
9. Average concentration of Pb^{2+}:	___________	
10. Temperature of saturated PbI_2 solution:	___________	___________
11. Average temperature of saturated PbI_2 solution:	___________	
12. Value of solubility product: $K_{sp} = [Pb^{2+}][I^-]^2$	___________	
13. Solubility of PbI_2 (g/100 g H_2O):	___________	

B. Spectrophotometric Method

Absorbance wavelength used for measurements: ___________ nm

	Measurement 1	Measurement 2
1. Temperature of saturated PbI_2 solution:	___________	___________
2. Average temperature of saturated PbI_2 solution:	___________	
3. Absorbance of saturated PbI_2 solution:	___________	___________
4. Average absorbance of saturated PbI_2 solution:	___________	
5. Measured concentration of I^- from calibration curve:	___________	
6. Calculated concentration of Pb^{2+} in saturated solution:	___________	
7. Value of solubility product: $K_{sp} = [Pb^{2+}][I^-]^2$	___________	
8. Solubility of PbI_2 (g/100 g H_2O):	___________	

DATA

9. List solubility values from 5 classmates:

   1. ______________

   2. ______________

   3. ______________

   4. ______________

   5. ______________

   Your own solubility value: 6. ______________

Calculations:

10. Average solubility from 6 measurements: ______________

11. Standard deviation of 6 measurements: ______________

C. Effect of Common Ions

1. Temperature of saturated PbI_2 solution: ______________

2. Absorbance wavelength measured: ______________

3. Absorbance of saturated PbI_2 solution containing 10^{-3} M $Pb(C_2H_3O)_2$: ______________

4. Measured concentration of I^- from calibration curve: ______________

5. Calculated concentration of Pb^{2+} contributed from PbI_2 only: ______________

6. Solubility of PbI_2 in 10^{-3} M $Pb(C_2H_3O)_2$ (g/100 g H_2O): ______________

D. Effect of Foreign Ions

1. Temperature of saturated PbI_2 solution: ______________

2. Absorbance wavelength measured: ______________

3. Absorbance of saturated PbI_2 solution containing 10^{-3} M $NaNO_3$: ______________

4. Measured concentration of I^- from calibration curve: ______________

5. Calculated concentration of Pb^{2+}: ______________

6. Solubility of PbI_2 in 10^{-3} M $NaNO_3$ (g/100 g H_2O): ______________

FINAL RESULTS

Solubility of PbI_2 (g/100 g H_2O): (You may not have data to fill all the enrties.)

	Titration Method	Spectrophotometric Method
In pure water:		
In 10^{-3} M $Pb(C_2H_3O)_2$:		
In 10^{-3} M $NaNO_3$:		

Calculations:

Name _______________________________________   Date _____________

EXPERIMENT 15

QUESTIONS

1.   In Part A, if the titration end point is missed when the KI solution is titrated with $Pb(NO_3)_2$, causing too much $Pb(NO_3)_2$ to be added, will the error in the solubility product be +, -, or 0?

____

2.   At constant temperature, the **solubility** of PbI_2 decreases if KI is added to the solution. Does the **solubility product** increase, decrease, or remain the same?

____

3.   The **solubility** of PbI_2 increases with increasing temperature. Does the **solubility product** increase, decrease, or remain the same?

____

4.   Assume that EPA regulations required that the concentration of Pb^{2+} in public drinking water supplies be limited to 0.050 mg L^{-1}. Your measured value for the solubility of PbI_2 should be considerably larger than this, indicating that the disposal of solid PbI_2 could present a water pollution problem. Calculate the amount of KI that must be added to each liter of a saturated solution of PbI_2 in order to lower the Pb^{2+} concentration to 0.050 mg L^{-1}.

_______________ **g KI must be added per liter of saturated PbI_2 solution.**

Calculations:

EXPERIMENT 16
Acid-Base Equilibria, pH, and Buffer Solutions

PRELABORATORY PREPARATION
1. Do the Prelaboratory Exercise and turn it in at the beginning of your laboratory period.
2. Read about the use of the pH meter, Section XII, Methods, page 25.
3. Review the use of the buret, Section IV, Methods, page 17.
4. Read the discussion of **reaction quotient**, Equation 15.3, in Experiment 15.

INTRODUCTION
Around 1887, Arrhenius proposed that the characteristic properties of acids (sour taste, reaction with some metals to release H_2 gas, and turning blue litmus red) are caused by **hydrogen ions, H^+**, in the solution, and that the characteristic properties of bases (bitter taste, slippery feel, and turning red litmus blue) are caused by **hydroxyl ions, OH^-**.

Arrhenius defined acids as compounds that, when dissolved in water, increase the concentration of H^+ in solution and bases as compounds that increase the concentration of OH^-.

Later, it became clear that hydrogen ions, as such, are not stable species in water but react quickly by attaching to another molecule. For example, when an acid, such as HCl, is dissolved in water, it does not simply dissociate to form H^+ and Cl^-, but instead, a proton[1] is transferred from an HCl molecule to a water molecule, as in Reaction 16.1.

$$HCl + H_2O \rightleftharpoons Cl^- + H_3O^+ \tag{16.1}$$

The ion H_3O^+ is called the **hydronium ion.**

In 1923, Brønsted and Lowry, each independently, proposed a more general theory of acids and bases:

An acid-base reaction always involves the transfer of a proton, H^+, between molecules. The acid is the proton donor and the base is the proton acceptor.

In Reaction 16.1, HCl is the acid and H_2O is the base.

Acid-Base Equilibria
Reaction 16.1 is an example of the general class of acid-base reactions, illustrated by Reaction 16.2. Let **HA** represent any monoprotic acid. It will react with any base, **B,** by transferring a proton to it:

$$HA + B \rightleftharpoons HB^+ + A^- \tag{16.2}$$

Water can act as either an acid or a base, donating or accepting a proton depending on its reaction partner. When an acid is dissolved in water, water serves as a base and accepts a proton:

$$HA + H_2O \rightleftharpoons H_3O^+(aq) + A^-(aq) \tag{16.3}$$

Two kinds of bases can be dissolved in water:

1. With bases such as ammonia (NH_3), or basic anions[2] from salts such as NaCN, water serves as an acid and donates a proton, forming OH^- in the process:

$$NH_3 + H_2O \rightleftharpoons NH_4^+(aq) + OH^-(aq) \tag{16.4}$$

(a) salt dissociates, forming a basic anion: $NaCN \xrightarrow{H_2O} Na^+(aq) + CN^-(aq)$

(b) anion accepts a proton from water: $CN^-(aq) + H_2O \rightleftharpoons HCN + OH^- \tag{16.5}$

---

[1] In this case, the species H^+ is called a proton instead of a hydrogen ion because it is not present as an ionic species in solution.

[2] The anion of a salt will act as a base if it forms a weak acid, such as HCN ($K_a = 4 \times 10^{-10}$), when it accepts a proton.

2. Basic salts such as NaOH dissociate in water to form OH⁻ directly:

$$NaOH \xrightarrow{H_2O} Na^+ + OH^-$$ (16.6)

Acids or bases that essentially dissociate completely in water, i.e. whose equilibria in Reactions 16.3 to 16.6 lie far to the right, are called *strong* acids and bases. Those that remain significantly undissociated, with equilibria to the left, are called *weak* acids and bases.

The equilibrium expression for Reaction 16.3, acid dissociation in water, is written:

$$K_a = \frac{[H_3O^+][A^-]}{[HA]}$$ (16.7)

where [HA] is the concentration of acid that remains undissociated at equilibrium. K_a is often called the **acid dissociation constant**. It is a constant quantity that depends on the temperature, and is tabulated for many acids in chemical data references. A parallel discussion of **base dissociation constants** will be found in your textbook.

Self-ionization of Water

Even pure water contains a small concentration of H_3O^+ and OH⁻ ions. If two electrodes are immersed in pure water at room temperature and connected to a battery, the water will conduct a very small, but measurable, current. This fact means that ions must be present in the water. The conductivity and, therefore, the concentration of ions, increases with temperature. The ions arise from Reaction 16.8, in which a proton, **H⁺**, is transferred from one water molecule to another.

$$H_2O + H_2O \rightleftharpoons H_3O^+(aq) + OH^-(aq)$$ (16.8)

In this case, one water molecule acts as an acid and the other as a base, so that water is both acid and base in the same reaction. Compounds, such as water, that can be either an acid or a base, are termed **amphoteric**.

The equilibrium constant expression for Reaction 16.8 is written:

$$K_w = [H_3O^+][OH^-]$$ (16.9)

K_w is an equilibrium constant for the self-ionization of water[3]. It sometimes is called the **water dissociation product** or water **ion-product constant**. Like acid and base dissociation constants, K_w is temperature dependent, increasing with temperature. To illustrate its temperature dependence, values for K_w at 25°C and 50°C are given below in Equations 16.10a, b:

$$K_w(25°C) = 1.00 \times 10^{-14} \ (mol/L)^2$$ (16.10a)

$$K_w(50°C) = 1.83 \times 10^{-13} \ (mol/L)^2$$ (16.10b)

The small values for K_w indicate that the equilibrium of Reaction 16.8 is far to the left, so that water may be considered to be a very weak acid and, also, a correspondingly very weak base. The fact that K_w is larger at 50°C than at 25°C shows that the degree of ionization in water increases with temperature.

---

[3] From the definition of an equilibrium constant, you might expect the equilibrium expressions for Reactions 16.3 and 16.8 to be written:

$$K_{16.3} = \frac{H_3O^+][A^-]}{[HA][H_2O]} \quad \text{and} \quad K_{16.8} = \frac{[H_3O^+][OH^-]}{[H_2O]^2}$$

instead of the expressions given in 16.7 and 16.9. However, the [H₂O] terms in the denominators are not included in 16.7 and 16.9. This is because [H₂O] is effectively constant at 55.5 mol/L in pure water and in all but the most concentrated aqueous solutions; there is no point including unnecessary constants in an equilibrium equation.

To illustrate this point, consider pure water at 25°C. The concentration of water molecules that have lost a proton is 1×10^{-7} mol/L and the concentration that have gained a proton is 1×10^{-7} mol/L. Given that the initial concentration of water molecules before any ionization occurred was 55.5 mol/L, the concentration remaining unchanged at equilibrium is $(55.5 - 2 \times 10^{-7}) \approx 55.5$ mol/L, essentially no change at all. Thus, the denominator terms for [H₂O] in the expressions for $K_{16.3}$ and $K_{16.8}$ above are constant, for all practical purposes.

EXAMPLE 1:

Use K_w to calculate values for $[H_3O^+]$ and $[OH^-]$ in pure water at 25°C and 50°C.

SOLUTION:

In pure water, the stoichiometry of Reaction 16.8 requires that $[H_3O^+] = [OH^-]$.
Let $x = [H_3O^+] = [OH^-]$. Then,

$$K_w = [H_3O^+][OH^-] = (x)(x) = x^2$$

At 25°C, by Equation 16.10a: $K_w(25°C) = x^2 = 1.00 \times 10^{-14} \ (mol/L)^2$

Solving for x gives: $x = \{1.00 \times 10^{-14} \ (mol/L)^2\}^{1/2} = 1.00 \times 10^{-7} \ mol/L.$

Since $x = [H_3O^+] = [OH^-]$:

$$[H_3O^+] = [OH^-] = 1.00 \times 10^{-7} \ mol/L, \text{ in pure water at 25°C.}$$

At 50°C, by Equation 16.10b: $K_w(50°C) = x^2 = 1.83 \times 10^{-13} \ (mol/L)^2$

Solving for x gives: $x = \{1.83 \times 10^{-13} \ (mol/L)^2\}^{1/2} = 4.28 \times 10^{-7} \ mol/L.$

Since $x = [H_3O^+] = [OH^-]$:

$$[H_3O^+] = [OH^-] = 4.28 \times 10^{-7} \ mol/L, \text{ in pure water at 50°C.}$$

Whenever $[H_3O^+] = [OH^-]$, the solution is said to be neutral, regardless of the actual value of the concentrations.

In general, $[H_3O^+]$ need not always be equal to $[OH^-]$. Excess OH^- or H_3O^+ can be added to pure water, perhaps by adding NaOH or HCl. In a multicomponent solution, all the separate equilibrium relations must be satisfied simultaneously. Therefore,

Equations 16.10 will represent the equilibrium condition between H_3O^+ and OH^- in any aqueous solution, not only in pure water.

Consider the behavior of the reaction quotient,[4] **Q**, for Reaction 16.8 if some NaOH is added to pure water. Before the base is added, $Q = K_w = 1.00 \times 10^{-14} \ (mol/L)^2$. Immediately after adding NaOH, the value of Q will exceed the value of K_w, indicating that equilibrium has been destroyed. To reach a new equilibrium, extra OH^- must react with H_3O^+, to form H_2O by the reverse reaction of 16.8. If the solution temperature is 25°C, a common room temperature, the reaction will continue until the concentrations of both ions are reduced to levels where $Q = K_w(25°C) = 1.00 \times 10^{-14} \ (mol/L)^2$. Then, equilibrium is reestablished, now shifted to the left of its former position. At the new equilibrium, $[H_3O^+][OH^-] = 1.00 \times 10^{-14} \ (mol/L)^2$, even though in this case $[OH^-] > [H_3O^+]$, and the solution is called **basic**. In solutions where $[H_3O^+] > [OH^-]$, the solution is called **acidic**.

pH Scale

In water solutions, the degree of acidity is defined in terms of the concentration of H_3O^+. The higher $[H_3O^+]$, the more acidic is the solution. Since $[H_3O^+]$ can vary from very small values $(<10^{-15})$ to relatively large values (>1), it is useful to express acidity in a way that compresses the scale of $[H_3O^+]$ to a manageable range. This is done by defining a quantity called **pH** as a measure of acidity:

$$pH = -\log_{10}[H_3O^+] \tag{16.11}$$

The "p" symbol means, in general: "take the negative logarithm to the base 10 of the quantity represented by the next symbol."

Thus, $pOH = -\log_{10}[OH^-]$ and $pK_w = -\log_{10}K_w$. By convention, concentration brackets, ionic charges, and waters of hydration are omitted, so that one writes "pH", not "$p[H_3O^+]$", and "pOH", not "$p[OH^-]$".

Note that taking the antilog of of Equation 16.11 gives:

$[H_3O^+] = antilog_{10}(-pH) = 10^{-pH} \ mol/L.$ Similarly, $[OH^-] = 10^{-pOH} \ mol/L.$

---

[4] See Equation 15.3, Experiment 15, for the definition of the reaction quotient.

In any equilibrium water solution at 25°C, it always is true that
$$[H_3O^+][OH^-] = 1.00 \times 10^{-14} \ (mol/L)^2$$
whether the solution is acidic, basic, or neutral. Therefore, at 25°C,
$$[H_3O^+][OH^-] = (10^{-pH})(10^{-pOH}) \ (mol/L)^2$$
$$= 10^{-(pH+pOH)} \ (mol/L)^2 = 1.00 \times 10^{-14} \ (mol/L)^2.$$
By equating the exponents of the last two equalities, we obtain:
$$pH + pOH = 14, \ (at \ 25°C) \hspace{3cm} (16.12)$$
The uses of Equations 16.11 and 16.12 are illustrated by Examples 2, a-e.

EXAMPLE 2:

a.   Find the pH, pOH, and $[OH^-]$ in an aqueous solution at 25°C where $[H_3O^+] = 6.4 \times 10^{-8}$ mol/L.

SOLUTION:

For $[H_3O^+] = 6.4 \times 10^{-8}$ mol/L: pH = -log(6.4×10^{-8}) = **7.2**.
From Equation 16.12:   pOH = 14 - 7.2 = **6.8**.
$[OH^-]$ may be found from: $[OH^-] = 10^{-pOH} = 10^{-6.8}$ = **1.6 x 10^7 mol/L**
Use Equation 16.10a to check the above calculations:
$K_w(25°C) = (6.4 \times 10^{-8}$ mol/L$)(1.6 \times 10^7$ mol/L$) = $ **1.1 x 10^{-14} (mol/L)^2,** as required. (The small discrepancy of
0.1×10^{-14} is due to rounding errors.)

b.   For pH = 4.7, find pOH, $[H_3O^+]$, and $[OH^-]$ in an aqueous solution at 25°C.

SOLUTION:

For pH = 4.7: $[H_3O^+] = 10^{-4.7}$ M = **2.0 x 10^{-5} mol/L**.
Also, pOH =14 - 4.7 = **9.3**, so that: $[OH^-] = 10^{-9.3}$ M = **5.0 x 10^{-10} mol/L**.
Use Equation 16.10a to check the above calculations:
$K_w(25°C) = (2.0 \times 10^{-5}$ mol/L$)(5.0 \times 10^{-10}$ mol/L$) = 10.0 \times 10^{-15}$ (mol/L)^2
= **1.0 x 10^{-14} (mol/L)^2,** as required.

c.   What is the pK_w in an aqueous solution at 25°C?

SOLUTION:

From 16.10a, $pK_w = -log_{10}K_w(25°C) = -log_{10}(1.00 \times 10^{-14})$ = **14.**

d.   What are the pH and pOH in a neutral aqueous solution at 25°C?

SOLUTION:

In a neutral solution at 25°C, $[H_3O]^+ = [OH^-] = 1.00 \times 10^{-7}$ mol/L. Therefore,
$$pH = pOH = -log_{10}(1.00 \times 10^{-7}) = 7.$$

e.   What are the pH and pOH in a neutral aqueous solution at 50°C?

SOLUTION:

In a neutral solution at 50°C, $[H_3O^+] = [OH^-] = 4.28 \times 10^{-7}$ mol/L, from Example 1. Therefore,
$$pH = pOH = -log_{10}(4.27 \times 10^{-6}) = 6.37$$

An inspection of the definition of pH shows that, as $[H_3O^+]$ is made larger, the pH decreases.
**The lower the pH, the higher the H_3O^+ concentration and the more acidic the solution; the higher
the pH, the more basic the solution.**
Example 2d. points out that a neutral solution at 25°C has pH = 7.00. Therefore, at 25°C:
pH < 7.00 indicates an acidic solution.
pH > 7.00 indicates a basic solution.
These relations are illustrated in Figure 16.1, which also gives typical pH ranges at 25°C of some common
substances. Notice, from Figure 16.1, that if $[H_3O^+] > 1.0$ M, then the pH must become negative. This is
entirely possible, although not common, because it becomes increasingly difficult to dissociate even
strong acids at such high $[H_3O^+]$ levels.

MEASUREMENT OF pH

The two most frequently used methods for experimentally measuring pH are with chemical indicators that change color at certain pH values, and with an electronic instrument called a pH meter. In this experiment, we will use both methods.

Chemical Color Indicators

A color indicator may be a weak acid or a weak base. Most of them are weak acids. It is due to the equilibrium behavior of weak acids and bases that they can be effective indicators of pH.

For a weak acid or base to serve as a color indicator, it must have different colors in its undissociated and dissociated forms.

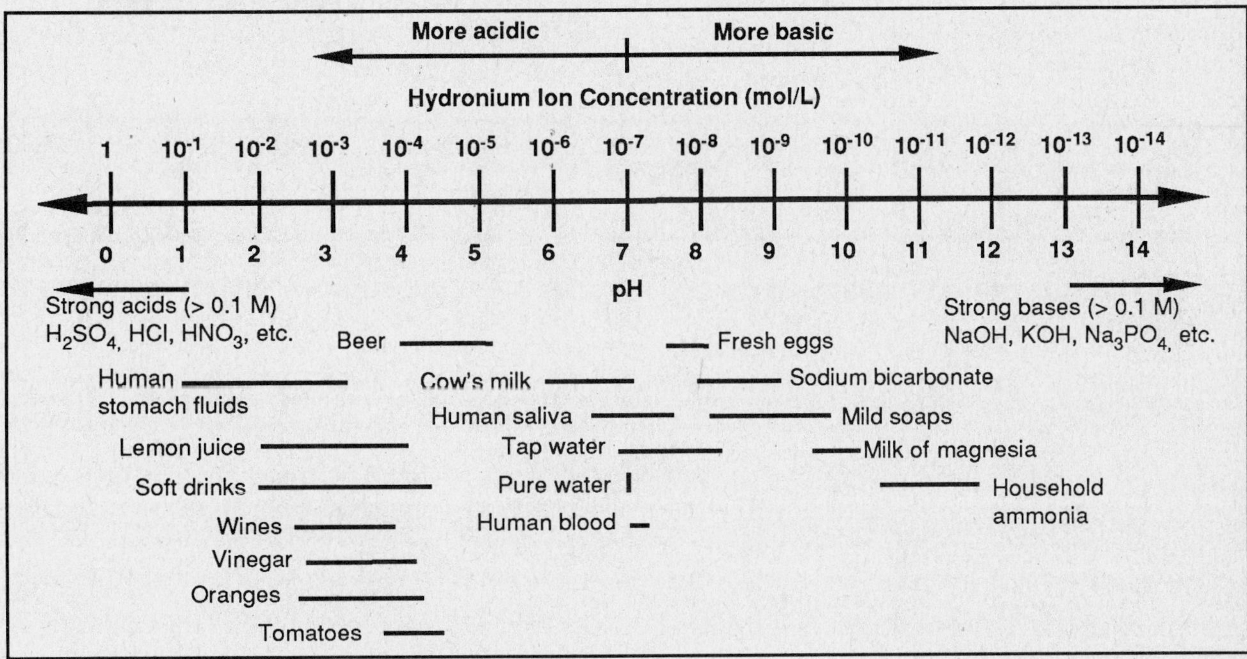

Figure 16.1. Chart showing typical ranges of hydronium ion concentration ($[H_3O^+]$) and pH for some common substances at 25°C.

Let **HIn** stand for a weak acid indicator compound. In water, some HIn molecules will dissociate, transferring a proton to water molecules. The dissociation equilibrium in water, may be written:

$$HIn \xrightleftharpoons{H_2O} H_3O^+ + In^-$$ (16.13)

Undissociated indicator molecule has color **A**, the acidic color.   Indicator anion has color **B**, the basic color.

If a small amount of indicator is added to an **acid** solution, where there already is an excess of H_3O^+, the high value for $[H_3O^+]$ very effectively holds the equilibrium of Reaction 16.13 far to the left, preventing any significant dissociation of the weak acid indicator. Then, the solution will have color **A**, the acidic color. Only when $[H_3O^+]$ drops below a certain value, characteristic for each weak acid, can the indicator dissociate sufficiently to change the solution to color **B**, the basic color.

Different indicators have different acid and base colors. Because each different indicator is a weak acid or weak base of different strength, a different value of $[H_3O^+]$ is required to shift the equilibrium of Reaction 16.13 from one side to the other. Accordingly, different indicators will exhibit their color change over different pH ranges. Some common indicators and their color change ranges are given in Table 16.1. Other indicators are listed in the various handbooks of chemical data.

Indicators may be in liquid solution or powder form for adding to solutions. Paper strips may be impregnated with litmus (a common indicator) for dipping into the solution to be tested. Multi-range pH paper is impregnated with several indicators to cover a wide pH range, each indicator changing color at a different pH. With multi-range papers (such as Hydrion), the pH is determined by dipping the paper in the solution to be tested and matching the color of the paper strip with a color chart. Multi-range indicators also are available in powder form, for dissolving to make indicator solutions.

Although chemical indicators change color over a range of pH values, they can measure a particular pH value to within 0.5-1.0 pH unit, by comparing the solution color intensity with a standard.

Table 16.1.   **Acid-Base Color Indicators[a]**

Indicator	Acid Color	Base Color	pH Range
o-cresol red	red	yellow	0 - 2
Methyl orange	red	yellow	3 - 4.4
Methyl red	red	yellow	4.4 - 6
Litmus	red	blue	4.5 - 8.3
Bromthymol blue	yellow	blue	6 - 8
Neutral red	red	yellow	7 - 8
Phenolphthalein	colorless	red	8 - 10
Thymolphthalein	colorless	blue	9.4 - 10.6
Alizarin yellow	yellow	violet	10 - 12
1,3,5-trinitrobenzene	colorless	red	12 - 13.4

[a] Roger G. Bates, Determination of pH, Theory and Practice, John Wiley and Sons, New York, 1964, pp. 138-139.

pH Meter

The electronic pH meter, properly used, is much more accurate and reproducible than color indicators. Ordinary student pH instruments are capable of measuring to ± 0.1 pH unit. The pH meter has almost entirely supplanted the use of color indicators in modern laboratories. Almost, but not quite, for sometimes a color indicator test is faster, easier, and quite adequate. One important advantage of the pH meter is that the electronic signal it produces can be used to automate analytical procedures and control industrial processes. It also has obvious advantages for color blind persons. The pH meter and its operation are described in Section XII, Methods, page 25.

PLAN OF EXPERIMENT
Part A: Acid and Base Solutions

You first measure the pH of several typical acid and base solutions, using color indicators and a pH meter. As shown below, the pH value is related to the percent dissociation of an acid. For a monoprotic acid:

$$\% \text{ dissociation} = \frac{[H_3O^+]}{[\text{Acid}]_0} \times 100\% = \frac{10^{-pH}}{[\text{Acid}]_0} \times 100\% \tag{16.14}$$

where $[\text{Acid}]_0$ is the initial acid concentration, before any dissociation takes place.

EXAMPLE 3:

a.   What is the percent dissociation of a 2.5×10^{-3} M HCl solution if the pH is measured to be 2.6? What does the percent dissociation indicate about the acid strength of HCl?

SOLUTION:

Since the pH was measured to be 2.6, $[H_3O^+] = 10^{-2.6}$. Therefore,

$$\% \text{ dissociation} = \frac{[H_3O^+]}{[\text{Acid}]_0} \times 100\% = \frac{10^{-2.6}}{2.5 \times 10^{-3}} \times 100\% = \frac{2.5 \times 10^{-3}}{2.5 \times 10^{-3}} \times 100\% = \textbf{100\%}$$

The large percent dissociation indicates that HCl is a strong acid.

b.   The pH of an 0.20 M solution of formic acid, HCOOH, is measured to be 2.2. What is the percent dissociation? Is HCOOH a strong or weak acid?

SOLUTION:

$$\% \text{ dissociation} = \frac{[H_3O^+]}{[\text{Acid}]_0} \times 100\% = \frac{10^{-2.2}}{0.20} \times 100\% = \frac{6.3 \times 10^{-3}}{0.20} \times 100\% = \mathbf{3.2\%}$$

The small percent dissociation indicates that formic acid is a weak acid.

Part B: Salt Solutions

You will measure the pH of several salt solutions. Some salt solutions are neutral, while others may be acidic or basic.

- **If an ion that is formed when a salt dissociates in water accepts a proton from water to form a weak acid and additional OH^-, the pH rises and the solution is basic.**

- **If an ion that is formed donates a proton to water to form H_3O^+, the pH falls and the solution is acidic.**

- **If an ion that is formed associates with OH^- to form a weak base, $[H_3O^+]$ becomes larger than $[OH^-]$, the pH falls and the solution is acidic.**

Examples 4, a-c, illustrate some of the ways that the pH can be influenced by the nature of the salt.

EXAMPLE 4:
a.   Should dissolving the salt NaCl in water affect the pH?

SOLUTION:

It depends on whether any of the ions that are formed will exchange protons with water or associate with OH^-. NaCl dissociates into Na^+ and Cl^- ions.
- **Neither of the ions formed contains a proton to donate to water.**
- **Cl^- could accept an H^+, as in Reaction 16.15:**

$$H_2O + Cl^- \rightleftharpoons HCl + OH^- \tag{16.15}$$

However, HCl is a strong acid and the equilibrium of 16.15 is *very far to the left.* Therefore, no OH^- or undissociated HCl is formed and no change in the $[H_3O^+] : [OH^-]$ ratio occurs.

Since Reaction 16.15 does not change the value of the $[H_3O^+] : [OH^-]$ ratio, it cannot affect the pH.

- **Na^+ could associate with OH^-, which is always present because of the self-ionization of water (Reaction 16.8), by Reaction 16.16:**

$$Na^+ + OH^- \rightleftharpoons NaOH \tag{16.16}$$

However, NaOH is a strong base and the equilibrium of Reaction 16.16 is *very far to the left.* Therefore, no undissociated NaOH is formed and no change in the $[H_3O^+] : [OH^-]$ ratio occurs.

Since Reaction 16.16 does not change the $[H_3O^+] : [OH^-]$ ratio, it cannot affect the pH.

Conclusion: Dissolving NaCl in water has no effect on the pH. NaCl is said to be a salt of a strong acid (because Cl^- does not accept protons from water) and a strong base (because Na^+ does not associate with OH^- in water).

b.   Should dissolving the salt Na_2S in water affect the pH?

SOLUTION:

     Na_2S dissociates into Na^+ and S^{2-} ions.
- **Neither of the ions formed contains a proton to donate to water.**
- **We saw from Part a, that Na^+ does not associate with OH^-.**
- **S^{2-} could accept two protons, as in Reactions 16.17a, b:**

$$H_2O + S^{2-} \rightleftharpoons HS^- + OH^- \tag{16.17a}$$

$$H_2O + HS^- \rightleftharpoons H_2S + OH^- \tag{16.17b}$$

Both HS^- and H_2S are weak acids which remain significantly undissociated at equilibrium and the equilibria of Reactions 16.17a, b are *to the right.* Therefore, additional OH^- is formed in the solution and the $[H_3O^+]$: $[OH^-]$ ratio is changed. Since Reactions 16.17a, b increase $[OH^-]$, they decrease the $[H_3O^+]$: $[OH^-]$ ratio and increase the pH.

Conclusion: Dissolving Na_2S in water increases the pH. Na_2S is said to be a salt of a weak acid (because S^{2-} accepts protons from water) and a strong base (because Na^+ does not associate with OH^- in water).

c.   Should dissolving the salt $CuCl_2$ in water affect the pH?

SOLUTION:

     $CuCl_2$ dissociates into Cu^{2+} and Cl^- ions.
- **We saw from Part a, that Cl^- does not accept protons from water.**
- **Cu^{2+} ions become hydrated in water solution, forming $Cu(H_2O)_4^{2+}$.**

The hydrated cupric ion acts as an acid and donates a proton to water.

$$Cu(H_2O)_4^{2+} + H_2O \rightleftharpoons H_3O^+ + Cu(H_2O)_3(OH)^+ \tag{16.18}$$

Additional $[H_3O^+]$ is formed in Reaction 16.18. This increases the $[H_3O^+]$: $[OH^-]$ ratio, lowering the pH.

Conclusion: Dissolving $CuSO_4$ in water forms the acidic hydrate $Cu(H_2O)_4^{2+}$ and lowers the pH.

Part C: Dissociation Constant of a Weak Acid

     You will determine the dissociation constant of an unknown weak acid. A weak acid, designated **HA,** dissolves in water according to Reaction 16.3. Its dissociation constant, $\mathbf{K_a}$, is given by Equation 16.7. The smaller the value of K_a, the weaker is the acid. The value of K_a can be found by measuring the pH. We will describe two methods for finding K_a from a pH measurement.

     Method 1 is useful for understanding the theory, but Method 2 is a simpler laboratory technique.

You will use Method 2 to make the measurement.

Method 1 for Determining $\mathbf{K_a}$:

     Make up a weak acid solution of known molarity. The degree of dissociation is indicated by the pH, see Equation 16.14 and Example 3. In general, $[\mathbf{H_3O^+}]_\mathbf{a}$, *formed by acid dissociation*, is large enough that $[\mathbf{H_3O^+}]_\mathbf{w}$, *from water dissociation,* may be neglected.

     $[\mathbf{H_3O^+}]_\mathbf{w}$ may be neglected, with respect to $[\mathbf{H_3O^+}]_\mathbf{a}$, if the pH of the solution is 5, or lower. Then $[\mathbf{H_3O^+}]_\mathbf{total} = [\mathbf{H_3O^+}]_\mathbf{a} + [\mathbf{H_3O^+}]_\mathbf{w} \approx [\mathbf{H_3O^+}]_\mathbf{a}$.

Under these conditions, $[H_3O^+] \approx [A^-]$, by Reaction 16.3. Then Equation 16.7 may be written:

$$K_a = \frac{[H_3O^+][A^-]}{[HA]} = \frac{[H_3O^+]^2}{[HA]_0 - [H_3O^+]} \tag{16.19}$$

where $[HA]$, the undissociated acid concentration at equilibrium, has been replaced by the equivalent quantity, $[HA]_0 - [H_3O^+]$.

     $[HA]_0$ is the initial acid concentration, before any dissociation occurs.

Equation 16.19 can be solved by knowing the initial concentration of acid added and determining $[H_3O^+]$ by measuring the pH.

The critical step experimentally is making up the acid solution carefully so that $[HA]_0$ is accurately known. Method 2 completely avoids this critical step.

Method 2 for Determining K_a:

This method does not require that you know $[HA]_0$.

Method 2 involves an approximation that requires the acid to have a K_a value of 10^{-3}, or less.

1. Dissolve an unweighed sample of a weak acid in water.
2. Accurately divide the solution into two equal parts.
3. Titrate one part to a phenolphthalein end point with standardized NaOH solution.

As OH^- is added, it associates with H_3O^+ to form water. This decreases $[H_3O^+]$, shifting Reaction 16.3 farther to the right and dissociating more acid.

The titration reaction is:

$$HA + OH^- \longrightarrow A^- + H_2O \tag{16.20}$$

At the titration end point, Reaction 16.20 has gone to completion and the acid is fully dissociated; all the acid has been converted to A^- and water. The solution composition is exactly the same as if the salt, NaA, made from the anion of the acid and the cation of the base, were dissolved in pure water.

The sample half portion that was titrated now contains only water and a concentration of A^- ion that is equal to the initial concentration of HA, i.e., $[A^-]$(after titration to end point) = $[HA]_0$

4. Now, mix the titrated half of the original sample with the untitrated half.

If you divided the sample accurately in two parts with reasonable care, $[A^-]$ in the titrated portion is equal to $[HA]$ in the untitrated portion, so that the solution made by mixing the two halves together contains equal concentrations of HA and A^-. These quantities, then, cancel out of the equilibrium expression.

$$K_a = \frac{[H_3O^+][A^-]}{[HA]} = [H_3O^+] \tag{16.21}$$

The result is that K_a is simply equal to $[H_3O^+]$, which can be determined by measuring the pH.

Part D: Buffer Solutions

When a salt of a weak acid is added to a solution of the same weak acid, the solution becomes resistant to changes in pH when additional acid or base is added. Such a solution is called a **buffer solution**, because it is "buffered" against pH changes.

A buffered acid solution has the same equilibrium expression, Equation 16.7, as a non-buffered solution. Solve Equation 16.7 for the hydronium ion concentration:

$$[H_3O^+] = K_a \frac{[HA]}{[A^-]}$$

Take $\log_{10}$ of both sides and multiply both sides by -1.

$$-\log_{10}[H_3O^+] = -\log_{10} K_a - \log_{10}\frac{[HA]}{[A^-]}$$

Substitute the symbols **pH** and **pK_a** for the first two terms:

$$pH = pK_a - \log_{10}\frac{[HA]}{[A^-]} \tag{16.22}$$

Equation 16.22 shows very clearly that the pH of a weak acid solution depends on two quantities: pK_a, which is a constant of the acid, and $\log_{10}$ of the ratio $[HA]/[A^-]$, which is an easily controlled variable in buffer solutions. Equation 16.22 is an especially useful form of the equilibrium expression for calculating the behavior of acid buffer solutions.

How to Make a Buffer Solution

The equilibrium reaction of the weak acid HA dissolved in water is

$$HA + H_2O \rightleftharpoons H_3O^+(aq) + A^-(aq) \tag{16.3}$$

where the equilibrium is to the left, because dissociation is incomplete for the weak acid. To make a buffer solution, the salt of HA must be added. Let **MA** stand for a univalent highly soluble salt of the weak acid HA. (The salt is chosen to be univalent only for simplicity in the following explanation. Multivalent salts would only require different ionic charges and stoichiometries.) The dissociation reaction of the salt is:

$$MA \xrightarrow{\ H_2O\ } M^+(aq) + A^-(aq) \tag{16.23}$$

Because the salt is highly soluble, it dissociates completely. Add the salt MA to a solution of the acid HA. When MA dissolves in the acid solution, Reaction 16.23 produces an excess of A^-, which drives the equilibrium of 16.3 to the left, so that essentially none of the acid, HA, remains undissociated. Because $[H_3O^+]$ was already small before MA was added, only a little A^- can be consumed in the equilibrium shift to the left. The nature of a buffer solution allows two important approximations to be valid:

1. Since essentially all of the A^- in solution comes from the salt, by Reaction 16.23, the concentration of A^- is essentially the same as the concentration of salt MA originally added:

$$[A^-] \approx [MA]_0 = \frac{\text{moles of MA originally added to solution}}{\text{final volume of solution}} \tag{16.24}$$

2. The concentration of undissociated acid in solution after adding the salt MA is essentially the same as the initial concentration, $[HA]_0$, before before adding the salt, because so little of the weak acid was dissociated originally.

$$[HA] \approx [HA]_0 = \frac{\text{moles of acid originally added to solution}}{\text{final volume of solution}} \tag{16.25}$$

Now, the pH of the buffer solution can be calculated from Equation 16.22, inserting $[HA]_0$ for $[HA]$, and $[MA]_0$ for $[A^-]$. K_a must be looked up in a chemical data table.

The buffering action of the solution can be understood by examining Equation 16.22. Because K_a is a fixed quantity, the only way for the pH to change is if the ratio $[HA]/[A^-]$ changes. Suppose a little strong acid were added to the buffer solution. It dissociates completely, forming H_3O^+. However, A^- in the solution immediately associates with the new H_3O^+, to form undissociated HA and H_2O. The end result is that $[HA]$ increases a little and $[A^-]$ decreases a little. If both concentrations were fairly large to start with, small changes in opposite directions will not change the ratio by very much. The $\log_{10}$ of the ratio will change much less than the ratio itself. (Test this claim for yourself with a quick calculation.) Thus, the pH change of a well-designed buffer solution, although *not* zero, will be small when a strong acid is added to it.

If a strong base is added to the solution, it takes a proton from H_3O^+, lowering $[H_3O^+]$. This shifts the equilibrium of 16.3 to the right, increasing $[A^-]$ a little and decreasing $[HA]$ a little. Once again, the pH change will be small.

The change in the ratio $[HA]/[A^-]$ is a minimum when the original ratio is unity and when the concentration values are large, compared to the added acid or base.

The ability of a buffer to consume acid or base is not unlimited. If enough acid or base is added, the buffering action becomes exhausted and the pH then changes rapidly. A measure of how much acid or base a buffer solution can consume without losing its buffering ability is called its **buffer capacity**. **Buffer capacity equals the moles of strong acid or base that cause a change of one pH unit in 1 liter of buffered solution. A high buffer capacity is attained by using large and equal concentrations of acid and salt.** The maximum buffer capacity for any given concentration of acid and salt is attained when $[HA] = [A^-]$; then $pH = pK_a$, see Equation 16.22. To obtain a buffer pH of any other value, the ratio $[HA]/[A^-]$ must be adjusted appropriately, at the expense of having a smaller buffer capacity. **The best way to make a buffer solution so that it has a particular pH, is to choose a weak acid with a pK_a as close to the desired pH as possible.**

Then, the ratio $[HA]/[A^-]$ will not have to deviate far from unity to attain the desired pH, preserving a buffer capacity near the optimum. Examples 5, 6, and 7 illustrate how to use Equation 16.22 for calculating the properties of buffer solutions.

EXAMPLE 5: Calculating the pH of a Buffer Solution

Calculate the pH of a buffer solution made by dissolving 0.250 mole of the weak acetic acid, $HC_2H_3O_2$, and 0.250 mole of its salt sodium acetate, $NaC_2H_3O_2$, in sufficient water to make 1 L of solution. In a data table, you can find that $K_a(HC_2H_3O_2) = 1.76 \times 10^{-5}$. Notice that this solution meets the requirements for a good buffer, by having the salt and acid concentrations equal and large.

SOLUTION:

Let $[HA] = [HC_2H_3O_2]$, and $[A^-] = [C_2H_3O_2^-]$. Then $[HA] = 0.250$ M and $[A^-] = 0.250$ M. Substitute these values, along with K_a, into Equation 16.22:

$$pH = pK_a - \log_{10} \frac{[HA]}{[A^-]} = -\log_{10}(1.76 \times 10^{-5}) - \log_{10}\frac{0.250}{0.250}$$

Because the $\log_{10}(1) = 0$, we have eliminated the last term in the pH equation by choosing $[HA] = [A^-]$.

$$pH = -\log_{10}(1.76 \times 10^{-5}) - 0 = \mathbf{4.75}$$

EXAMPLE 6: Making a Buffer Solution that Has a Particular pH

How many grams of $NaC_2H_3O_2$ must be added to 1 L of 0.250 M $HC_2H_3O_2$ to make a buffer solution of pH = 5.00. Assume the addition of the solid salt does not change the solution volume.

SOLUTION:

$$pH = pK_a - \log_{10}\frac{[HA]}{[A^-]} = 5.00$$

In this case we know the desired pH and wish to calculate $[A^-]$. To rearrange the equation, first take the antilog of both sides.

$$[H_3O^+] = K_a \frac{[HA]}{[A^-]} = 1.00 \times 10^{-5}$$

Now solve for $[A^-]$:

$$[A^-] = (1.76 \times 10^{-5})\left(\frac{0.250 \text{ M}}{1.00 \times 10^{-5}}\right) = 0.440 \text{ M}$$

The solution requires 0.440 mol L^{-1} of the salt, in order to have a pH = 5.00. Because our solution is only 500 mL, we must add $0.440/2 = 0.220$ moles of salt to the 500 mL of solution. The molar mass of $NaC_2H_3O_2$ is 82.0. Therefore, $(82.0 \text{ g mol}^{-1})(0.220 \text{ mol}) = \mathbf{18.0 \text{ g } NaC_2H_3O_2 \text{ must be added.}}$

EXAMPLE 7: Calculating the pH Change Caused by Adding a Strong Acid to a Buffer Solution

Calculate the change in pH caused by adding 1.00×10^{-2} moles of concentrated HCl to the buffer solution of Example 5. Assume that the added volume is so small that the volume of the buffer solution does not change significantly.

SOLUTION:

Adding the HCl introduces 1.00×10^{-2} moles of H_3O^+, which associates with acetate ion to form undissociated acetic acid.

$$C_2H_3O_2^- + H_3O^+ \rightleftharpoons HC_2H_3O_2 + H_2O$$

Because $[C_2H_3O_2^-]$ is large, the equilibrium is far to the right and essentially all of the added H_3O^+ reacts. The new concentrations of the buffer components become:

$$[C_2H_3O_2^-]_{new} = [C_2H_3O_2^-]_{old} - 1.00 \times 10^{-2} = 0.250 - 0.010 = 0.240 \text{ M}$$
$$[HC_2H_3O_2]_{new} = [HC_2H_3O_2]_{old} + 1.00 \times 10^{-2} = 0.250 + 0.010 = 0.260 \text{ M}$$

Using these values in Equation 16.22:

$$pH_{new} = pK_a - \log_{10}\frac{[HA]}{[A^-]} = -\log_{10}(1.76 \times 10^{-5}) - \log_{10}\frac{0.260}{0.240}$$

$$pH_{new} = 4.75 - \log_{10}(1.083) = 4.75 - 0.035 = 4.71$$

The old pH value was 4.75, so the change is a decrease of 0.04 pH units. The effect of the buffer may be appreciated by comparing the answer above with the pH change that would occur if the same amount of acid were added to 1 L of pure water. In pure water, the pH would change from 7 to 2, a decrease of 5 full pH units.

SAFETY

1. Wear approved eye protection.

2. Concentrated acids and bases evolve considerable heat when they react with water. If they are not diluted properly, evolved heat can cause localized boiling and possible spattering of reagent, with danger for everyone nearby.

    TO DILUTE AN ACID OR BASE:

    ALWAYS SLOWLY POUR THE ACID OR BASE INTO WATER, WITH CONSTANT STIRRING.

    This procedure dilutes the acid or base rapidly, limiting the rate of reaction and minimizing the temperature rise.

    NEVER POUR WATER INTO CONCENTRATED ACID OR BASE.

3. The acids and bases used in this experiment can cause severe skin burns. Their vapors are corrosive and, in some cases, toxic. Contact with your eyes and mucous membranes are especially dangerous. They also can destroy clothing. Be very careful when handling these chemicals. Avoid contact with skin or clothing. Do not deliberately inhale the vapors. It is wise to wear a lab coat or apron.

    IF ACID OR BASE DOES COME INTO CONTACT WITH YOUR EYES, SKIN, OR CLOTHING,

    IMMEDIATELY RINSE THE AFFECTED AREAS WITH COPIOUS AMOUNTS OF WATER.

    A little water is worse than none, for it speeds up the reaction. Your laboratory will have special eye wash devices. Know the locations of the eyewash stations and how to use them.

    DO NOT ASSUME YOU CAN NEUTRALIZE AN ACID SPILL BY ADDING A STRONG BASE. YOU ONLY WILL MAKE MATTERS MUCH WORSE.

4. Clean up all spills immediately with large amounts of rinse water, followed by a covering of baking soda ($NaHCO_3$) for acids, or sodium bisulfate ($NaHSO_4$) for bases.

5. The measuring electrodes of the pH meter are very delicate and easily broken. Handle them carefully and avoid bumping them against anything.

PROCEDURE
Work with a partner for Parts **A** and **B**.

A. Acidic and Basic Solutions
1. Obtain about 40 mL of 1.0 M HCl stock solution. Prepare by dilution about 25 mL each of 1.0 M HCl, 0.10 M HCl, and 0.010 M HCl. Put small quantities of the acid solutions in small, labelled test tubes. Have at least 4 test tubes for each concentration.

2. Using pH paper and indicator solutions made available to you, determine the pH of each of the three solutions. First test with pH paper, and then observe the colors obtained with several indicator solutions. Enter your observations into the table on the data sheet.
    Note that the indicator color tells you on which side of the color change range the solution pH lies.

Try to bracket high and low values of the solution pH with different indicators.

3.   After completing the pH tests on the acid solutions, measure the solutions with a pH meter for comparison. Record the results.

4.   Prepare 1.0 M, 0.10 M, and 0.010 M solutions of NaOH, $HC_2H_3O_2$, and NH_4OH. Determine the pH of these solutions with pH paper, indicator solutions, and a pH meter. Record the data.

B. Salt Solutions
Use a pH meter or pH paper to measure the pH of the following salt solutions:
0.1 M NaCl, 0.1 M $NaC_2H_3O_2$, and 0.1 M Na_2CO_3.
Record and explain the results.

C. Analysis and Dissociation Constant of a Weak Acid
1.   Obtain a sample of an unknown weak acid and dissolve it into 100 mL distilled water in a 250 mL Erlenmeyer flask.

2.   Divide this solution accurately into 2 equal parts, using another 250 mL Erlenmeyer flask.

3.   Titrate one of the solutions to a phenolphthalein end point with 0.2 M NaOH. Do not overtitrate. Add the NaOH slowly, swirling the solution in the flask while you titrate. When you near the endpoint, indicated by a pink color in the solution that disappears on swirling, add the NaOH drop by drop, swirling the solution after each drop. Stop when you obtain a permanent pink color.

4.   Record the volume of NaOH required. You need this to calculate the moles of acid in the unknown sample.

5.   Follow the procedure of **Method 2,** in the **Introduction, Part C.**, to determine the acid dissociation constant. You will mix the titrated half of the original sample, with the untitrated half. Then pK_a = pH.

6.   Use pH paper or a pH meter to measure the pH of the combined solutions. Record the pH and calculate K_a.

7.   Save the final solution for **Part D.**

D. Buffer Solutions
The solution from **Part C** is a buffer solution. It was made by mixing a solution of a weak acid with a solution of the salt of the acid. Even though the salt solution was made by titrating an acid solution to the end point with a base, it nevertheless is identical to a solution made by dissolving the solid salt in water. When the titrated and untitrated solutions were mixed, they made a buffer with $[HA]/[A^-]$ = 1, which has maximum buffer capacity for its concentration.

1.   Observe the effect of the buffering action by comparing the change in pH of the buffer solution with that of water, when an acid and base are added, as follows: (Use pH paper or a pH meter.)
    a. Prepare two 25 mL volumes of buffer solution, and two 25 mL volumes of water
        in test tubes.
    b. Add 5 drops of 0.1 M HCl to one buffer sample and to one water sample.
    c. Record the pH values of each sample.
    d. Add 5 drops of 0.1 NaOH to each of the other buffer and water samples.
    e. Record the pH values of each sample.

Name ____   Date _____________

EXPERIMENT 16
PRELABORATORY EXERCISE

1. The pH of a solution was measured to be 8.5. Calculate $[H_3O^+]$ and $[OH^-]$.

$[H_3O^+]$ = _____________ ; $[OH^-]$ = _____________

2. Calculate the percent dissociation and K_a of a weak acid, HA, knowing that a 10^{-2} M solution of the acid has pH = 5.0.

% dissociation = _____________ ; K_a = _______________

3. **a.** What is the pH of a buffer solution made by dissolving 0.100 moles of benzoic acid, C_6H_5COOH, and 0.100 moles of its salt, sodium benzoate, C_6H_5COONa, in sufficient water to make 1 L of solution. $K_a(C_6H_5COOH) = 6.46 \times 10^{-5}$.

pH = _____________

b. If 1.00×10^{-3} mol of nitric acid, HNO_3, is added to 1 L of the buffer solution of **part a**, what *change* in pH will occur?

Change in pH = _____________

Calculations:

Name ____   Date ____________

EXPERIMENT 16
DATA
(Observe significant figures in all calculations.)

A. Acidic and Basic Solutions
1.   Enter, in the table below, the indicator solutions used and the colors observed. Also, enter the pH values indicated by pH paper and a pH meter.

Kind of Solution	Indicator solution used	(Enter indicator color for each solution.) Molarity of Acid or Base Solution		
		1.0 M	0.1 M	0.001 M
HCl				
	pH from pH paper ⟶			
	pH from pH meter ⟶			
NaOH				
	pH from pH paper ⟶			
	pH from pH meter ⟶			
$HC_2H_3O_2$				
	pH from pH paper ⟶			
	pH from pH meter ⟶			
NH_4OH				
	pH from pH paper ⟶			
	pH from pH meter ⟶			

DATA

2. Calculate the percent dissociation in each of the three acetic acid ($HC_2H_3O_2$) solutions. Also, calculate K_a by **Method 1** for each of the acetic acid solutions.

	[$HC_2H_3O_2$]		
	1.0 M	0.1 M	0.01 M
% dissociation			
K_a			

3. Explain any trends you observe in % dissociation or K_a for different concentrations of acetic acid.

Calculations:

B. Salt Solutions

	0.1 M NaCl	0.1 M NaC$_2$H$_3$O$_2$	0.1M Na$_2$CO$_3$
Observed pH:			

What conclusions can you draw from the above results?

C. Analysis and Dissociation of a Weak Acid by Method 2

1. Unknown acid No.: _______________

2. Initial buret reading: _______________

3. Final buret reading: _______________

Name _______________________________________   Date _____________

EXPERIMENT 16

DATA
(Observe significant figures in all calculations.)

4. Volume of NaOH used: _______________

5. Moles of NaOH added in titration: _______________

6. Moles of acid in sample: _______________

7. pH of combined solution made by mixing
titrated and untitrated solutions: _______________

8. pH of mixed solution: _______________

9. K_a of unknown acid: _______________

Calculations:

D. Buffer Solutions: Test of Buffering Action

	pH in buffer	pH in water
1. Initial solution (pH was measured in part C):	_________	_________
2. After adding 5 drops of 0.1 M HCl:	_________	_________
3. After adding 5 drops of 0.1 M NaOH:	_________	_________

Observations:

EXPERIMENT 17
Titration Analysis of Weak Acid Solutions: Potassium Hydrogen Phthalate, Acetic Acid in Vinegar, and Citric Acid in Fruits

PRELABORATORY PREPARATION
1. Do the Prelaboratory Exercise and turn it in at the beginning of your laboratory period.
2. Review the definitions of molarity, normality and equivalent mass in Chemical Units, page xi.
3. If necessary, review the theory of acid-base reactions in Experiment 16.
4. Review the use of the buret, Section IV, Methods, page 17.
5. Review vacuum filtering technique, Section IX, Methods, page 23.

INTRODUCTION

In this experiment, the percent acetic acid in vinegar and percent citric acid in fruit juice are measured by the method of **titration**. Titration analysis is a technique for measuring the concentration of an unknown solution by making it react stoichiometrically with a standard reagent, which has an accurately known concentration. The standard reagent is added incrementally until the **equivalence point** is reached, where stoichiometric quantities of unknown sample and standard reactants have been added to the solution. Then, using the known concentration of the standard reagent, the concentration of the reagent being analyzed can be determined.

In an acid–base titration, the equivalence point is reached when the molar ratio of acid and base reagent in the titrated solution are the same as their ratio in the stoichiometric neutralization equation.

The solution being analyzed is called the **titrand**, and the reagent being titrated into the titrand is called the **titrant**. The titrant must be **standardized** by having its molar concentration accurately determined. If the volume of titrant added to the titrand during the analysis is carefully measured, the number of moles of titrant needed to neutralize the titrand and reach the equivalence point can be calculated. By knowing the moles of titrant, you can calculate the moles of acid originally in the solution.

The equivalence point of a titration is determined by monitoring the titration with a pH meter or by adding a chemical color indicator to the solution. During a titration of an acid with a base, the pH increases continually. With each drop of added base, the rate of increase is slow when the moles of acid and base are not near the equivalence ratio. As the reagent mole ratio nears the equivalence ratio, the rate of change of pH increases rapidly. The rate of change of pH is a maximum when equivalence is reached. This is how the state of equivalence is recognized. The indicated equivalence point sometimes is called the **titration end point.**

Titrations of weak and strong acids with a strong base are similar in that the neutralization reactions go to completion in both cases. They differ in that the equivalence point for strong acids occurs at pH = 7, where the solution is neutral, while the equivalence point for weak acids occurs at a pH > 7, where the solution is basic. This is why the term **equivalence point** is more suitable than **neutralization point**.

Phenolphthalein indicator, which turns from colorless to red at about pH = 9, (see Table 16.1, Experiment 16) is suitable for all titrations in this experiment. Choosing an indicator with a color change at such a high pH insures that enough base is added so that all three protons in citric acid are neutralized in the titration with NaOH.

Both acetic and citric acids are weak acids, meaning that they are only partially ionized in aqueous solution. During a titration however, their dissociation equilibria continue to shift until they have completely reacted with the standard reagent. The characteristics of weak acids are discussed in Experiment 16. The other weak acid, potassium hydrogen phthalate, is used in the calibration of the standard NaOH reagent.

EXAMPLE 1:

50 mL of acetic acid, $HC_2H_3O_2$ are titrated to the equivalence point with 23.4 mL of 0.226 M standardized NaOH solution. What is the concentration of the acetic acid solution? The neutralization reaction is:

$$HC_2H_3O_2 + NaOH \rightleftharpoons NaC_2H_3O_2 + H_2O \qquad (17.1)$$

SOLUTION:

The stoichiometric ratio of reagents is 1:1. Therefore, at the equivalence point, the moles of NaOH added are equal to the moles of $HC_2H_3O_2$ originally present.

moles of NaOH added = (volume of NaOH) × (molarity of NaOH)

$$= (23.4 \times 10^{-3} \text{ L})(0.226 \text{ mol/L}) = 5.29 \times 10^{-3} \text{ mol NaOH}$$

Therefore the total moles of $HC_2H_3O_2$ originally present was 5.29×10^{-3} moles.

The original concentration of the 50 mL of $HC_2H_3O_2$ solution was:

Concentration = (moles of $HC_2H_3O_2$)/(volume of $HC_2H_3O_2$)

$$[HC_2H_3O_2] = (5.29 \times 10^{-3} \text{ mol})/(50 \times 10^{-3} \text{ L}) = \textbf{0.106 mol/L}$$

EXAMPLE 2:

A 500 mL sample of citric acid ($H_3C_6H_5O_7$) is completely titrated with 44.6 mL of the same standard NaOH solution used in Example 1. What is the concentration of the citric acid solution?

SOLUTION:

Citric acid is a triprotic weak acid. Each molecule contains 3 acidic protons which can be neutralized, one after the other, by titrating with a base. Complete titration means that all 3 protons have reacted with NaOH. The overall complete neutralization reaction is:

$$H_3C_6H_5O_7 + 3 NaOH \rightleftharpoons Na_3C_6H_5O_7 + 3 H_2O \qquad (17.2)$$

The stoichiometric ratio of reagents is 1:3. Therefore, at the equivalence point, the moles of $H_3C_6H_5O_7$ originally present are equal to 1/3 the moles of NaOH added.

Moles of NaOH added = (volume of NaOH) × (molarity of NaOH)

$$= (40.6 \times 10^{-3} \text{ L})(0.226 \text{ mol/L}) = 9.18 \times 10^{-3} \text{ mol NaOH}$$

Therefore the total moles of $H_3C_6H_5O_7$ originally present was

$$(1/3)(9.18 \times 10^{-3}) \text{ moles} = 3.06 \times 10^{-3} \text{ moles}.$$

The original concentration of the 500 mL of $H_3C_6H_5O_7$ solution is:

Concentration = (moles of $H_3C_6H_5O_7$)/(volume of $H_3C_6H_5O_7$)

$$[H_3C_6H_5O_7] = (9.18 \times 10^{-3} \text{ mol})/(500 \times 10^{-3} \text{ L}) = \textbf{0.0184 mol/L}$$

Because it is triprotic, 1/3 mole of citric acid can provide 1 mole of protons, so that its **equivalent** mass is equal to 1/3 of its molar mass. See the next section below for the definition of *equivalent mass*.

MOLARITY (M), NORMALITY (N), and EQUIVALENTS

Molarity (M) = moles of solute per liter of solvent.

If 1.0 L of a solution contains 1.0×10^{-2} moles NaOH and 3.0×10^{-3} moles KCl, the solution is 1.0×10^{-2} M in NaOH and 3.0×10^{-3} M in KCl.

Normality (N) = equivalents of solute per liter of solution.

One equivalent (eq) of an acid = mass of the acid that contributes one mole of protons in an acid-base reaction.

One equivalent (eq) of a base = mass of the base that reacts with one mole of protons in an acid-base reaction.

One molecule of HCl can contribute one proton; therefore, one mole of HCl can contribute one mole of protons. Thus, one equivalent of HCl (the mass that contributes one mole of protons) is equal to one molar mass of HCl, or 36.46 g. Therefore, the normality of an HCl solution is equal to its molarity.

One molecule of citric acid ($H_3C_6H_5O_7$) can contribute three protons; therefore one mole of $H_3C_6H_5O_7$ can contribute three moles of protons. Thus, one equivalent of $H_3C_6H_5O_7$ is equal to one-third molar mass of $H_3C_6H_5O_7$, or 1/3 × 192.13) = 64.04 g.

The normality of an $H_3C_6H_5O_7$ solution is three times its molarity. A solution of $H_3C_6H_5O_7$ that is 1.00 M is 3.00 N. A 2.00 N solution of $H_3C_6H_5O_7$ must contain 2/3 moles of $H_3C_6H_5O_7$ per liter of solution.

Normality is always equal to, or greater than, molarity.

Acetic Acid in Vinegar

Vinegar is a common "kitchen acid" that is at least 5% by mass acetic acid in water solution. Acetic acid may be the product of commercial synthesis or a "natural" process in which fruit or grain is allowed to ferment to ethanol (see the Introduction to Experiment 28), which then is oxidized to acetic acid. (Wine, exposed to air for too long, turns to vinegar by the same process.)

$$CH_3CH_2OH + O_2 \longrightarrow CH_3COOH + H_2O \qquad (17.3)$$
$$\text{ethanol} \qquad\qquad\qquad \text{acetic acid}$$

Wine, cider, and rice vinegars are named after the original fermenting material. Federal regulations require that solutions labeled "vinegar" contain at least 5% acetic acid.

Citric Acid in Fruits

Citric acid, $H_3C_6H_5O_7$, is another common "kitchen acid." It is responsible for the sour taste of citrus fruits. It is a triprotic acid, each molecule containing 3 acidic protons which can be transferred, one after the other, to water molecules, or neutralized, one after the other, by titrating with a base.

PLAN OF EXPERIMENT

You first standardize a solution of base to be used as the titrant. Then you use your standardized titrant to analyze several acid solutions. Typical pH ranges for vinegar and citric acid in fruit are shown in Figure 16.1, Experiment 16.

The titrant base is NaOH. It is very difficult to dissolve a known mass of solid NaOH into a known volume of water to obtain a standard solution of NaOH with an accurately known concentration, because NaOH is extremely hygroscopic (absorbs water from the atmosphere very strongly). This makes it almost impossible to weigh solid NaOH accurately, because a significant mass of water is always present. NaOH can be standardized, however, by titrating it to an end point against an acid of accurately known concentration. The acid serves as a primary standard for calibrating the NaOH, which becomes a secondary standard.

A primary standard is a substance that can be readily prepared to give reproducible and accurate values of some quantity to be measured, in this case the concentration of acid. Potassium hydrogen phthalate, $KHC_8H_4O_4$, is useful as a primary acid standard. It is monoprotic, containing 1 mole of acidic protons per mole of $KHC_8H_4O_4$. It can be highly purified, does not oxidize readily, can be dried easily to a constant mass, and has a large molar mass, which permits high accuracy when weighing conveniently sized samples. These characteristics make $KHC_8H_4O_4$ a good primary standard that can be made into a solution with an accurately known concentration.

One drawback to the use of $KHC_8H_4O_4$ as a primary standard is that it is a weak acid, which causes it to have a titration end point that is less sharp than desirable. The end point sharpness can be maximized by insuring that both the titrant and titrand solutions are free of CO_2, which forms carbonate ion in basic solution.

Distilled water used to prepare the $KHC_8H_4O_4$ and NaOH solutions should be boiled to drive off any dissolved CO_2.

The neutralization reaction between NaOH and $KHC_8H_4O_4$ is:

$$NaOH + KHC_8H_4O_4 \rightleftharpoons KNaC_8H_4O_4 + H_2O \qquad (17.4)$$

Once you have standardized the base titrant, you can measure the concentrations of unknown acid solutions, including vinegar and citric acid from fruits.

SAFETY

1. Wear approved eye protection.

2. NaOH can cause severe eye, skin, and clothing damage. If contact accidentally occurs, rinse the affected parts with large amounts of water.

   KNOW THE LOCATIONS OF THE EYEWASH DEVICES.

   Clean up any NaOH spills immediately with large amounts of water, followed by a covering of sodium bisulfate ($NaHSO_4$).

PROCEDURE

The standardized NaOH left over after this experiment may be used in Experiment 18, if no more than 1 week intervenes. After 1 week, the NaOH solution should be recalibrated against $KHC_8H_4O_4$.

A. Standardization of NaOH Titrant with a Primary Standard of $KHC_8H_4O_4$

1. Prepare about 500 mL of approximately 0.1 M NaOH. Estimate the mass of $KHC_8H_4O_4$ that will react with about 20 mL of the NaOH solution. (Molar mass = 204.2 g mol^{-1}.)
Before proceeding, check your estimate with your instructor.

2. Weigh accurately a piece of glassine weighing paper. Then place on it an amount of $KHC_8H_4O_4$ that gives an accurate mass close to your estimate. Carefully transfer the sample into a 250 mL beaker. Calculate the number of equivalents of $KHC_8H_4O_4$ in your sample, using the stoichiometry of Reaction 17.4.

3. Add about 50 mL of boiled and cooled distilled water and 2-3 drops of phenolphthalein indicator to the $KHC_8H_4O_4$ solution.

4. Titrate with the NaOH solution. Make two determinations.

B. Acetic Acid Content of Vinegar

1. Pipet 2.00 mL of a vinegar sample into a 250 mL beaker. Add a measured volume (about 50 mL) of distilled water (it need not be boiled) and 2-3 drops of phenolphthalein indicator.

2. Titrate with the standardized NaOH solution. Make 2 determinations.

C. Citric Acid Content of Citrus Fruits

1. Select a citrus fruit, cut it in half, and squeeze its juice into a 250 mL beaker.

2. Vacuum filter to remove solid material.

3. Pipet 2.00 mL of the filtered juice into a 250 mL beaker.

4. Add a measured volume (about 250 mL) of distilled water and 2-3 drops of phenolphthalein indicator.

5. Titrate with the standardized NaOH solution. The color change occurs after all 3 acidic protons in citric acid have been neutralized. Make 2 determinations.

6. Compare your results with those of your labmates to estimate the variability of citric acid content among different kinds of citrus fruits and among different samples of the same kind of fruit.

Name ____   Date _____________

EXPERIMENT 17
PRELABORATORY EXERCISE

1.   A total of 23.24 mL of 0.1150 M NaOH was required to reach a phenolphthalein end point when titrating a 5.03 g sample of vinegar. The density of the vinegar was 1.0136 g mL^{-1}.

    a. How many equivalents of acetic acid were present in the sample.   ___________

    b. What is the mass percentage of acetic acid in the sample?   ___________

Calculations:

Name ____   Date ____________

EXPERIMENT 17

DATA
(Observe significant figures in all calculations.)

A. Molarity of NaOH Solution

		Determination 1	Determination 2
1.	Mass of glassine weighing paper:	__________	__________
2.	Total mass of sample plus weighing paper:	__________	__________
3.	Mass of $KHC_8H_4O_4$ sample:	__________	__________
4.	Moles of $KHC_8H_4O_4$ in sample:	__________	__________

Titration with NaOH

5.	Initial buret reading:	__________	__________
6.	Final buret reading:	__________	__________
7.	Volume of NaOH titrant:	__________	__________
8.	Molarity of NaOH titrant:	__________	__________
9.	Average molarity of NaOH solution:	__________	
10.	Relative average deviation:	__________	

B. Acetic Acid Content of Vinegar

1.	Vinegar brand name:	__________	
2.	Molarity of NaOH titrant:	__________	

		Determination 1	Determination 2
3.	Initial volume of vinegar, before adding water:	__________	__________
4.	Volume of water added to vinegar:	__________	__________
5.	Total vol. of vinegar solution, after adding water:	__________	__________
6.	Initial buret reading:	__________	__________
7.	Final buret reading:	__________	__________
8.	Volume of NaOH titrant added	__________	__________
9.	Moles of acetic acid in sample	__________	__________
10.	Normality of vinegar solution	__________	__________
11.	Av. normality of vinegar solution	__________	
12.	Relative av. deviation of normality	__________	
13.	Mass of acetic acid per mL of vinegar	__________	
14.	The density of vinegar is 1.01 g mL^{-1}. Calculate the mass percent of acetic acid in your vinegar sample.	__________	

DATA

C. Citric Acid Content of Fruits

1. Citrus fruit used: _______________

2. Molarity of NaOH titrant: _____________

Determination
(indicate units on your entries)

		1	2
1.	Mass of glassine weighing paper:	_____________	_____________
2.	Total mass of sample plus weighing paper:	_____________	_____________
3.	Initial volume of filtered citrus fruit juice, before adding water:	_____________	_____________
4.	Volume of water added to fruit juice:	_____________	_____________
5.	Total volume of citrus fruit juice solution, after adding water:	_____________	_____________
6.	Initial buret reading:	_____________	_____________
7.	Final buret reading:	_____________	_____________
8.	Volume of NaOH titrant added:	_____________	_____________
9.	Moles of citric acid in sample:	_____________	_____________
10.	Normality of fruit juice solution:	_____________	_____________
11.	Average normality of fruit juice solution:	_____________	
12.	Relative average deviation of normality:	_____________	
13.	Equivalents of citric acid in sample:	_____________	

14. The density of citric acid is 1.01 g mL^{-1}. Calculate the mass percent of citric acid in your juice sample.   _____________

15. Using your results and those of your classmates, make a table comparing the citric acid content of several different fruits.

Fruit	Mass percent citric acid in juice (give the observed range)

Calculations:

Name ____   Date ____________

EXPERIMENT 17
QUESTIONS

1.   Give the effect (+, 0, -) on the determination of the moles of acid in a sample, if the following experimental errors occur:

a.   A drop of standardized NaOH adheres to the inside of the titrand container and is never washed down into the solution.

____

b.   The buret used was not clean and some NaOH adhered to the buret walls above the meniscus level.

____

c.   The vinegar or juice solution was so deeply colored that the initial color change of the indicator was not observed and the acid was overtitrated.

____

EXPERIMENT 18
Dissociation Constants of a Polyprotic Acid: A Determination by pH Titration

PRELABORATORY PREPARATION
1. Do the Prelaboratory Exercise and turn it in at the beginning of your laboratory period.
2. Review the definitions of **normality** and **equivalent mass** in the Chemical Units section.
3. Review the use of the pH meter, Section XII, Methods, page 25.
4. If you need a review of the principles of acid-base reactions, read the Introduction to Experiment 16.

INTRODUCTION
In this experiment, you titrate an unknown polyprotic acid using a pH meter to continuously monitor changes in pH during the titration. A graph of pH versus volume of titrant is used to determine the acid dissociation constants in a particularly simple way.

The titration of the triprotic acid H_3PO_4 (phosphoric acid) can serve as an example of the method. H_3PO_4 dissociates in water in three successive steps, each of which establishes its own independent equilibrium and has its own equilibrium constant.

$$H_3PO_4 + H_2O \rightleftharpoons H_3O^+ + H_2PO_4^- ; \qquad K_{a1} = \frac{[H_3O^+][H_2PO_4^-]}{[H_3PO_4]} = 7.5 \times 10^{-3} \tag{18.1}$$

$$H_2PO_4^- + H_2O \rightleftharpoons H_3O^+ + HPO_4^{2-} ; \qquad K_{a2} = \frac{[H_3O^+][HPO_4^{2-}]}{[H_2PO_4^-]} = 6.2 \times 10^{-8} \tag{18.2}$$

$$HPO_4^{2-} + H_2O \rightleftharpoons H_3O^+ + PO_4^{3-} ; \qquad K_{a3} = \frac{[H_3O^+][PO_4^{3-}]}{[HPO_4^{2-}]} = 4.4 \times 10^{-13} \tag{18.3}$$

The acid species H_3PO_4, $H_2PO_4^-$, and HPO_4^{2-} are all present at the same time in aqueous solution. Their equilibrium constants show that H_3PO_4 is the strongest acid, $H_2PO_4^-$ the next strongest, and HPO_4^{2-} the weakest. HPO_4^{2-} is nearly as weak an acid as is water.

In a titration with a strong base, ionization occurs stepwise, the strongest acid, H_3PO_4, reacts first, leaving the weaker acids virtually intact until the strongest acid is completely reacted. Then the second strongest acid, $H_2PO_4^-$, reacts, and finally the weakest acid, HPO_4^{2-}, reacts. In other words, Reaction 18.1 goes to completion first, before Reactions 18.2 and 18.3 can begin. Then 18.2 is completed, and, finally, 18.3 is completed.

The titration of a triprotic acid with a strong base will have 3 equivalence points, one for each of the 3 neutralization reactions.

In practice, HPO_4^{2-}, the acid in 18.3, is too weak an acid for feasible titration in aqueous solution, because its dissociation constant, K_{a3}, is so small. $K_{a3}(H_3PO_4) = 4.4 \times 10^{-13}$ is so close to $K_w(\text{water}) = 1.00 \times 10^{-14}$, that the pH change at the equivalence point can scarcely be distinguished from the titration of pure water.

In general, for the titration curve of a polyprotic acid to show separate, well-defined equivalence point breaks, the individual K_a values all must be larger than about 10^{-9} and differ from one another by at least a factor of 10^3.

K_{a1} and K_{a2} for phosphoric acid fulfill these requirements, but K_{a3}, at 4.4×10^{-13}, is too small, and so the third equivalence point is not observed. Figure 18.1 is a typical titration curve obtained for phosphoric acid.

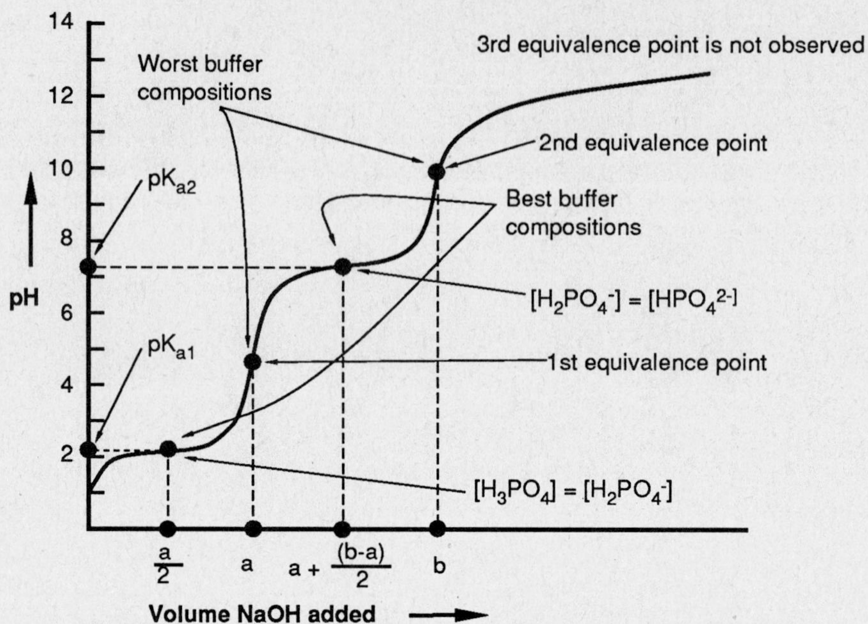

Figure 18.1.   A typical titration curve for the titration of H_3PO_4 with NaOH. a = volume of NaOH added at 1st equivalence point and b = volume added at 2nd equivalence point. The measured dissociation constants are: $pK_{a1} = 2.1$, $K_{a1} = 7.9 \times 10^{-3}$; $pK_{a2} = 7.2$, $K_{a2} = 6.3 \times 10^{-8}$.

Notice, in Figure 18.1, that the rate of pH change is smallest where the volume of NaOH added equals **a/2** and **{a + (b-a)/2}**. These points are halfway to the 1st and 2nd equivalence points respectively, where the buffer capacity of the titrand solution is greatest. The rate of pH change is greatest at the equivalence points, **a** and **b**, where the buffer capacity is smallest.[1]

> **Phosphoric acid contains one equivalent (1 mole of acidic protons) per mole up to the first equivalence point, two equivalents per mole up to the second equivalence point, and three equivalents per mole for a complete titration.**

The term **normality**, (symbol: **N**), indicates the number of moles of acidic protons that are neutralized in a titration to any particular equivalence point. Thus, a 1 M phosphoric acid solution is 3 N for complete titration of all 3 protons, but only 1 N for titration to just the first equivalence point.

How to Determine K_a Values from the Titration Curve

Using a titration curve to determine acid dissociation constants is based on the equilibrium constant equations, 18.1-18.3. When the titration first begins, the only important reaction is 18.1. Reaction 18.2 will not begin until Reaction 18.1 is essentially complete, and 18.3 will not begin until 18.2 is essentially complete. During the initial stages of the titration, $[H_2PO_4^-]$ is increasing and $[H_3PO_4]$ is decreasing, in Reaction 18.1.

> **Eventually, they must become equal, i.e., at some point in the titration, $[H_2PO_4^-]/[H_3PO_4] = 1$.**

The point where $[H_2PO_4^-]/[H_3PO_4] = 1$ occurs half-way to the first equivalence point, when 1/2 of the NaOH moles needed to complete the neutralization of the first acidic proton have been titrated into the solution.

---

[1] If you have studied buffer solutions or have done Experiment 16, you know that the most effective buffer solution is one containing equal concentrations of a weak acid and a salt of that acid. At the halfway points of the phosphoric acid titration, the titrand solution is equivalent to one made from equal concentrations of the acid and its sodium salt. Therefore, at the halfway points the buffering action of the titrand solution will be a maximum and the pH changes are a minimum for each increment of added base. Figure 18.1 illustrates this directly. The slope of the **pH vs. added base** curve is smallest at the titration points halfway to equivalence, where the buffer capacity is a maximum, and largest at the equivalence points, where the buffer capacity is a minimum.

At this point, we have for K_{a1}:

$$K_{a1} = \frac{[H_3O^+][\cancel{H_2PO_4^-}]}{[\cancel{H_3PO_4}]} = [H_3O^+]_{1/2\ ep} \ ; \ \ or\ pK_{a1} = pH_{1/2\ ep} \qquad (18.4)$$

Thus, a measurement of $[H_3O^+]_{1/2\ ep}$, when 1/2 of the first acidic protons have been neutralized, gives K_{a1} directly. Similarly, a measurement of $[H_3O^+]_{1/2\ ep}$, halfway between the first and second equivalence points, gives K_{a2} directly. The halfway points are easily read from the graph of the titration curve. K_{a3} cannot be measured with this technique because, after the second equivalence point is passed, the titration curve rises continuously toward about pH = 13.

PLAN OF EXPERIMENT

You titrate an unknown polyprotic acid and determine its dissociation constants from the titration curve. Only two equivalence points will be observed, either because the unknown acid is diprotic, or because its third (or higher) dissociation constants are too small. By using a standardized titrant, you can calculate the equivalent mass of the acid for each equivalence point. If you know the density of the acid and can assume the acid to be diprotic, you can calculate its molar mass.

SAFETY

1.   Wear approved eye protection.

2.   NaOH is a strong base and is very corrosive to eyes, skin and clothing. The weak acid unknowns may

be strong irritants, especially to your eyes. Handle all the reagents carefully and avoid contact.

3.   The measuring electrodes of the pH meter are very delicate and easily broken. Handle them carefully

and avoid bumping them against the beaker or stirrer.

PROCEDURE

(Use NaOH standard solution left over from Experiment 16. If it is more than 1 week old, recalibrate it against $KHC_8H_4O_4$. If necessary, prepare 200 mL of fresh standard according to the procedure of part A, Experiment 16.)

1.   Standardize the pH meter as described in Section XII, Methods, page 25.
Be sure to rinse the electrodes with distilled water carefully each time before placing them in the buffer solution or in the acid sample solution.

2.   Obtain a sample of unknown polyprotic acid and record its number and molarity on the Data sheet. Accurately pipet a 25 mL sample of the unknown acid into each of three 250 mL beakers.

3.   Titrate each portion separately. Before beginning each titration, record the initial pH of the acid solution. During each titration, stir continuously with a magnetic stirrer, stirring motor, or, by swirling the beaker a little *very carefully so you do not bump the electrodes.*
The 1st titration will serve as a practice run. Titrate the first portion rapidly and qualitatively, to obtain a general idea of where the equivalence points are located. This will save time with the 2nd and 3rd quantitative titrations.

1st Titration (rapid and qualitative):
1.   Add titrant in 3-5 mL increments until the pH begins to rise rapidly. Then, pass through the equivalence point with 1 mL increments.

2.   During the 1 mL steps, record the buret level and pH after each increment. When the pH once more is changing slowly, return to 3-5 mL increments until the second equivalence point is approached.

3.   Use 1 mL increments again through the second region of rapid pH change. During the 1 mL steps, record the buret level and pH after each increment. Titrate well past the second equivalence point, until the pH changes slowly again.

2nd and 3rd Titrations (careful and quantitative):

1. Using the first titration as a guide, titrate with 3-5 mL increments, to just before the NaOH volume where the pH will begin changing rapidly.

Record the buret level and pH after adding each increment of NaOH.

Then, pass through the equivalence point dropwise, measuring at approximately 0.1 mL intervals. Record the buret level and pH after each small increment.

2. After the equivalence point is passed, resume measuring at larger intervals again until the next equivalence point approaches, and then use 0.10 mL intervals again through the second equivalence point. Continue adding titrant well past the second equivalence point, until the pH increases only slightly with added titrant. Record the buret level and pH after each increment.

Switch the pH meter to standby between titrations. Do not turn it off.

4. Graph the 2nd and 3rd titration curves separately. Determine pK_{a1} and pK_{a2} for your unknown acid from each curve and average their values. Enter all results on the Data sheet.

Name ____   Date _____________

EXPERIMENT 18
PRELABORATORY EXERCISE

1. **a.** What is the normality of a 0.5 M NaOH solution? _____________

b. What is the molarity of a 0.02 N NaOH solution? _____________

c. What is the normality of a 0.1 M H_2CO_3 acid solution with respect to a complete titration of all ionizable hydrogens? _____________

d. What is the molarity of a 0.1 N H_2CO_3 solution where the normality was measured by a complete titration? _____________

2. The aqueous solution dissociation reactions of sulfurous acid, H_2SO_3, are:

$$H_2SO_3 + H_2O \rightleftharpoons H_3O^+ + HSO_3^- \; ; \qquad K_{a1} = \frac{[H_3O^+][HSO_3^-]}{[H_2SO_3]}$$

$$HSO_3^- + H_2O \rightleftharpoons H_3O^+ + SO_3^{2-} \; ; \qquad K_{a2} = \frac{[H_3O^+][SO_3^{2-}]}{[HSO_3^-]}$$

A 100 mL sample of H_2SO_3 was titrated with 0.1000 N NaOH. Some measured values of the pH during the titration are given below.

pH:	1.4	1.8	7.1	7.6	10.1	11.2
mL NaOH:	0	50	100	150	200	300

(1st equiv. pt.)   (2nd equiv. pt.)

a. Calculate K_{a1} and K_{a2}. K_{a1} = _______________ ; K_{a2} = _______________

b. What is the molarity of the H_2SO_3 solution?   **Molarity** = _____________

c. What is the normality of the H_2SO_3 solution with respect to a complete titration?   **Normality** = _____________

Calculations:

Name ____   Date _____________

EXPERIMENT 18

DATA
(Observe significant figures in all calculations.)

1.   Unknown acid No.:   ________________

2.   Grams of unknown acid per liter of solution: ________________

3.   Normality of NaOH standard: ________________

4.   **Titration data:**

1st titration		2nd titration		3rd titration	
mL NaOH added	Measured pH	mL NaOH added	Measured pH	mL NaOH added	Measured pH
0		0		0	

For approx. 1st titration:

Record the mL NaOH where approach to eq. pts. begins:

Approach to:
1st equiv, pt.   ______

2nd equiv, pt.   ______

For 2nd titration:

pK_{a1} = ______ pK_{a2} = ______

Av. pK_{a1} = ___________

Av. pK_{a2} = ___________

For 3rd titration:

pK_{a1} = ______ pK_{a2} = ______

DATA

4. Equivalent Mass of Unknown Acid

	Titration 2	Titration 3
Initial volume of unknown acid:	__________	__________
mL NaOH at 1st equiv. pt.:	__________	__________
mL NaOH at 2nd equiv. pt.:	__________	__________
Normality of unknown acid, with respect to the 2nd equiv. pt.:	__________	__________
Average normality (2nd equiv. pt.):	__________	
Equivalent mass of unknown acid at 1st equiv. pt.:	__________	
Equivalent mass of unknown acid at 2nd equiv. pt.:	__________	
Molar mass of unknown diprotic acid:	__________	

Calculations:

Name ____   Date ______________

EXPERIMENT 18
(Additional graph paper is at the back of this book.)

EXPERIMENT 19
Oxidation-Reduction Titration:
Analysis of an Oxalate

PRELABORATORY PREPARATION
1. Do the Prelaboratory Exercise and turn it in at the beginning of your laboratory period.
2. Review the use of the buret, Section IV, page 17, and the technique for transferring powders, Section II, page 12, Methods.
3. Review the definitions of **equivalent mass** and **normality** in the Chemical Units section.
4. If necessary, review the discussion of oxidation-reduction reactions in Experiment 13.

INTRODUCTION
Oxidation-reduction reactions, often called redox reactions, involve the transfer of electron charge from one atom to another.

The atom donating the electron charge is said to be *oxidized*. **Its oxidation number increases (changes in the positive direction), which corresponds to electron loss.**

The atom accepting the electron charge is said to be *reduced*. **Its oxidation number decreases (changes in the negative direction), which corresponds to electron gain.**

One atom can be oxidized and donate electrons, only if another atom accepts the electron and is reduced. The two atoms that exchange electrons in an oxidation-reduction reaction, or the two molecules in which they reside, are called the **oxidation-reduction couple**. In oxidation-reduction reactions, usually only a few of all the atoms involved undergo changes in oxidation number. We will identify these atoms by calling them **redox atoms**.

Because an electron cannot leave one atom without becoming attached to another atom, every oxidation-reduction reaction must have at least two redox atoms: one that is oxidized and has an increase in oxidation number, and another that is reduced, with a decrease in oxidation number.

The oxidation number of an atom defines its **oxidation state**. An element in its pure natural state has oxidation number = 0. Many elements can have several different oxidation states. Manganese, for example, can exist in oxidation states 0, +2, +3, +4, +6, and +7. The oxidation states greater than zero correspond to the loss of 2, 3, 4, 6, and 7 electrons, respectively. With so many possible oxidation states, manganese compounds can undergo many different oxidation-reduction reactions. In Example 1, the reaction of this experiment serves to illustrate oxidation states and oxidation number changes.

EXAMPLE 1:
Potassium permanganate ($KMnO_4$) and sodium oxalate ($Na_2C_2O_4$) react in a solution acidified with sulfuric acid[1], according to Reaction 19.1a, b:

Molecular equation:

$$2\ KMnO_4 + 5\ Na_2C_2O_4 + 8\ H_2SO_4 \longrightarrow$$
$$2\ MnSO_4 + 10\ CO_2 + 5\ Na_2SO_4 + K_2SO_4 + 8\ H_2O \qquad (19.1a)$$

Ionic equation, with oxidation numbers assigned. Rules for assigning oxidation numbers are in your textbook.

$$2\ MnO_4^- + 5\ C_2O_4^{2-} + 16\ H^+ \longrightarrow 2\ Mn^{2+} + 10\ CO_2 + 8\ H_2O \qquad (19.1b)$$

Oxid. numbers: +7, -2   +3, -2   +1   +2   +4,-2   +1,-2

---

[1] H_2SO_4 is used because the anion, SO_4^{2-}, does not react with any of the reactants or products of Reaction 19.1. HCl cannot be used because MnO_4^- oxidizes Cl^- to $1/2\ Cl_2$, and HNO_3 cannot be used because Mn^{2+} reduces NO_3^- to NO_2^-.

Mn and **C** are the only atoms that change their oxidation numbers. Therefore, they are the redox atoms in Reaction 19.1. Manganese changes from +7 in MnO_4^-, to +2 in Mn^{2+}, and carbon changes from +3 in $C_2O_4^{2-}$, to +4 in CO_2. In Reaction 19.1, Mn(+7) is reduced to Mn(+2), and, at the same time, C(+3) is oxidized to C(+4).

The absolute value of the change in oxidation number experienced by a redox atom is equal to the number of *equivalents* in one mole of the molecule containing the redox atom. This leads to the following definitions:

a. **One equivalent mass of a redox atom is equal to its atomic mass divided by the absolute value of its change in oxidation number.**

b. **When the redox atom is part of a molecule, one equivalent mass of that molecule is equal to its molar mass divided by the absolute value of the oxidation number change of its redox atom.**

c. **The equivalent mass of a molecule is the mass that contains one equivalent mass of its redox atom.**

EXAMPLE 2:

In Reaction 19.1, Mn (the redox atom in $KMnO_4$ and $MnSO_4$) changes its oxidation number from +7, in $KMnO_4$, to +2, in $MnSO_4$. The absolute value of the oxidation number change is 5 units. Therefore, there are 5 equivalents in each mole of $KMnO_4$, and the equivalent mass of $KMnO_4$, in *this* reaction, is its molar mass ($158.04 \text{ g mol}^{-1}$) divided by 5.

$$\text{Equiv. mass of } KMnO_4 \text{ (in reaction 19.1)} = \frac{158.04 \text{ g mol}^{-1}}{5 \text{ eq. mol}^{-1}} = 31.61 \text{ g eq.}^{-1}$$

Notice that the equivalent mass of a reagent depends on the particular reaction in which it is involved, because the nature of the reaction determines the oxidation number changes.[2]

The **normality** of a solution is the number of equivalents of reagent per liter of solution. A solution that is 0.1000 N $KMnO_4$ *and used in Reaction 19.1* contains:

$$0.1000 \text{ eq L}^{-1} \times 31.61 \text{ g eq}^{-1} = \textbf{3.161 g L}^{-1} \textbf{ } KMnO_4$$

PLAN OF EXPERIMENT

It is very difficult to prepare and keep pure potassium permanganate solutions, because permanganate ion is so strong an oxidizer that it will react with most trace impurities, and, in time, even with the water solvent itself. It is best to make up the solution initially to an approximate concentration, and then standardize with a reagent of known concentration. Only freshly standardized $KMnO_4$ solutions should be used for quantitative titrations. You will prepare and standardize a solution of $KMnO_4$, by titrating it against a $Na_2C_2O_4$ solution of known concentration. The standardized permanganate solution then is used to analyze a sample that contains an unknown amount of oxalate ion.

Permanganate ion (MnO_4^-) is intensely purple in color, and all other reagents and products in Reaction 19.1 are colorless. Thus, MnO_4^- can serve as the end point indicator.

As MnO_4^- is titrated into a colorless oxalate solution, Reaction 19.1 forms Mn^{2+}, which also is colorless. As long as any oxalate remains, with which permanganate can react, the purple color of added titrant will disappear. When all the oxalate has reacted, the next drop of MnO_4^- is in excess and the solution turns faintly pink. An excess of one drop of MnO_4^- gives a visible pink color.

The permanganate endpoint color is not permanent. After the endpoint is reached, the permanganate ion, now in excess, slowly reacts with Mn^{2+}, forming MnO_2, and causing the endpoint color to fade gradually. A pink color that persists after stirring the solution for 30 sec should be regarded as the true endpoint.

---

[2]To illustrate how the equivalent mass depends on the particular reaction, we consider the reaction whereby $KMnO_4$ reacts slowly with water (see Plan of Experiment):

$$4 MnO_4^- + 2 H_2O \longrightarrow 4 MnO_2 + 3 O_2 + 4 OH^- \tag{19.2}$$

In Reaction 19.2, the oxidation number of Mn changes from +7 in MnO_4^-, to +4 in MnO_2. The absolute value of the oxidation number change is 3 units. Thus, the equivalent mass of $KMnO_4$, *in Reaction 19.2*, is:

$$(158.04 \text{ g mol}^{-1})/(3 \text{ eq mol}^{-1}) = 52.68 \text{ g eq}^{-1}$$

quite different from its equivalent mass in Reaction 19.1.

As the endpoint is approached, the pink permanganate color will last longer and longer. You must titrate more slowly near the endpoint to avoid overtitrating. It is good practice, near the end point, to record the buret reading after each drop of added titrant. The true endpoint is the first faint pink color that lasts 30 sec. If one drop causes an intense color, the drop just before it corresponded to the correct endpoint.

The reaction is slow at room temperature, so the titrand solution must be heated to about 80°C to ensure that equilibrium is maintained during the titration. Even at an elevated temperature, the reaction will start slowly and the pink color of each drop of titrant may persist a few seconds at the beginning of titration. However, the Mn^{2+} formed by the reaction acts as a catalyst, and the reaction rate increases as the concentration of Mn^{2+} builds up. Reactions that are catalyzed by one of the products formed are called autocatalytic.

At the very beginning of the titration, it is important to add the titrant very slowly. If it is added rapidly, brown MnO_2 may precipitate, formed by a reaction that competes with the initially slow oxalate reaction.

If only a little MnO_2 is formed, a faint brown color appears. This is not serious if there is enough oxalate present to reduce MnO_2 to Mn^{2+}. Simply stop titrating and stir until the brown color disappears. If MnO_2 persists, the titration is ruined and you must start over. After a few mL of MnO_4^- solution have been added, and the catalyzed oxalate reaction rate has increased, the rate of titration may be increased.

SAFETY

1. Wear approved eye protection.

2. H_2SO_4 is very corrosive. Handle it carefully.

3. Sodium oxalate is toxic. Do not touch it except with a spatula and pipet oxalate solutions with a

    pipetting bulb. **NEVER PIPET BY MOUTH.**

4. Permanganate solutions are intensely colored and will leave a permanent stain if spilled on clothing.

    Handle the solutions carefully.

PROCEDURE
A. Preparation of Permanganate Solution
1. Obtain 500 mL of an approximately 0.1 N $KMnO_4$ solution in a beaker.

2. Cover the solution with a watchglass and boil it gently for 5 min to oxidize contaminants before standardization. Then remove it from the heat and allow to cool.

3. When the beaker has cooled sufficiently to be picked up by hand, filter the solution into a clean, glass-stoppered bottle, through a sintered glass filter or through a wad of glass wool (about 5 cm in diameter) placed in a funnel. Organic matter readily reduces MnO_4^-, and for this reason, filter paper cannot be used.

The purpose of the filtering procedure is to remove manganese dioxide, MnO_2, with which $KMnO_4$ usually is contaminated.

Aqueous solutions of permanganate are not stable because MnO_4^- slowly oxidizes water, a reaction greatly accelerated by the presence of solid MnO_2. Unless MnO_2 is removed, permanganate decomposition can occur in the buret, during the experiment.

B. Standardization of Permanganate Solution
1a.   *If a standardized solution of sodium oxalate is available,* pipet 25.00 mL of it into each of two labelled 250 mL beakers (or Erlenmeyer flasks). (Otherwise, go to **1b.**) Record the molarity of the standard oxalate solution and calculate the number of moles of oxalate in each of your 2 samples. Dilute each solution to about 100 mL with distilled water.

1b.   *If no standard oxalate solution is available*, obtain about 2g of solid sodium oxalate and dry it (if necessary) on a watch glass in an oven at 110°C for 30 min. After drying, let the sample cool to room temperature. Transfer the cool dry sample to a piece of weighing paper and weigh sample and paper accurately. Transfer two samples, of about 0.25 g each, of the sodium oxalate into 250 mL beakers (or Erlenmeyer flasks). The transfer may be done in several steps to get a mass close to 0.25 g into the container. Weigh the remaining sample and paper after each transfer, to obtain the mass of sample transferred by difference.

Mark each beaker with the mass of its sample. Add about 100 mL distilled water to dissolve the solid sodium oxalate samples.

2.   Rinse a buret with unstandardized potassium permanganate solution, then fill it nearly to the top with the same solution. Run a little permanganate through the tip into an extra beaker, to clear out any bubbles. Then read the starting level to 0.01 mL and record it on the Data sheet.

The dark color of the permanganate solution makes it difficult to see the bottom of the meniscus. Instead, read the volume from the top of the liquid level.

3.   Acidify the oxalate solutions by slowly and carefully adding 25 mL of 6 M H_2SO_4, with continuous stirring.

4.   Prepare to heat one oxalate solution with a gas burner or hotplate. Locate the solution beaker below the buret so you can titrate the heated solution directly. Before starting the titration, heat the titrand solution to about 80°C, but not above 90°C. Then begin the titration and maintain the titrand temperature between 70°C and 85°C throughout.

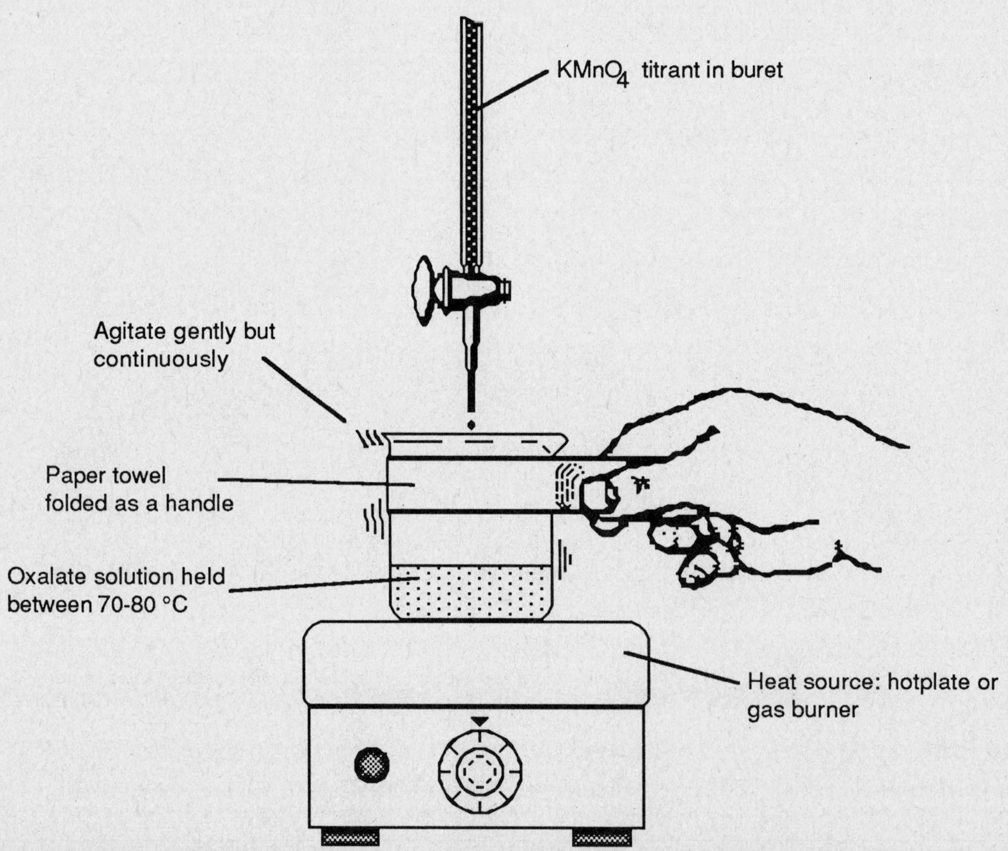

Figure 19.1.   Apparatus arrangement for titrating warmed oxalate solution.

Titration Technique

Use the following method for titrating:

Constantly swirling the beaker (holding it with a folded towel), start the titration slowly, by adding 20 single drops of titrant, 2 sec apart (or until color fades between drops), to build a small concentration of Mn^{2+} for catalyzing the reaction. Then add about 5 mL of permanganate solution all at once and wait a few seconds, while still swirling, for the pink color to disappear. Then add another 5 mL, and continue in the same manner.

As the endpoint nears, the pink color will persist longer and longer. When you observe that the pink color lifetime is lengthening, you must add the titrant slowly again, drop by drop with constant swirling, observing the color carefully after each drop. Wait until the color disappears before adding the next drop. **Near the endpoint, record the titrant level to 0.01 mL after adding each drop, as insurance against overtitrating. When a single drop causes the color to last for at least 30 seconds, the endpoint has been reached. If one drop gives a permanent intense color, the drop before it corresponded to the correct endpoint.**

7.   Record the volume of titrant added and calculate the normality of the permanganate titrant solution.

8.   Do a repeat determination.

C. Analysis of an Unknown Oxalate

1.   Obtain a sample containing an unknown amount of oxalate. Dry it, if necessary, as in part B.

2.   Transfer the dry cool sample to a piece of weighing paper, weigh accurately, and record the mass.

3.   Transfer about 0.25-0.50 g of the unknown sample into each of two labelled 250 mL beakers. Weigh the remaining sample and weighing paper after each transfer to obtain the transferred mass by difference. Proceed with the analysis as in part B. Make two determinations.

4.   Calculate the moles and mass percentage of oxalate in your unknown.

Name ____   Date _____________

EXPERIMENT 19
PRELABORATORY EXERCISE

1.   What mass of $KMnO_4$ is required to make 0.500 mL of 0.500 N solution?   _________

2.   A sample weighing 0.2144 g was analyzed for oxalate ($C_2O_4^{2-}$) by titrating with potassium permanganate standard solution. It required 12.4 mL of 0.104 N $KMnO_4$ to obtain the first sign of pink color that persisted for 30 sec.

    a. Calculate the moles of oxalate in the sample.   _____________

    b. Calculate the mass percent oxalate in the sample.   _____________

Calculations:

Name ____   Date ____________

EXPERIMENT 19

DATA
(Observe significant figures in all calculations.)

Determination
1   2

B. Permanganate Standardization

1a.   Molarity of ready-made standard oxalate solution: ______   ______

Volume of ready-made oxalate solution used as titrand: ______   ______

Moles of oxalate in titrand solution: ______   ______

1b.   Mass of solid oxalate sample and weighing paper: ______   ______

Mass of solid sodium oxalate sample: ______   ______

Moles of oxalate in titrand solution: ______   ______

2.   Final $KMnO_4$ buret reading: ______   ______

3.   Initial $KMnO_4$ buret reading: ______   ______

4.   Volume of $KMnO_4$ used: ______   ______

5.   Normality of $KMnO_4$ standard solution: ______   ______

6.   Average normality of $KMnO_4$ standard solution: ______

C. Analysis of Unknown Oxalate Sample

1.   Unknown No. ______

2.   Mass of solid unknown sample and weighing paper: ______   ______

3.   Mass of solid unknown sample: ______   ______

4.   Final $KMnO_4$ buret reading: ______   ______

5.   Initial $KMnO_4$ buret reading: ______   ______

6.   Volume of $KMnO_4$ used: ______   ______

7.   Moles of oxalate ion in unknown sample: ______   ______

8.   Average moles of oxalate in unknown: ______

9.   Mass percent oxalate ion in unknown: ______

Calculations:

Name _______________________________________   Date ____________

EXPERIMENT 19
QUESTIONS

1.   Oxalic acid is slowly decomposed by sulfuric acid:

$$H_2C_2O_4 \xrightarrow{H_2SO_4} H_2O + CO + CO_2$$

If the acidified oxalate solution (see Procedure Part B, step 3) is allowed to stand for several hours before titration, what error (+, -, or 0) might be introduced:

    a.   in the normality of the standardized permanganate solution, in part B?   _______

    b.   in the percent oxalate in the unknown?   _______

2.   Why not use HCl or HNO_3 to acidify the oxalate solution, in order to avoid the problem mentioned in question 1?

3.   Why does the pink color of the endpoint eventually disappear?

4.   Why is it important to begin the titration very slowly?

EXPERIMENT 20
Oxidation-Reduction Titration:
Iodometric Determination of Cu^{2+}

PRELABORATORY PREPARATION
1. Do the Prelaboratory Exercise and turn it in at the beginning of your laboratory period.
2. Review the definitions of **equivalent mass**, **normality**, and **molarity** in the Chemical Units section.
3. If necessary, review the discussion of oxidation-reduction reactions in Experiment 13.

INTRODUCTION
This experiment is an example of an indirect analysis where the compound to be analyzed is made to react with another compound, in order to form a product which can be analyzed quantitatively more easily and accurately than the original compound of interest. The basis of this analysis for Cu^{2+} is a reaction of Cu^{2+} with an excess of iodide ion, I^-, in which Cu^{2+} is quantitatively reduced to insoluble copper(I) iodide:

$$2\,Cu^{2+} + 4\,I^- \longrightarrow 2\,CuI(s) + I_2 \qquad (20.1)$$

Molecular iodine, I_2, is not very soluble in water, but it reacts quickly with the iodide ion that is present, to form soluble, brown-colored, triiodide ion, I_3^-:

$$I_2 + I^- \longrightarrow I_3^- \qquad (20.2)$$

The actual analysis is made by titrating the triiodide ion with standardized sodium thiosulfate solution, $Na_2S_2O_3$ (photographer's "hypo"). Thiosulfate ion reduces I_3^- to $3\,I^-$ and, in turn, is oxidized to the tetrathionate ion:

$$I_3^- + 2\,S_2O_3^{2-} \longrightarrow 3\,I^- + S_4O_6^{2-} \qquad (20.3)$$

The titration is complete when all of the I_3^- (and, hence, all of the I_2 formed in Reaction 20.1) is converted to I^-. The endpoint is identified by a starch indicator which is intensely blue in the presence of a trace of I_3^-, and turns colorless when all the I_3^- is consumed by Reaction 20.3.

EXAMPLE:
A 140.32 g sample of copper ore is analyzed for copper content by dissolving it in nitric acid and analyzing the solution for Cu^{2+} by means of Reactions 20.1-20.3. An excess of KI is dissolved into the solution containing Cu^{2+}, to provide iodide ion for Reaction 20.1. The I_3^- which then results from Reaction 20.2 is titrated to a starch solution endpoint with 12.38 mL of 0.1036 N $Na_2S_2O_3$ solution. Calculate the mass percent copper in the ore sample.

SOLUTION:
The volume in liters of titrant, multiplied by titrant normality, equals equivalents of thiosulfate ion used:

$$\text{equiv thiosulfate} = (0.01238\ L)(0.1036\ \text{equiv}\ L^{-1}) = 1.283 \times 10^{-3}$$

The equivalent mass of a reactant in an oxidation-reduction reaction is the mass that donates or receives one mole of electrons. In Reaction 20.3, the equivalent mass of $S_2O_3^{2-}$ is equal to its molar mass, and the equivalent mass of I_3^- is equal to 1/2 of its molar mass. This can be deduced from the stoichiometry, which shows that 2 moles of $S_2O_3^{2-}$ titrant reduce 1 mole of I_3^- to 3 moles of I^-. Each I_3^- ion must receive 2 electrons from the titrant, to provide the necessary negative charge for the 3 I^- product ions:

$$I_3^- + 2\,e^- \longrightarrow 3\,I^-$$

Therefore, each mole of $S_2O_3^{2-}$ donates one mole of electrons, and each 1/2 mole of I_3^- accepts one mole of electrons.

In an indirect analysis, the moles of titrant used to measure the reaction product must eventually be related back to the amount of unknown in the original sample. In this experiment, the amount of $S_2O_3^{2-}$ used to titrate I_3^- must be related back to the amount of Cu^{2+} in the original sample. This can be done either in terms of equivalents or in terms of moles. We will do it both ways, in order to show how the concept of equivalents can simplify your calculations.

Calculation Using Equivalents

In any reaction, or consecutive sequence of reactions, one equivalent of any reactant always will react with one equivalent of any other reactant. Therefore, the number of equivalents of Cu^{2+} in the sample is equal to the number of equivalents of $S_2O_3^{2-}$ used in the titration of I_3^-, even though these reactants are involved in different steps of the overall process. We saw above that 1.283×10^{-3} equivalents of $S_2O_3^{2-}$ were used in the titration, so that

$$\text{equivalents of } Cu^{2+} = \text{equivalents of } S_2O_3^{2-} = 1.283 \times 10^{-3}$$

In Reaction 20.1, each Cu^{2+} ion accepts 1 electron, to form CuI. Therefore, the equivalent mass of Cu^{2+} *in this reaction* is equal to its molar mass:

$$1.283 \times 10^{-3} \text{ equivalents of } Cu^{2+} = 1.283 \times 10^{-3} \text{ moles of } Cu^{2+} \text{ in the ore sample.}$$

Atomic mass $Cu = 63.5$ g mol^{-1}. Therefore,

$$\text{mass Cu in ore} = (63.5 \text{ g mol-1})(1.283 \times 10^{-3} \text{ mol}) = 8.14 \times 10^{-2} \text{ g}$$

$$\text{percent Cu in ore} = \frac{8.14 \times 10^{-2} \text{g Cu in sample}}{140.3 \text{ g total sample}} \times 100\% = \mathbf{5.80 \times 10^{-2}\%}$$

Calculation Using Molar Masses:

Examination of Reactions 20.1-20.3 shows that:

a. 2 moles of Cu^{2+} forms 1 mole of I_2.

b. 1 mole of I_2 forms 1 mole of I_3^-.

c. 1 mole of I_3^- reacts with 2 moles of $S_2O_3^{2-}$.

d. Therefore, the presence of 2 moles of Cu^{2+} in the original solution, requires 2 moles of $S_2O_3^{2-}$ to titrate its product, I_3^-.

The moles of Cu^{2+} equals the moles of $S_2O_3^{2-}$ titrant used, so that:

$$\textbf{moles } Cu^{2+} = \textbf{moles } S_2O_3^{2-} = \mathbf{1.283 \times 10^{-3}}$$

Atomic mass $Cu = 63.5$ g mol^{-1}, so that mass Cu in ore $= (63.5$ g mol-1$)(1.283 \times 10^{-3}$ mol$) = 8.14 \times 10^{-2}$ g

$$\text{percent Cu in ore} = \frac{8.14 \times 10^{-2} \text{g Cu in sample}}{140.3 \text{ g total sample}} \times 100\% = \mathbf{5.80 \times 10^{-2}\%}$$

Both approaches lead to the same answer. Use the method that seems easiest for you.

PLAN OF EXPERIMENT

You first prepare a standard solution of Cu^{2+}, which is used, in turn, to standardize the $Na_2S_2O_3$ solution. This is necessary because anhydrous $Na_2S_2O_3$ cannot be used as a primary standard, since it readily becomes hydrated in moist air. The presence of an unknown amount of water in the $Na_2S_2O_3$ solid makes it difficult to obtain an accurately known mass of $S_2O_3^{2-}$. Thus, a known solution of Cu^{2+}, made by dissolving a known mass of $Cu(s)$ in nitric acid, is used as the primary standard. Before the copper solution can be used to standardize the thiosulfate solution, it must be boiled, in a hood, to expel excess nitric acid and any oxides of nitrogen which might have formed. After being standardized, the thiosulfate solution is used to analyze one or more unknowns containing Cu^{2+}.

Use of the starch indicator can be a little tricky. If the starch is added to a solution containing too high a concentration of I_3^-, it may partially decompose and not change color reliably. The starch indicator must be added when most of the I_3^- has been consumed by Reaction 20.3.

Add starch shortly before the end point is reached, when the brown color of the titrand, due to I_3^-, has become very faint. Adding the starch produces an intense blue color, which disappears at the end point.

The blue color may reappear upon standing, but the first disappearance of color is the true end point.

SAFETY

1. Wear approved eye protection.

2. Fumes of nitric acid are toxic and corrosive. Use it only in a hood and avoid inhaling its vapors.

PROCEDURE

A. Preparation of Cu^{2+} Primary Standard Solution

1. Weigh accurately an amount of Cu metal between 1 and 2 g, using weighing paper or a weighing bottle. Transfer the sample to a 250 mL Erlenmeyer flask.

2. **DO ALL PARTS OF THIS STEP IN A HOOD.** Add 15 mL 6 M HNO_3 to the copper. Heat the solution on a steam bath or in a water bath until the copper is dissolved. Continue heating carefully to evaporate the excess acid. It is important not to evaporate away too much of the nitric acid.

> **When evaporating the nitric acid, periodically remove the flask from the heat and allow it to cool for 2 minutes. Watch for crystallization to occur during cooling. When you observe crystallization, stop the evaporation process.**
> **DO NOT EVAPORATE PAST THE POINT WHERE CRYSTALLIZATION FIRST OCCURS. Excessive evaporation produces insoluble oxy- and hydroxy-nitrates which will interfere with the analysis.**

3. Remove the flask from the hood and add 50 mL distilled water. Then neutralize the excess acid by adding, dropwise and with stirring, 7.5 M NH_4OH until a trace of $Cu(OH)_2$ precipitate forms, or until a deep blue color just appears, indicating the presence of the $Cu(NH_3)_4^{2+}$ complex. Avoid adding excess NH_4OH.

4. Acidify the solution with 5 mL glacial acetic acid (concentrated, 99%). Stir the solution. Rinse the stirring rod with distilled water from a wash bottle, collecting the rinse water into the flask so that no copper is lost.

5. Quantitatively transfer the solution to a 100 mL volumetric flask, using a small funnel. Rinse the Erlenmeyer flask well, collecting the rinse water into the volumetric flask. You must collect all of the dissolved copper. Finally, rinse the funnel interior, collecting the rinse water in the volumetric flask.

6. Add distilled water, swirling the flask, to the measured mark on the flask neck. Mix the contents thoroughly.

7. Calculate the molarity and normality of the Cu^{2+} solution and record the values on the Data sheet.

B. Standardization of the Thiosulfate Solution
1. Use solid $Na_2S_2O_3$ (anhydrous or hydrated) to prepare 0.5 L of approximately 0.1 N solution, and store in a glass stoppered bottle.

2. Pipet 20.00 mL of Cu^{2+} primary standard solution into a clean, dry, 250 mL Erlenmeyer flask. Then add 10 mL 2.0 M KI.

3. Titrate with the 0.1 N $Na_2S_2O_3$ solution. At the appropriate time, when the brown color is nearly gone, add 2-3 mL starch indicator solution. The solution will turn blue.

4. Continue titrating to the disappearance of the blue color. Record the volume of titrant used on the Data sheet.

5. Do a repeat determination.

6. Calculate and record the average normality of the $Na_2S_2O_3$ solution.

C. Analysis of Unknown Cu^{2+} Solution
1. Obtain an unknown solution containing Cu^{2+}. Pipet 20.00 mL into a clean, dry, 250 mL Erlenmeyer flask.

2. Add 10 mL 5.0 M acetic acid and 10 mL 2.0 M KI.

3. Titrate with standard $Na_2S_2O_3$ solution, as in steps 3 and 4 of part B.

4. Do a repeat determination.

5. Calculate the molarity of Cu^{2+} in the unknown solution. Also, assume that the unknown is a solution of $Cu(NO_3)_2$ and calculate the concentration of copper nitrate in the solution.

Name _______________________________________   Date ____________

EXPERIMENT 20
PRELABORATORY EXERCISE

1.   In Reactions 20.1, 20.2, and 20.3, determine the number of equivalents per mole for:

a. Cu^{2+} ____________________   d. I_3^- ____________________

b. I^- ____________________   e. $S_2O_3^{2-}$ ____________________

c. I_2 ____________________

2.   24.8210 g of dry $Na_2S_2O_3$ is dissolved and diluted to 1.000 L in a volumetric flask.
a. What is the calculated normality of the solution with respect to Reaction 20.3?

____

b. Recall that $Na_2S_2O_3$ is not suitable for a primary standard. Is the error in the
calculated normality likely to be **+**, **-**, or **random**. (Circle the correct answer.) Explain your answer.

3. When analyzing for Cu^{2+} by means of Reactions 20.1-20.3, 16.42 mL of 0.1171 N $Na_2S_2O_3$ were used
to titrate the sample. How many moles of Cu^{2+} were in the sample?

____

Calculations:

Name ____   Date ____________

EXPERIMENT 20

DATA
(Observe significant figures in all calculations.)

Determination
1   **2**

A. Preparation of Cu^{2+} Primary Standard Solution

1. Mass of Cu metal: _________   _________

2. Volume of Cu^{2+} solution::   _________   _________

3. Molarity of Cu^{2+} solution: _________   _________

4. Average molarity of Cu^{2+} solution: _________

5. Average normality of Cu^{2+} solution: _________

B. Standardization of the $S_2O_3^{2-}$ Solution

1. Mass of solid $Na_2S_2O_3$:   _________   _________

2. Volume of $Na_2S_2O_3$ solution: _________   _________

3. Approximate molarity of $Na_2S_2O_3$ solution: _________   _________

4. Approximate normality of $Na_2S_2O_3$ solution: _________   _________

5. Volume of Cu^{2+} primary standard solution: _________   _________

6. Volume of $Na_2S_2O_3$ titrant used: _________   _________

7. Measured normality of $Na_2S_2O_3$ solution: _________   _________

8. Av. measured normality of $Na_2S_2O_3$ solution: _________

C. Analysis of Unknown Cu^{2+} Solution

1. Unknown No.   _________

2. Volume of unknown: _________   _________

3. Volume of $Na_2S_2O_3$ titrant used: _________   _________

4. Moles of Cu^{2+} in unknown: _________   _________

5. Average moles of Cu^{2+} in unknown: _________

Name ___________________________________   Date _________________

EXPERIMENT 20
QUESTIONS

1.   Why don't you add the starch indicator to the titrand right at the beginning of the titration?

2.   The half-reactions involved in this reaction are:
   a. $Cu^{2+} + I^- + e^- \longrightarrow CuI(s)$
   b. $S_4O_6^{2-} + 2\,e^- \longrightarrow 2\,S_2O_3^{2-}$
   c. $I_3^- + 2\,e^- \longrightarrow 3\,I^-$

On the basis of your titrations, which half-reaction has the highest reduction potential? ______; the lowest reduction potential? ______
Explain your answers.

EXPERIMENT 21
Spectrophotometric Determination of Manganese

PRELABORATORY PREPARATION
1. Do the Prelaboratory Exercise and turn it in at the beginning of your laboratory period.
2. Read "Chemical Analysis with an Absorption Spectrometer," Section XIII, Methods, page 28.

INTRODUCTION
In a solution containing a colored compound, the intensity of color can be used to measure the concentration of the compound; the more intense the color, the higher the concentration. The color of a solution is an indication of how much light is absorbed at different wavelengths, when white light is passed through the solution. A **spectrophotometer** is an instrument for measuring light absorption at particular wavelengths. The absorption is expressed as a numerical value that can be related directly to the concentration of colored substance.

When a spectrophotometer is used to measure the amount of light absorbed in a sample, the measured value includes loss of light from several causes unrelated to the sample concentration, such as light lost by reflection and scattering from the surfaces of optical components in the instrument and light absorbed in the glass of the sample cell and instrument lenses. Therefore, it always is necessary to calibrate the spectrophotometer by measuring the net light absorption when the sample cell contains standard solutions of known concentrations. This procedure compensates for instrumental artifacts.

Substances that are weakly colored or colorless in solution can be measured by converting them, quantitatively, into compounds that are more strongly colored. Such is the case with manganese in this experiment, where manganese is put into solution as the Mn(II) ion, Mn^{2+}.

Mn(II) ion is almost colorless. It is easily oxidized, however, to Mn(IV) in water solution, where it forms the intensely purple species MnO_4^- (permanganate ion).

The very intense color (large **absorbance,** see Section XIII, Methods) means that the analysis can be very sensitive because the light absorption will be relatively large, even with small amounts of manganese in the sample. This method finds wide application in the analysis of metal alloys and ores where manganese is a minor constituent.

PLAN OF EXPERIMENT
Two unknowns containing manganese are analyzed, one a solution of Mn^{2+} and the other a piece of steel containing manganese as an alloying element. The steel is dissolved in acid to put its manganese into solution in the form of Mn^{2+}. KIO_4 is added to the Mn^{2+} solutions to supply IO_4^-, which serves as the oxidant for oxidizing Mn^{2+} to permanganate ion, MnO_4^-:

$$2\,Mn^{2+} + 5\,IO_4^- + 9\,H_2O \longrightarrow 2\,MnO_4^- + 5\,IO_3^- + 6\,H_3O^+ \tag{21.1}$$

Solution pH must be 4, or lower, to prevent formation of MnO_2, which would contribute an unpredictable amount to the measured light absorption. The solution also must be free of chloride ion, which would be oxidized to chlorine by permanganate and interfere with the analysis.

The concentration of MnO_4^- is measured by comparing its light **absorbance** at 530 nm with that of calibration standards. Sample concentration is directly proportional to sample absorbance over a wide range, so that a graph of absorbance versus concentration is linear. A calibration graph can be used to convert the measured absorbance of an unknown sample to concentration of MnO_4^-. From the concentration of MnO_4, the amount of Mn^{2+} in the sample solution can be calculated, using the stoichiometry of Reaction 21.1.

Observe that the absorbance scale on the spectrophotometer is not linear, but logarithmic. On a logarithmic scale, the reading error is greatest at high absorbances because of scale compression. Measurement error is also relatively large at the low end of the scale because the signal is small. **The most reliable measurements are obtained when absorbance values lie in the middle portion of the scale, between 0.15 and 1.0.**

In Procedure **Parts A, B,** and **C,** you prepare all of the solutions for analysis. In **Part D,** you calibrate the spectrophotometer and, in **Part E,** you measure the unknowns.

SAFETY

1. Wear approved eye protection.

2. Concentrated nitric acid, HNO_3, is extremely corrosive and its fumes are toxic. Handle it very carefully, avoiding skin and clothing contact. Do not inhale its vapors. Dissolve the steel sample with nitric acid in a fume hood.

PROCEDURE

A. Preparation of a Standard Mn^{2+} Solution

1.   The stock Mn^{2+} solution will be dispensed from burets marked **"STOCK SOLUTION"** and labelled with the exact concentration. Record the concentration on your Data sheet. **Be careful not to use the burets marked "UNKNOWN No. XX"**

2.   Use a clean, dry 400 mL beaker to obtain your stock solution. Before dispensing the stock solution, record the initial buret reading on your Data sheet. Then carefully dispense about 20 mL into your beaker and record the final buret reading and the exact volume of stock solution obtained on the Data sheet.

3.   Add 80 mL of distilled water to the 20 mL of stock solution. Then, *very carefully,* add 100 mL of 6M HNO_3. Use a graduated cylinder to measure these volumes.

4.   Add about 0.5 g of solid potassium periodate, KIO_4. Use a triple-beam balance to weigh the KIO_4.

5.   Now you must boil the solution without losing any material by spattering. **Do not add boiling chips, because they will make later quantitative transfer to a volumetric flask difficult.**
Instead, place a glass stirring rod with a roughened or broken tip in the solution to assist smooth boiling. Lean the rod in the pouring lip of the beaker and cover the beaker with a watch glass. The rod and watch glass can be easily rinsed into the beaker after boiling to collect all the sample.

Boil the solution gently for about 3 min, or until a purple color appears. If the color has appeared after 4 min, stop heating and wait until the boiling stops. Then add a little more KIO_4 and resume heating gently until color appears.

6.   After the purple color appears, cool the solution and transfer the contents of the beaker quantitatively to a 250 mL volumetric flask. Rinse the underside of the watch glass cover and the stirring rod carefully with distilled water, collecting the rinse water in the beaker. Then transfer the rinse water to the volumetric flask. Rinse the beaker well, adding the rinse water to the volumetric flask.

7.   Dilute the solution in the volumetric flask with distilled water exactly to the index mark and mix thoroughly. Calculate the concentration of Mn^{2+} in this standard solution and record it on your Data sheet.

8.   Pour about 100 mL of this solution into a clean, dry 150 mL Erlenmeyer flask. Stopper this flask and label it **"Standard."**

B. Preparation of Unknown Solution

1.   The unknown solutions will be dispensed from burets marked **"UNKNOWN No. XX"**. Your instructor will tell you which unknown to use. Label a clean, dry 400 mL beaker with your unknown number and record this number on your Data sheet.

2.　　Before dispensing your unknown, record the initial buret reading on your Data sheet. Then carefully dispense 20 mL of unknown solution into the labelled 400 mL beaker. Record the final buret reading and the exact volume of unknown obtained on the Data sheet.

3.　　Repeat steps 3-8 of **Part A**, using the unknown solution. Label the final flask **"Unknown No. _"**. This unknown solution will be analyzed in **Part E.**
　　Be sure all flasks are correctly labelled so that your standard and unknown solutions do not get mixed up.

C. Preparation of Steel Sample Containing Manganese

1.　　Obtain a steel sample weighing 0.5-1.0 g. Weigh it carefully, to 0.1 mg, into a clean, dry 250 mL beaker. Record the sample number and mass on your Data sheet.

2.　　*In a hood,* carefully add 30 mL of 6 M HNO_3 to the steel and boil gently until the steel is dissolved. Some solid residue of carbon and carbides will remain.

3.　　Remove heat, let the boiling stop, and wait another 3 min. Then, *very slowly and carefully to avoid frothing,* add 1 g of ammonium peroxydisulfate, $(NH_4)_2S_2O_8$. The peroxydisulfate ion will oxidize the solid carbon and carbide residue to CO_2 gas.

4.　　Boil the solution gently for about 10 min to destroy excess peroxydisulfate.
　　If the solution is pink or brown, it indicates that MnO_2 has formed. If this occurs, remove heat and let the solution cool for about 3 min. Then add, *slowly to avoid foaming,* **about 0.1 g of ammonium hydrogen sulfite or sodium hydrogen sulfite and boil for another 5 min. Then remove heat.**

5.　　Add 50 mL cold distilled water and 10 mL concentrated phosphoric acid, H_3PO_4, to remove the color of Fe^{3+} by forming a complex ion.

6.　　Let the solution cool to room temperature. Add, *slowly and carefully,* about 0.5 g of solid potassium periodate, KIO_4. Boil gently for about 3 min, or until the purple color appears.

7.　　Repeat steps 6-8 of **Part A** to prepare the solution for analysis. Label the flask **"Steel Sample No. _"**. This solution will be analyzed in **Part E.**

D. Calibration of Spectrophotometer

1.　　Prepare 3 calibration solutions from your standard solution as follows:
　　a. Label 3 clean, dry 25 mL volumetric flasks: *1, 2,* and *3.*
　　b. Pipet accurately 1, 5, and 10 mL portions of your standard into these flasks. Record on your Data sheet the size of the portion in each labelled flask.
　　c. Fill each flask with distilled water exactly to the index mark and mix well.

2.　　Measure the absorbance at 530 nm of the 3 accurately diluted solutions and, also, of your undiluted standard solution. Follow the procedure in **Section XIII, Methods,** carefully. Use distilled water as the blank. Contamination and/or bubbles on the cell walls can cause measuring errors. Carefully observe the following precautions:
　　• **Before each measurement, including the water blank, rinse the spectrophotometer sample cell 2 or 3 times with a small quantity of the solution to be measured.**
　　• **To remove bubbles on the cell walls, press a piece of parafilm over the top of the cell with your finger to seal it, and** *slowly* **invert and rotate the cell to bring all the bubbles to the liquid surface.**

3.　　Measure and record the absorbances. Calculate the concentration of Mn^{2+} originally present in each calibration sample and plot a calibration curve with **absorbance** on the **y-axis** versus **mol/L** on the **x-axis.**

E. Measurement of Unknowns
a. Unknown Solution
1.   Rinse the sample cell 2 or 3 times with a small quantity of the unknown solution from **Part B.** Measure and record the absorbance at 530 nm of your unknown. Determine its concentration from your calibration curve and record this value. Repeat this measurement 3 times, emptying and refilling the sample cell with fresh unknown each time.

b. Steel Sample
1.   Rinse the sample cell 2 or 3 times with a small quantity of the unknown solution made from the steel sample in **Part C.** Measure and record the absorbance at 530 nm of this unknown solution. Determine its concentration from your calibration curve and record this value. Repeat this measurement 3 times, emptying and refilling the sample cell with fresh unknown each time.

If this solution is too concentrated in MnO_4^-, the spectrophotometer reading will be off-scale or beyond your calibration scale. In this case, accurately dilute the sample sufficiently to make a satisfactory measurement. It will be important to know the exact amount of dilution.

Name ____　　Date ____________

EXPERIMENT 21
PRELABORATORY EXERCISE

1.　　A solution contains 0.20 mg/m L Mn^{2+}. What minimum mass of KIO_4 must be added to 20 mL of the solution in order to completely oxidize the Mn^{2+} to MnO_4^-? See Equation 21.1 for the reaction.

Mass of KIO_4 needed = ______________

2. a.　　A standard solution of Mn^{2+} was oxidized quantitatively to MnO_4^- with periodate ion and carefully diluted to make 4 solutions with known concentrations of MnO_4^-. These solutions were then used to calibrate a spectrophotometer. The measured absorbances at 530 nm, and the corresponding known concentrations were:

Sample No.	1	2	3	4
$[MnO_4^-]$ (mol L^{-1})	1.11×10^{-4}	2.22×10^{-4}	2.96×10^{-4}	4.40×10^{-4}
Absorbance at 530 nm	0.26	0.52	0.69	1.04

Graph the calibration curve and use it to answer **part b.** Use the graph paper on the next page.

b.　　A steel sample weighing 2.474 g was dissolved in HNO_3 and the Mn^{2+} in the resulting solution was oxidized to MnO_4^-. The solution then was diluted to 100 mL and analyzed in the calibrated spectrophotometer, giving an absorbance reading of **0.62.** What was the mass percent of manganese in the steel sample?

Mass % Mn in steel sample = ______________

Calculations:

PRELABORATORY EXERCISE

Name ____   Date ____________

EXPERIMENT 21

DATA
(Observe significant figures in all calculations.)

A. Preparation of Standard Solution

1. Concentration of stock Mn^{2+} solution: ____________

2. Initial buret reading of stock solution: ____________

3. Final buret reading of stock solution: ____________

4. Volume of stock solution obtained: ____________

5. Concentration of standard Mn^{2+} solution: ____________

B. Preparation of Unknown Solution

1. Unknown solution number: ____________

2. Initial buret reading of unknown solution: ____________

3. Final buret reading of unknown solution: ____________

4. Volume of unknown solution obtained: ____________

C. Preparation of Steel Sample Containing Manganese

1. Unknown steel sample number: ____________

2. Mass of 250 mL beaker: ____________

3. Mass of beaker plus steel sample: ____________

4. Mass of steel sample: ____________

Calculations:

D. Calibration of Spectrophotometer
1. Volume of standard solution pipetted into volumetric flask:

No. 1: __________;   No. 2: __________;   No. 3: __________

	Absorbance Measured at 530 nm	Calculated Concentration
2. Standard solution:	____________	____________
3. Standard solution with 25/10 dilution:	____________	____________
4. Standard solution with 25/5 dilution:	____________	____________
5. Standard solution with 25/1 dilution:	____________	____________

6. Graph the calibration curve using the graph paper at the end of the experiment.

DATA

E. Measurement of Unknowns

	Determination		
	1	2	3

a. Unknown solution No. __________

   1. Absorbance of unknown solution: __________ __________ __________

   2. Concentration of Mn^{2+} in unk. solution: __________ __________ __________

   3. Average conc. of Mn^{2+} in unk. solution: __________

   4. Standard deviation: __________

b. Steel sample No. __________

   1. Absorbance of steel sample: __________ __________ __________

   2. Concentration of Mn^{2+} in solution: __________ __________ __________

   3. Average conc. of Mn^{2+} in solution: __________

   4. Standard deviation: __________

   5. Mass of Mn in steel sample: __________

   6. Mass percent of Mn in steel sample: __________

Calculations:

Name ____   Date ____________

EXPERIMENT 21
QUESTIONS

1.   Why should you try to adjust sample concentrations so that measured absorbances fall between 0.15 and 1.0 on the spectrophotometer absorbance scale?

2.   Explain why it is necessary to calibrate the spectrophotometer before measuring any unknowns.

3.   In **step 5** of **Part C**, why is it necessary to remove the solution color imparted by the presence of ferric ion, Fe^{3+}?

DATA

Name ____   Date ____________

EXPERIMENT 21

DATA

DATA

EXPERIMENT 22
Ion-Exchange Chromatography:
Determining the Concentration of an Anion

PRELABORATORY PREPARATION
1. Do the Prelaboratory Exercise and turn it in at the beginning of your laboratory period.
2. Review the technique of decantation, Section VIII, Methods, page 20.
3. Review the use of the buret, Section IV, Methods, page 17.
4. Review the definitions of equivalent mass, normality, and molarity in "Chemical Units", page xi.

INTRODUCTION
Like other types of chromatography, ion-exchange chromatography is a method for separating the components of a chemical mixture. The method selectively removes ions from a solution by causing them to exchange positions with other ions attached to the surface of a solid polymeric resin. In ion-exchange chromatography, a solution containing ions is made to flow through a column of porous beads made from a solid resin. The resin is a long-chain polymer that has a large number of either acidic or basic functional groups (depending on the type of resin) attached at intervals along the main chain. These functional groups are locations where ion-exchange processes can occur and are sometimes called active sites.

An acidic active site is one where an H^+ is readily attached, such as a carboxylic acid group on the resin, denoted by Resin-COO$^-$, which can form Resin-COO$^-$H$^+$.

A basic site is one where OH- is readily attached, such as a quarternary amine group on the resin, denoted by Resin–N(CH$_3$)$_3^+$, which can form Resin–N(CH$_3$)$_3^+$OH$^-$.

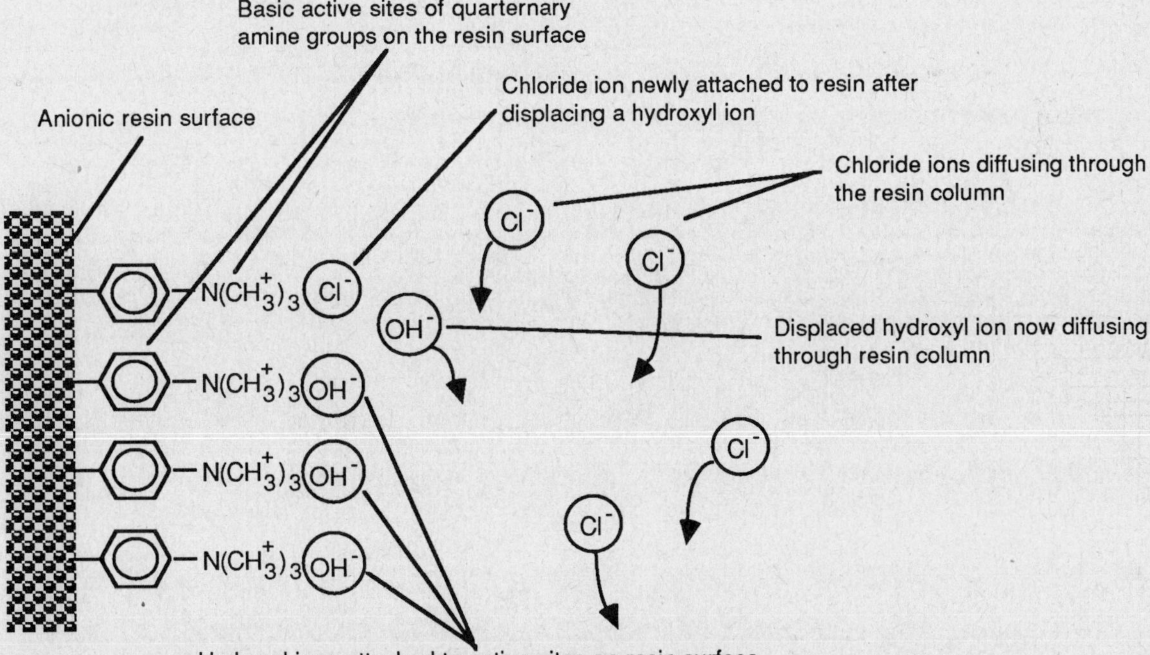

Figure 22.1. Diagram of the surface composition of an ion-exchange resin, illustrating how chloride ions can displace hydroxyl ions at the active sites.

Ions attached to the active sites can be made to exchange positions with ions in solution. In this experiment, an unknown solution containing either Cl^- or CO_3^{2-} anions is made to exchange with OH^- attached to the resin. Figure 22.1 illustrates how this process can remove chloride ions from solution and release hydroxyl ions into solution. Resins with acidic active sites are called cationic resins, because they serve as sites for cation-exchange. Similarly, resins with basic active sites are called anionic resins.

The active sites have different attractive strengths for different ions, and this selective attraction can serve as a means for ion separation and quantitative analysis. The exchange process is reversible and strongly attached resin ions, call them A^-, can be replaced by more weakly bound solution ions, call them B^-, simply by maintaining a high concentration of B^- in contact with the resin. The exchange reaction between the hydroxide form of the anionic resin of Figure 22.1 and a chloride ion solution may be written:

$$\text{Resin-N(CH}_3)_3^+\text{OH}^- + Cl^- \longrightarrow \text{Resin-N(CH}_3)_3^+Cl^- + OH^- \tag{22.1}$$

The equilibrium constant for Reaction 22.1 has the usual form:

$$K = \frac{[\text{Resin-N(CH}_3)_3^+Cl^-][OH^-]}{[\text{Resin-N(CH}_3)_3^+OH^-][Cl^-]}$$

PLAN OF EXPERIMENT

A vertical ion-exchange column is prepared with an anionic resin, so that an anion solution can be poured in at the top, allowed to percolate downward through the column, and exit into a flask at the bottom. The resin is prepared for performing a quantitative analysis of chloride ion by first soaking the resin in sodium hydroxide, in order to saturate all of the active sites with bound OH^-. The hydroxyl ion-saturated resin is always handled as a slurry. It is thoroughly washed, to remove excess unbound hydroxyl ions, and then poured carefully into the column tube so as to attain a uniform density with no air pockets.

The analysis is made by pouring the unknown solution containing Cl^- or CO_3^{2-} into the top of the column. As the solution passes down through the column, an exchange reaction such as 22.1 occurs, in which the unknown anion becomes bound to the resin and OH^- is released into solution.

When an unknown anion becomes bound to the resin, one OH^- is released for each negative charge on the anion, in order to preserve electroneutrality.

- **In the case of a chloride ion solution, each Cl^- that becomes bound to the resin will release one OH^-.**

- **In the case of a carbonate ion solution, each CO_3^{2-} that becomes bound to the resin will release two OH^- ions.**

The displaced OH^- then passes freely through the rest of the column and exits into the collection flask at the bottom.

At the end of the experiment, the number of hydroxyl ions contained in the solution in the collection flask is equal to the number of chloride ions or twice the number of carbonate ions that were in the unknown solution. The number of hydroxyl ions in the collection flask is measured by titrating the solution to the equivalence point with a standardized acid to determine the hydroxyl ion concentration. The hydroxyl ion concentration is then related back to the anion concentration in the unknown solution.

EXAMPLE:

A 20.0 mL sample of a KCl solution was passed through an anion-exchange column that had been saturated with OH^-. The column then was thoroughly washed to insure that all of the released hydroxyl ions were collected in the effluent. The effluent was titrated to the equivalence point with 12.4 mL of 0.112 N HCl. What was the concentration of KCl in the sample?

SOLUTION:

The sample solution contained K^+ and Cl^- ions. As the solution diffused through the column, Cl^- ions displaced OH^- ions at the active sites on the resin surface. One OH^- ion was displaced for each Cl^- ion in the solution. The OH- ions passed down the column and into the collection flask with the effluent. The equivalents of HCl needed to titrate the OH^- in the effluent was:

$$0.0124 \text{ L} \times 0.112 \text{ equiv./L} = 1.39 \times 10^{-3} \text{ equivalents HCl}$$

- The equivalents of HCl are equal to the equivalents of OH^- in the effluent.
- The equivalents of OH^- in the effluent are equal to the equivalents of Cl^- in the sample.

Therefore, we know that there were 1.39×10^{-3} equivalents of Cl^- in the 20 mL sample that was passed through the column. Each equivalent of Cl^- in the sample corresponds to 1 mole of KCl in the sample solution. Thus, the concentration of KCl in the sample solution was:

$$[KCl] = (1.39 \times 10^{-3} \text{ moles}/20.0 \text{ mL})(1000 \text{ mL/L}) = \mathbf{6.95 \times 10^{-2} \text{ mol/L}}$$

Several precautions must be observed to make certain that the unknown anion is quantitatively exchanged for hydroxyl ion:

1.   Anion-exchange resins are never stored in the hydroxide form because they rapidly decompose, especially if dry. Such resins are usually stored in the chloride form. Therefore, new resin must be loaded with OH^- by a thorough exchange process which involves repeated soakings in NaOH solution. The hydroxide form must be kept as a wet slurry at all times.

2.   It is important that the exchange column be long enough to insure that all of the unknown anions come into contact with fresh resin and exchange with hydroxyl ions, before reaching the bottom of the column.

3.   The resin must be distributed in the column very uniformly. The solution always will follow the path of least resistance and any non-uniformities in packing of the resin, such as air pockets or low-packing-density channels, will have the effect of shortening the effective column length. During and after filling the column, the resin always must be covered with water to prevent drying. Dry resin will shrink, cause voids in the packing, and begin to decompose.
   Always maintain 2 or 3 cm of water over the top of the resin in the column to prevent the formation of packing voids and the decomposition of the resin.

SAFETY
   Wear approved eye protection.

PROCEDURE

1.   Obtain about 50 mL of unknown solution in a clean beaker and record its number on the Data sheet.

2.   The anion resin has already been prepared by soaking in NaOH. Obtain about 70 mL of the resin slurry in a 250 mL beaker.

3.   Wash the resin by decantation with water. Add distilled water until the beaker is about half-full, swirl the resin two or three times, allow the resin to settle, and decant most of the wash water into another beaker. Test the wash water with phenolphthalein indicator and then discard the wash water. Add new water and repeat this process until the wash water gives a colorless test to phenolphthalein.
   A colorless test with phenolphthalein indicates a wash water having a pH less than 8 (see Table 16.1, Experiment 16). This occurs when all excess OH^- has been washed from the column.

After attaining a colorless test with phenolphthalein the column is saturated with *bound* OH^-, which can be released only by exchange with another anion.

4.   Add about 5 mL distilled water to a 50 mL buret and insert a plug of glass wool just above the stopcock.

5.   Transfer the resin slurry to the buret. The length of the resin column should be about 15 cm.
   To prevent drying and the formation of air pockets, be certain the top of the resin beads are always 2 to 3 cm below the surface of the liquid in the buret. Be sure to leave at least 20 mL of free space above the resin to accommodate your unknown sample.

6.   Add about 5 mL of water to the buret and drain the same amount from the bottom. Test the effluent with indicators; it should be acidic with respect to phenolphthalein (pH < 8; acidic: colorless; basic: red) and basic with respect to methyl orange (pH > 4.4; acidic: red; basic: yellow). If it is not, the resin has not been adequately washed. Consult your instructor before proceeding.

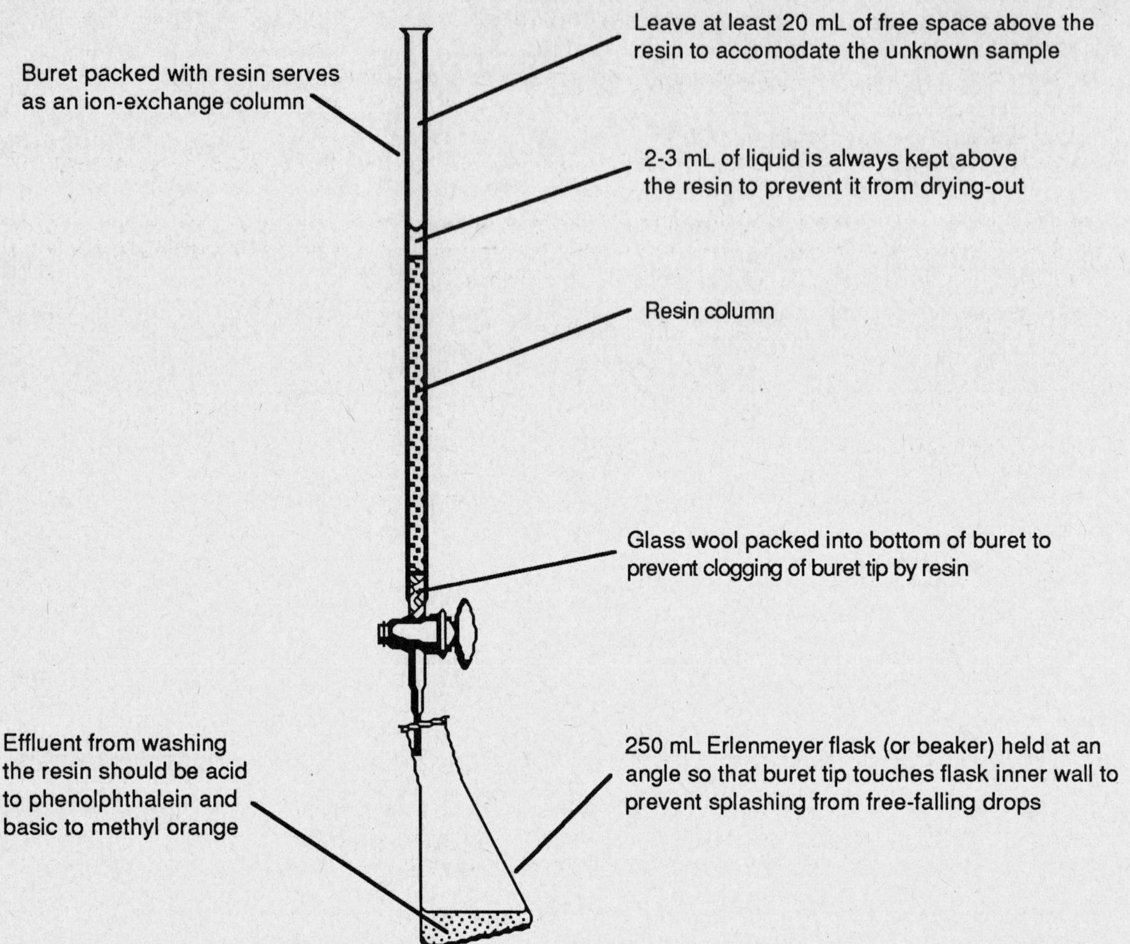

Figure 22.2. Ion-exchange column ready to receive unknown sample.

7.   Accurately pipet 20.0 mL of your unknown solution into the top of the buret.

8.   Prepare to collect effluent from the buret by placing a 250 mL Erlenmeyer flask under the tip. Avoid loss from splashing of free-falling drops by holding the buret tip against the inner side of the flask.

9.   Slowly drain (not faster than 2 mL/min) about 20 mL of liquid from the bottom of the buret into the Erlenmeyer flask, until 2-3 mL of liquid remains above the top of the resin beads.
Do not let the liquid level drop below the top of the resin beads at any time.

10.   Wash the unknown through the column by adding a 25 mL portion of distilled water at the top and then draining 25 mL from the bottom (not faster than 5 mL/min) into the same flask containing the first effluent from the unknown. Repeat this washing procedure once again.

11.   Wash the column a third time with 25 mL of distilled water, but this time collect the effluent separately in a small clean beaker.

Name ____   Date ________________

EXPERIMENT 22
PRELABORATORY EXERCISE

1.   25 mL of a solution of Na_2CO_3 of unknown concentration was passed through an ion-exchange column that had been saturated with OH^-. The column then was thoroughly washed to insure that all of the released OH^- was collected in the effluent. The effluent was titrated to the equivalence point with 15.6 mL of 0.175 N HCl. Calculate the concentration of the Na_2CO_3 solution in moles per liter.

   $[Na_2CO_3]$ = ______________________

Calculations:

Name ____   Date ________________

EXPERIMENT 22

DATA
(Indicate the correct units for measured quantities.)

1. Unknown solution number: ________________

2. Exact normality of standard HCl: ________________

Determination
	1	2

3. Exact volume of unknown solution added to column: _________ _________

4. Final buret reading in titration: _________ _________

5. Initial buret reading in titration: _________ _________

6. Volume of acid used to titrate OH^-: _________ _________

7. Equivalents of acid used: _________ _________

8. Assume unknown contained NaCl.

Moles of NaCl in volume added to column: _________ _________

Concentration of NaCl in unknown: _________ _________

Average concentration of NaCl in unknown: _________

g/mL of NaCl in unknown: _________

9. Assume unknown contained Na_2CO_3.

Moles of Na_2CO_3 in volume added to column: _________ _________

Concentration of Na_2CO_3 in unknown: _________ _________

Average concentration of Na_2CO_3 in unknown: _________

g/mL of Na_2CO_3 in unknown: _________

Calculations:

Name _______________________________________   Date ___________________

EXPERIMENT 22
QUESTIONS

1.   What is the error (+, -, or 0) in the calculated concentration of the unknown solution if:

   a.   the buret containing the resin is not accurately calibrated?   _____

   b.   the pipet used for adding the unknown delivered only 19.91 mL?   _____

   c.   the buret containing the resin had anions adsorbed on the glass
        from a previous experiment?   _____

   d.   air pockets or channels formed in the resin column?   _____

   e.   the unknown solution is passed through the resin column faster
        than recommended?   _____

   f.   the unknown solution is passed through the resin column slower
        than recommended?   _____

EXPERIMENT 23
Reaction Enthalpies Determined By Calorimetry

PRELABORATORY PREPARATION
Do the Prelaboratory Exercise and turn it in at the beginning of your laboratory period.

INTRODUCTION
Chemical reactions are always accompanied by energy changes. If the internal energy of a chemical system decreases during a reaction, the system must release the excess energy to the surrounding environment; if the internal energy increases, the additional energy must be absorbed from the surrounding environment. The chemical system and the surrounding environment can exchange energy by means of heat transfer and by emission and absorption of electromagnetic radiation. Energy transfer by electromagnetic radiation is measured by spectroscopy. Energy transfer by heat flow is measured by calorimetry, which is the subject of this experiment.

Calorimetric measurements are important to many fields. To give just a few examples, calorimetry is used to measure the calories in foods, the energy content of fuels (coal, oil, natural gas, gasoline, alcohol, etc.), and the conversion efficiency of solar collectors. Calculations based on calorimetry allow the determination of how high a pressure is needed to convert amorphous carbon to diamonds, how much useful work can be obtained from a particular engine with a particular fuel, what pressure-volume-temperature conditions are the most effective for efficiently converting atmospheric nitrogen to nitrate fertilizers, and what materials might be suitable for fabricating the heat shields that protect space shuttles when they reenter the atmosphere. Calorimetry is the basic technique for studying any process where heat energy is of interest.

Both the *direction* and *amount* of heat energy exchanged between a chemical system and the surroundings are important measured properties for understanding and predicting the behavior of chemical systems.

- **If a chemical reaction releases heat energy to the environment, the system ends up with lower internal energy than it started; such a reaction is said to be *exothermic*.**

- **If a chemical reaction absorbs heat energy from the environment, the system ends up with higher internal energy than it started; such a reaction is said to be *endothermic*.**

When a chemical reaction releases or absorbs heat energy, at least one of the parameters of pressure, volume, or temperature must change. The *amount* of heat energy released or absorbed depends on which parameters are constant and which change.

- **When pressure is allowed to change and volume and temperature are held constant, the heat energy change of a reaction is called the *reaction energy*, written ΔE.**

- **When volume is allowed to change and pressure and temperature are held constant, the heat energy change of a reaction is called the *reaction enthalpy*, written ΔH.**

In general, the reaction energy and reaction enthalpy have different values for the same reaction. In this experiment, we will study the heat energy changes of several reactions under constant pressure conditions.[1] Consequently, the quantity measured will be the **reaction enthalpy**.

---

[1] The pressure is constant at atmospheric pressure in this experiment because the reaction vessel is not hermetically sealed. The loosely fitting lid allows the pressure inside the calorimeter to equilibrate with the external pressure.

Reaction enthalpy is defined as $\Delta H = H_f - H_i$, where H_f is the enthalpy of the system in its final state, after reaction, and H_i is the enthalpy of the system in its initial state.

- **If the reaction is exothermic, the system loses heat energy to the surroundings and its final enthalpy is less than its initial enthalpy; then $H_f < H_i$ and ΔH is a negative quantity.**

- **If the reaction is endothermic, the system gains heat energy from the surroundings and its final enthalpy is greater than its initial enthalpy; then $H_f > H_i$ and ΔH is a positive quantity.**

In this experiment, you will determine the magnitude and sign of ΔH for several reactions using an apparatus called a calorimeter. A calorimeter is simply a thermally well-insulated reaction container. Suppose two different solutions that react chemically are mixed together in a calorimeter. If the reaction is exothermic, the reacting molecules will release heat to their surroundings and the reaction solution will get hotter. If the calorimeter confines all of the heat, without any leakage to the outside, the temperature increase of the solution can be measured and ΔH determined. If the reaction absorbs heat (is endothermic), the solution will become cooler, and a measurement of the temperature decrease allows ΔH to be determined.

CALORIMETRIC CALCULATIONS
Heat Units:
The quantity of heat energy is measured in **joules**[2] (symbol: **J**).

To raise the temperature of 1 g of water by 1°C, 4.1840 J of heat energy must be added to the water. If 4.1840 J of heat energy are removed from 1 g of water, the temperature of the water will decrease by 1°C.

Specific Heat:
The specific heat of a substance is equal to the number of joules required to raise the temperature of 1 g of the substance by 1°C.

From the definition of the joule, above, *the specific heat of water is 4.1840 J g^{-1} deg^{-1}.*

If the specific heat of a substance is known, it is easy to calculate how many joules must be gained or lost in order to change its temperature by a specified amount:

joules = specific heat $\times$ grams of substance $\times$ temperature change

EXAMPLE 1:
How many joules must be added in order to heat 25 g of water from 18°C to 33°C?

SOLUTION:
Temperature change $= \Delta T =$ final temp. - initial temp $= 33 - 18 = 15°C$

joules added $= (4.1840 \text{ J g}^{-1} \text{ deg}^{-1})(25 \text{ g})(15 \text{ deg}) = \mathbf{1.57 \times 10^3}$ **J**

EXAMPLE 2:
The specific heat of iron is 0.460 J g^{-1} deg^{-1}. A 1000 g piece of iron is heated to 500°C and then left to cool. If it loses 21 kJ to the surroundings during the first 30 s, what is its temperature after 30 s of cooling?

SOLUTION:

$$\text{joules lost} = \text{sp. ht.} \times \text{mass} \times \Delta T$$

$$21 \times 10^3 \text{ J} = 0.46 \text{ J g}^{-1} \text{ deg}^{-1} \times 1000 \text{ g} \times \Delta T$$

$$\Delta T = \frac{21 \times 10^3}{(0.46)(1000)} = 45.5 \text{ deg}$$

$$T(\text{after 30 s}) = 500 - 45.5 = \mathbf{454.5°C}$$

---

2 An older unit for heat energy is the **calorie,** which is still found in some textbooks. It takes 1 calorie to raise the temperature of water by 1°C. Therefore, **1 calorie = 4.1840 J.**

Heat Capacity:
   The heat capacity, C, of a sample is equal to its specific heat multiplied by its mass.

For example, the heat capacity of the 1000 g iron block in Example 2 above is:
   $C = 0.46 \ J \ g^{-1} \ deg^{-1} \times 1000 \ g = \textbf{460 J deg}^{-1}$.
Notice that specific heat is a property of a particular material and is independent of the mass of any given sample, while heat capacity *always* is related to a given sample.
   The heat capacity of a particular sample is equal to the number of joules needed to raise the temperature of the entire mass of sample by 1°C.

   A value for heat capacity can be specified for any sample (solid, liquid, or gas) or any collection of samples, even if the collection is inhomogeneous in composition. The heat capacity for a system made up of several different kinds of parts is equal to the sum of the separate heat capacities of each of the parts. In general:
   C(entire system) = {total heat energy added or lost}/ΔT

EXAMPLE 3:
   Suppose you want to heat a beaker containing 500 mL of a nitric acid solution, along with 25 g of copper metal pieces that are to be dissolved. A thermometer and a mechanical stirrer also are immersed in the solution. All parts of the system, including the beaker, must be heated together. Using an electrical heater (so we can measure the energy accurately) we find that 11.9×10^3 J are required to raise the temperature of the entire apparatus and sample by 3°C. The heat capacity of the entire system is:
   $C(\text{entire system}) = 11.9 \times 10^3 \ J/3°C = \textbf{3.97} \times \textbf{10}^3 \ \textbf{J deg}^{-1}$

PLAN OF EXPERIMENT
   In this experiment, you construct a simple but effective calorimeter, Figure 23.1, and measure the enthalpy changes of several different reactions in it. You determine the amount of heat energy absorbed or released in a reaction by measuring the temperature change of the solution in which the reactants are dissolved.
   In order to use the temperature change to calculate the heat energy, you must determine the heat capacity of the entire system.
   The system includes everything that changes temperature.
The parts of the calorimeter system included in the measured heat capacity are the reagent solution, immersed portions of the thermometer and stirrer, and the interior walls of the calorimeter container.

Procedure for Finding the Heat Capacity of the Calorimeter
   The method is based on the principle that *energy must be conserved*. A measured amount of cool water is placed in the calorimeter along with a thermometer and stirrer and allowed to come to thermal equilibrium, meaning that the temperature no longer changes. Then, a measured amount of warm water is added to the system and a new thermal equilibrium is reached.
   The energy lost by the warm water must be equal to the energy gained by the cool water plus the calorimeter, stirrer, and thermometer.
The calculations involved in determining the heat capacity of the calorimeter apparatus are illustrated in Example 4.

EXAMPLE 4:
   Fifty mL of water are placed in a calorimeter that also contains a thermometer and stirrer. After a few minutes of gentle stirring, the water temperature becomes steady at 19.7°C.
   Then 50 mL of water that had been warmed to 42.2°C is added to the cool water in the calorimeter and the water mixture is stirred gently until the temperature becomes constant again at a final temperature of 30.6°C. Follow the calculation steps below to find the heat capacity of the calorimeter. It is necessary to know the density and specific heat of water.
   The density of water is 1.0 g mL^{-1}, so that 1 mL weighs 1 g.
   The specific heat of water is 4.184 J g^{-1} deg^{-1}.

Step 1: Calculate the temperature change, ΔT_1, and energy lost, ΔH_1, for the warm water when it cooled to the final temperature of the system.

ΔH_1 = sp. ht. of warm water $\times$ g of warm water $\times \Delta T_1$

$\Delta H_1 = (4.184$ J g^{-1}deg-1)(50 g)(30.6 - 42.2)deg

$\Delta H_1 = -2.43 \times 10^3$ J

Step 2: ΔH_1 must also be equal to the energy gained by the cool water plus the calorimeter interior, thermometer, and stirrer, when these increased in temperature.

The temperature increase of the cool water and immersed calorimeter parts, ΔT_2, is

$$\Delta T_2 = 30.6 - 19.7 = +10.9 \text{ deg}$$

Notice that the temperature changes of the warm and cool water portions are not equal and opposite, even though the masses of hot and cool water are equal. This is because some of the heat energy was absorbed by the calorimeter apparatus.

The energy gained by *only* the cool water, ΔH_2, is easy to determine with a calculation like that in Step 1:

ΔH_2 = sp. ht. of cool water $\times$ g of cool water $\times \Delta T_2$

$\Delta H_2 = (4.184$ J g^{-1}deg^{-1})(50 g)(10.9 deg)

$\Delta H_2 = +2.28 \times 10^3$ J

Step 3: Determine the energy gained, ΔH(cal), by the calorimeter apparatus by recognizing that it must be equal to the difference between the energy lost by the warm water and that gained by the cool water.

- **The warm water lost 2.43×10^3 J.**
- **The cool water gained 2.28×10^3 J.**
- **The energy not accounted for is: $(2.43 \times 10^3 - 2.28 \times 10^3) = 0.15 \times 10^3$ J**
- **This energy must have been gained by the calorimeter apparatus to warm it to 30.6°C.**

The initial temperature of the calorimeter apparatus was the same as for the cool water, 19.7°C. Therefore, the temperature change for the calorimeter apparatus was the same as for the cool water:

$$\Delta T(\text{cal}) = 30.6 - 19.7 = \mathbf{10.9°C}$$

The heat capacity of the calorimeter apparatus is:

$$C = \{\text{energy gained}\}/\Delta T = \{0.15 \times 10^3 \text{ J}\}/10.9 \text{ deg} = \mathbf{13.8 \text{ J deg}^{-1}}$$

Procedure for Finding the Enthalpy of Reaction

The steps involved in finding the enthalpy of a reaction are:

1. Cause the reaction to occur in a calibrated calorimeter.
2. Measure the temperature change that occurs.
3. Determine the heat capacity of the reacting system.
4. Add the heat capacities of the calorimeter and the reacting system to obtain the heat capacity of the entire system.
5. Calculate the energy change in the surroundings required to give the observed temperature change for the given sample.
6. Calculate the energy change in the surroundings required to give the observed temperature change for one mole of sample. The energy change in the reactants is equal in magnitude and opposite in sign to the energy change in the surroundings. Therefore, the molar reaction enthalpy is:

$$\Delta H \text{ (reactants)} = -\Delta H \text{ (surroundings for 1 mole reactant)}$$

The method is illustrated in Example 5, using the same calorimeter that was calibrated in Example 4.

EXAMPLE 5:

450 mL of pure water at room temperature were placed into the clean, dry calorimeter of Example 3 and allowed to come to thermal equilibrium with the calorimeter interior.

At thermal equilibrium, the water temperature was 20.2°C.

Then, 10.14 g of calcium bromide, $CaBr_2$, also at room temperature, were added to the calorimeter and allowed to dissolve in the water, with gentle stirring. The temperature was measured at regular intervals until it stopped changing significantly.

At the new thermal equilibrium, the temperature was 17.4°C.

Determine the molar enthalpy of solution of $CaBr_2$.

SOLUTION:

Step 1: $CaBr_2$ was dissolved in a calorimeter with a calibrated heat capacity of 13.8 J/deg.

Step 2:   The temperature change was: $\Delta T = T_{final} - T_{initial} = 17.4 - 20.2 = 2.8°C$.

Step 3:   The heat capacity of the 450 g water in the calorimeter is:
$$C(H_2O) = \text{sp. ht.}(H_2O) \times g(H_2O) = 4.184 \text{ J/(g deg)} \times 450 \text{ g} = 1.88 \times 10^3 \text{ J/deg}.$$

The specific heat of solid $CaBr_2$ is about 0.28 J/(g deg), so that the heat capacity of the 10.14 g $CaBr_2$ added to the calorimeter is
$$C(CaBr_2) = 0.28 \text{ J/(g deg)} \times 10.14 \text{ g} = 2.8 \text{ J/deg}$$

The heat capacity of the $CaBr_2$ is insignificant compared to that of the water and may be neglected.

Therefore, the heat capacity of the reacting system is just that of the water.
$$C(\text{reacting system}) = 1.88 \times 10^3 \text{ J/deg}$$

Step 4:   Adding the heat capacities of the calorimeter and the reacting system to obtain the heat capacity of the entire system gives:
$$C(\text{entire syst.}) = C(H_2O) + C(\text{cal}) = 1.88 \times 10^3 \text{ J/deg} + 13.8 \text{ J/deg} = 1.89 \times 10^3 \text{ J/deg}$$

The heat capacity of the calorimeter is just barely significant compared to the heat capacity of the water and makes a very small contribution.

Step 5:   The energy change in the surroundings required to give the observed temperature change with the given sample is:
$$\Delta H(\text{surroundings}) = C(\text{entire system}) \times \Delta T = (1.89 \times 10^3 \text{ J/deg})(-2.8 \text{ deg}) = 5.29 \times 10^3 \text{ J}$$

Step 6:   The molar mass of $CaBr_2$ is 199.9 g/mole, so the $CaBr_2$ sample contained
$$10.14/199.9 = 0.0507 \text{ moles } CaBr_2 \text{ in the sample}$$
The amount of energy required from the surroundings for dissolving 1 mole of $CaBr_2$ is:
$$\Delta H(\text{surroundings for 1 mole reactant}) = -5.29 \times 10^3 \text{ J}/0.0507 \text{ moles} = -104 \text{ kJ/mol}$$
The molar enthalpy of solution for $CaBr_2$ is the negative of the energy above.
$$\Delta H_{sol}(CaBr_2) = -\Delta H(\text{surroundings for 1 mole reactant}) = +104 \text{ kJ/mol} \text{ (endothermic)}$$

The handbook value for the molar enthalpy of solution of $CaBr_2$ is 102 J/mol. Therefore the percent error in this example experiment is:
$$\{104 - 102 \times 10^3\}/\{102 \times 10^3\} \times 100\% = 1.96\%$$

SAFETY

1.   Wear approved eye protection.

2.   Use care when handling the acid and base solutions.

3.   Calcium metal will react with moisture on your fingers. Handle it only with tongs or tweezers.

PROCEDURE

Work with a partner. Use two thermometers for this experiment.

A. Construction of Calorimeter

1.   Make a calorimeter, as in Figure 23.1, by nesting two foam plastic cups of the type used for hot and cold drinks. A cover is desirable and may be made from a sheet of corrugated cardboard or foam plastic. There must be two holes through the cover, one for a thermometer and one for a stirrer.

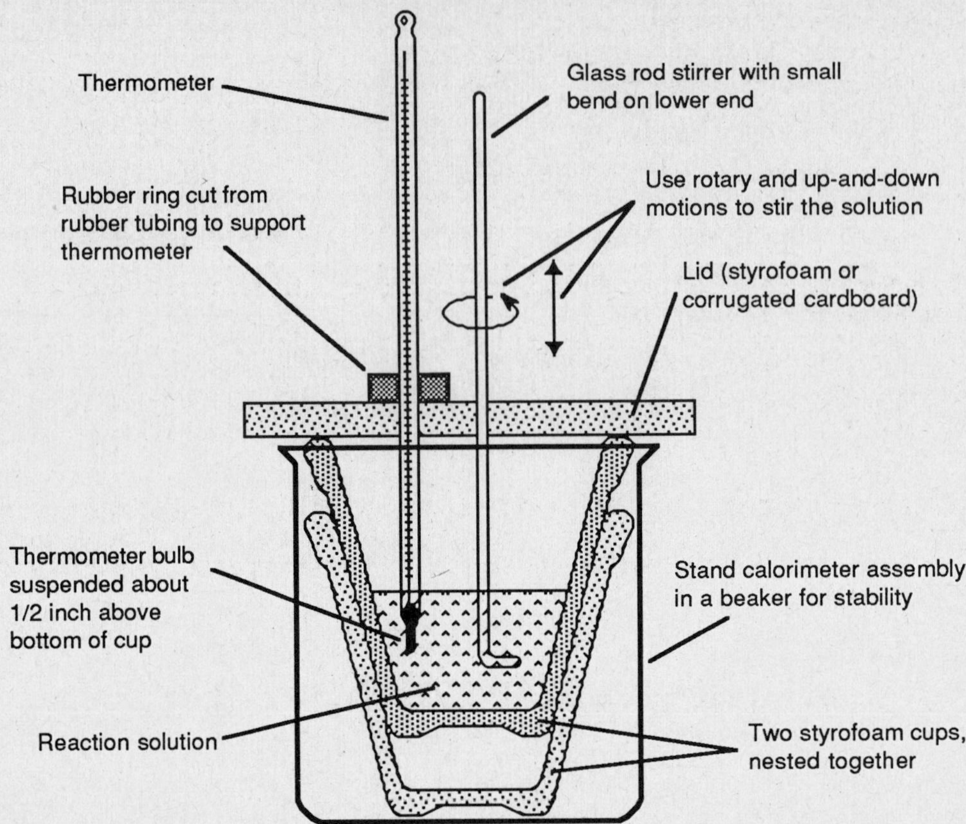

Figure 23.1. Calorimeter construction using foam plastic cups.

2.   Support the thermometer as shown in Figure 23.1 with a rubber ring cut from a piece of rubber tubing. Suspend the thermometer with its bulb 1/4 to 1/2 inch from the bottom of the inner cup. You will need a second thermometer for Part B, steps 5 and 6. The stirrer will work best if a short length at the bottom is bent at a right angle.

B. Calibration of the Calorimeter

1.   Measure 50 mL of water into an Erlenmeyer flask and begin heating it over a low burner flame.

2.   Weigh the clean, dry calorimeter to 0.1 g using a triple-beam balance.

3.   Pour 50 mL of cold water into the calorimeter and reweigh.

4.   Place the lid, with thermometer and stirrer, over the calorimeter. Measure the temperature of the cold water in the calorimeter. Do not record this temperature.
     Leave the thermometer in place in the calorimeter. Use your second thermometer for all temperature measurements outside of the calorimeter.

5.   Turn off the flame under the Erlenmeyer flask and let the heated water cool until it is about 20°C hotter than the cold water, using your second thermometer for measuring the temperature.

6.   Re-measure and record the temperatures of both the cold and the hot water.

7.   Immediately pour the hot water into the calorimeter, replace the lid and begin stirring gently.
     Be ready to record calorimeter temperatures immediately.

8.   Record temperature readings every 30 sec for about 5 min until the temperature becomes constant. Continue stirring throughout the measuring period.

9.   Weigh the calorimeter and contents to determine the mass of hot water added.
     Note that the measured heat capacity of the calorimeter includes the heat gained by the thermometer and stirrer. The heat capacity will probably be between 20 and 30 J/deg.

C. Measurement of ΔH for the reaction:

$$H^+ + OH^- \longrightarrow H_2O \quad \text{(ionic equation)}$$

Sodium hydroxide and hydrochloric acid are mixed directly in the calorimeter. The molecular equation for the reaction is: $HCl + NaOH \longrightarrow H_2O + NaCl$

1.   Add 50.0 mL of 2.0 M HCl solution to the calorimeter.

2.   Measure 50.0 mL of 2.0 M NaOH solution into a small beaker or Erlenmeyer flask. Allow the solutions to stand until their temperatures are the same at room temperature.
When measuring, be certain the thermometer bulb is completely immersed, but is not touching the side or bottom of the calorimeter.
Use 2 thermometers if available, with a rubber ring around them for support, as in Figure 23.1. If you do not have two thermometers, move one thermometer quickly between the solutions when comparing temperatures, drying the bulb quickly with a piece of paper towel to avoid transferring any solution.

3.   When both solutions are at the same temperature, record it.

4.   Rapidly pour the NaOH solution into the HCl solution. Put the lid with stirrer on the calorimeter and immediately begin gently stirring. Insert the thermometer quickly, while stirring constantly, and begin to take temperature readings at 10 s intervals. After the maximum temperature is passed (about 1.5-2 min) continue to record the temperature at 30 s intervals for about 5 min more.

5.   Plot the temperature-time data as in Figure 23.2 and determine the extrapolated high temperature.

6.   Repeat the process for a second determination.

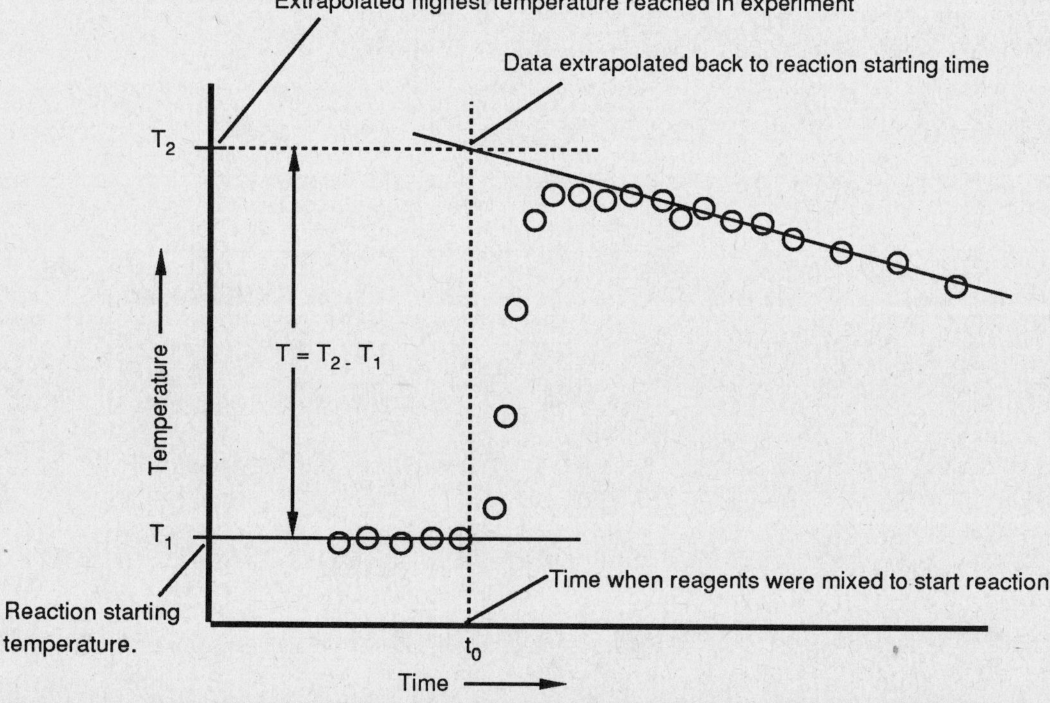

Figure 23.2. Graph of temperature versus time data for an exothermic reaction. After reaching a maximum value, the temperature slowly declines because of heat leakage out of the calorimeter. The linear part of the curve following the temperature maximum must be extrapolated back to the reaction starting time in order to determine the temperature that would have been reached if the reaction and new thermal equilibrium had taken place instantaneously with no heat loss. The temperature change due to the reaction is $\Delta T = T_2 - T_1$, where T_2 is the extrapolated high temperature and T_1 is the initial temperature.

D. Measurement of ΔH for the reaction:

$$Ca + 2H^+ \longrightarrow Ca^{2+} + H_2 \quad \text{(ionic equation)}$$

Calcium metal is added directly to hydrochloric acid in the calorimeter. The molecular equation for the reaction is: $Ca(s) + 2HCl \longrightarrow 2CaCl_2 + H_2$

1.   Add 50.0 mL of 1.0 M HCl solution to the calorimeter.

2.   Obtain a sample of metallic calcium. Be sure to handle it only with tongs or tweezers. The calcium probably has an oxide coating. Before weighing it, remove the heaviest parts of oxide crust by gently scraping the calcium with a small spatula.

3.   Weigh accurately about 0.5 g Ca in a weighing bottle on an analytical balance.

4.   Record the temperature of the HCl solution in the calorimeter.
5.   Add the Ca to the HCl solution and stir gently.

6.   While stirring constantly, record the temperature-time data as before and plot it as in Figure 23.2.

7.   Repeat the process for a second determination.

E. Determining the sign of ΔH for the reaction:

$$Ca + 2H_2O \longrightarrow Ca^{2+} + 2OH + H_2 \quad \text{(ionic equation)}$$

The molecular equation for the reaction is: $Ca(s) + 2H_2O \longrightarrow 2Ca(OH)_2 + H_2$

1.   Add an estimated 0.5 g of Ca to 100 mL distilled water in the calorimeter.

2.   Take enough temperature readings to tell whether the temperature is increasing or decreasing.

Name ____   Date _____________

EXPERIMENT 23
PRELABORATORY EXERCISE

1.   A calorimeter was calibrated to determine its heat capacity. 100 mL water were put into the calorimeter and the temperature became constant at 18.3°C. Then, 100 mL of warmer water at 35.6°C were added. The equilibrium temperature, after mixing, was 26.8°C. Calculate the heat capacity of the calorimeter.

Heat capacity of calorimeter: _______________

2.   The following chemical reaction between compound **A** and compound **B** was made to take place in the calorimeter of Question 1:

50.0 mL of 0.100 M **A** reacted completely with 50.0 mL of 1.00 M **B**. Before mixing, both solutions had come to the same room temperature of 21.4°C. After the reaction was over, the final extrapolated equilibrium temperature was 27.7°C.

Calculate the enthalpy of reaction per mole of **A**, assuming that all the **A** has reacted. Assume the specific heat of the reacting solution is the same as for pure water.

Calculations:

Name _______________________________________   Date ____________

EXPERIMENT 23
DATA

Record temperature-time data here.

Part B		Part C				Part D			
		1st determination		2nd determination		1st determination		2nd determination	
t (sec)	T (°C)	t (sec)	T (°C)	t (sec)	T (°C)	t (sec)	T (°C)	t (sec)	T (°C)

DATA

Because all your solutions are dilute, you can assume that the specific heat and density of all solutions used are equal to the corresponding values for water when making your calculations.

A. Calibration of Calorimeter

1. Mass of calorimeter: ___________________

2. Mass of calorimeter + cold water: ___________________

3. Mass of cold water: ___________________

4. Temperature of cold water just before mixing: ___________________

5. Temperature of hot water just before mixing: ___________________

6. Final equilibrium temperature after mixing: ___________________

7. Mass of calorimeter + total water: ___________________

8. Mass of hot water added: ___________________

9. Calibrated heat capacity of calorimeter (J/deg): ___________________

Calculations:

C. Measurement of ΔH for the reaction:

$$H^+ + OH^- \longrightarrow H_2O \quad \text{(ionic equation)}$$

	Determination 1	Determination 2
1. Volume of 2.0 M HCl:	_____________	_____________
2. Volume of 2.0 M NaOH:	_____________	_____________
3. Total volume of reaction mixture:	_____________	_____________
4. Number of moles of water formed:	_____________	_____________
5. Initial temperature:	_____________	_____________
6. Final extrapolated temperature after reaction:	_____________	_____________
7. Number of joules released:	_____________	_____________
8. Calibrated heat capacity of calorimeter (from Part B.):	_____________	
9. Enthalpy of reaction per mole of water formed:	_____________	_____________
10. Average enthalpy of reaction per mole of water formed:	_____________	

Calculations:

Name ____   Date ____________

EXPERIMENT 23

DATA

D. Measurement of ΔH for the reaction:

$$Ca + 2H^+ \longrightarrow Ca^{2+} + H_2 \quad \text{(ionic equation)}$$

		Determination	
		1	**2**
1.	Volume of 2.0 M HCl:	_______________	_______________
2.	Volume of 2.0 M NaOH:	_______________	_______________
3.	Total volume of reaction mixture:	_______________	_______________
4.	Mass of Ca:	_______________	_______________
5.	Moles of Ca:	_______________	_______________
6.	Volume of 1.0 M HCl:	_______________	_______________
7.	Number of moles of H+ in the solution:	_______________	_______________
8.	Initial temperature:	_______________	_______________
9.	Final extrapolated temperature after reaction:	_______________	_______________
10.	Number of joules released:	_______________	_______________
11.	Calibrated heat capacity of calorimeter (from Part B.):	_______________	
12.	Enthalpy of reaction per mole of Ca reacting:	_______________	_______________
13.	Average enthalpy of reaction per mole of Ca reacting:	_______________	

Calculations:

E. Determining the sign of ΔH for the reaction:

$$Ca + 2H_2O \longrightarrow Ca^{2+} + 2OH^- + H_2 \quad \text{(ionic equation)}$$

1.   Is the reaction exothermic or endothermic?   _______________________
Describe the observations that led to this conclusion.

Name ____   Date _____________

EXPERIMENT 23
QUESTIONS

1.   **a.** Use your results to Parts B and C to calculate the enthalpy of reaction per mole of Ca reacting for the reaction.

$$Ca + 2\,H_2O \longrightarrow Ca^{2+} + 2\,OH^- + H_2$$

Calculations:

Does the sign of ΔH agree with your measurement in Part E?_____________

b. Explain why it might be better to calculate the reaction above, as you did in Part a, than to measure it directly in your apparatus.

2.   Suppose 2.0 M ammonium hydroxide, $(NH_4)OH$, were substituted for NaOH in Part C.

a. Would this change the number of joules released? _____________
Explain.

b. Would it change the ΔH calculation in question 1a? _______
Explain.

Name _______________________________________   Date ____________

EXPERIMENT 23
DATA

DATA

EXPERIMENT 24
Intermolecular Forces:
Preparation of a Soap and Comparison of the Properties of
Soaps and Synthetic Detergents

PRELABORATORY PREPARATION

Do the Prelaboratory Exercise and turn it in at the beginning of your laboratory period.

INTRODUCTION

You can't wash an oil spot out of clothing using water alone because oil does not dissolve in water. In fact, oil and water actually repel one another, so that oil, in the presence of water, will adhere even more strongly to clothing. If you add soap or detergent to water, you can dissolve oil from clothing and rinse it away in the water solution. This experiment demonstrates how the chemical structure of soaps and detergents allows them to be effective cleaning agents for oils and greases.

Most soaps are soluble sodium or potassium salts of weak organic acids. The most common commercial soap is sodium stearate, $NaC_{18}H_{35}O_2$, is . It dissolves in water, forming the sodium and stearate ions:

$$NaC_{18}H_{35}O_2 \xrightarrow{H_2O} Na^+ + C_{18}H_{35}O_2^- \tag{24.1}$$

The stearate anion, $C_{18}H_{35}O_2^-$, is the active cleaning agent. It has an ionic charged end that dissolves well in water, owing to attraction by polar water molecules, and a non-polar hydrocarbon end with a long hydrocarbon chain, that dissolves well in oils and greases. The charged end is called **hydrophilic** (water-loving) and the hydrocarbon tail is called **hydrophobic** (water-fearing).

non–polar oil-soluble end   **stearate anion**   ionic water-soluble end
(hydrophobic)   (hydrophilic)

If water containing dissolved soap is rubbed over an oil spot, the anion hydrocarbon chains are strongly attracted into the oil, while the ionic ends remain bound to water molecules. In this way, soap anions act as linkages, joining oil molecules to water molecules. Thermal motion of the water molecules pulls the ionic ends of the soap anions back and forth, exerting a constant tugging action on the oil molecules. Eventually, the oil spot is broken up into small droplets and dispersed into the water. Many soap anions will have their hydrocarbon "tails" dissolved into the same oil droplet. Figure 24.1 is a visualization of how water molecules cluster around the head of a soap anion and help it pull oil away from an oil deposit and into water suspension. In this figure, the hydrocarbon chain is illustrated as a tail attached to a negatively charged head.

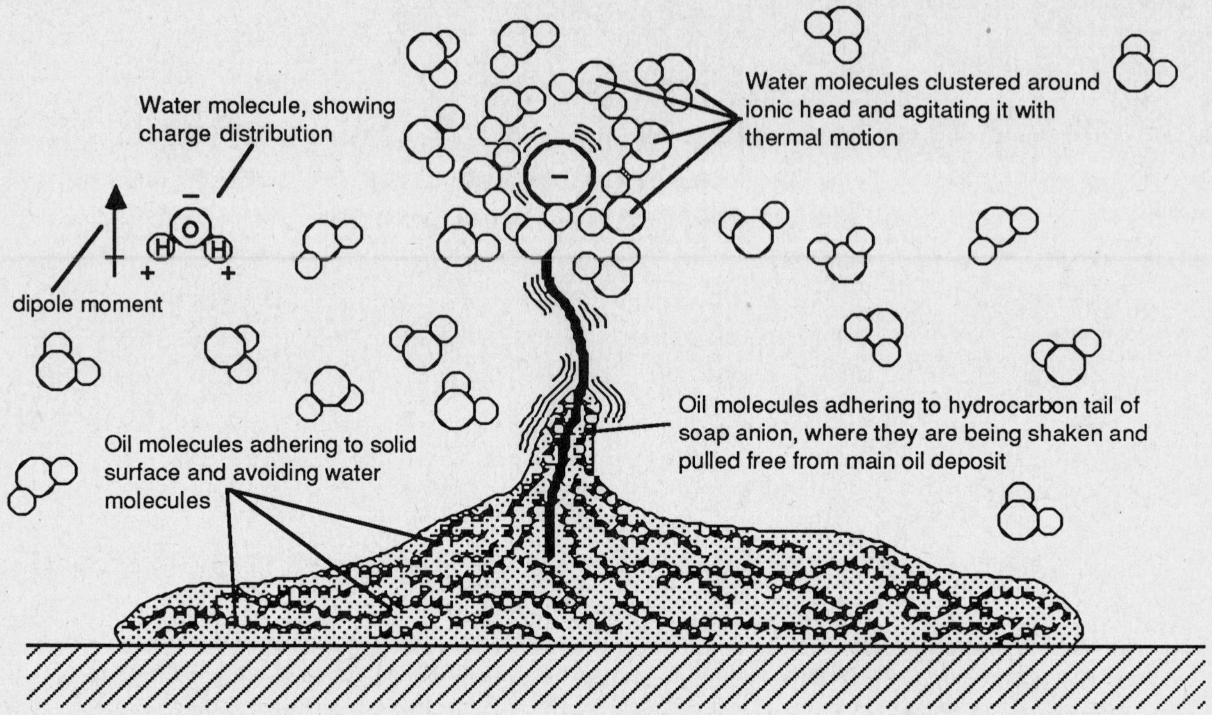

Figure 24.1. Detailed visualization of how water molecules clustered around the ionic head of a soap anion can agitate and pull oil molecules free from a surface into suspension.

The repulsion between adjacent ionic "heads" forces the entire oil and soap structure into a spherical shape called a **micelle** (see Figure 24.2). Adjacent micelles repel one another because they all have a negatively charged outer surface and, therefore, remain dispersed as an **emulsion.** The soap is called an **emulsifying agent.**

The formation of micelles and their dispersion into water to form an emulsion is illustrated in Figure 24.2. Notice that soap anions tend to form micelles even if no oil is present. This happens because the hydrocarbon ends attract while the charged ends repel one another. When a soap micelle makes contact with an oil surface, the soap micelle opens up as the hydrocarbon tails are drawn into the oil.

The cleansing action, or **detergency**, of soaps such as sodium stearate depends upon the fact that they ionize readily in water, as in Reaction 24.1, forming the anion charged at one end and with a long, non-polar hydrocarbon chain on the other end. You can imagine what would happen if the ionic end lost its charge—the soap would lose its ability to form micelles and, hence, its cleaning ability. This is just what happens in hard or acidic water. Hard water contains metal cations, such as Ca^{2+} and Mg^{2+}, that react with the charged end of soap anions to form molecules that do not ionize and are insoluble as well. Having lost their charged end, the soap molecules cannot pull oil and grease into a water emulsion. This is the main problem with soaps, because such cations are quite common in domestic water supplies. Soap will not clean in hard water until most of the metal cations have been precipitated by reacting with the soap, as in Reaction 24.2. This forms the insoluble gray "curd" commonly called "bathtub ring."

For convenience in writing chemical equations, we will indicate a hydrocarbon chain by the general symbol **R**, and write an ion such as stearate as shown below, in order to clearly show the charged (**COO-**)

$$R\text{-}\overset{\displaystyle O}{\overset{\|}{C}}\text{-}O^{-}$$

and non-polar (**R**) parts of the ion. Then, the reaction of a soap anion with a magnesium ion, one of the cations that make water hard, can be written as:

$$Mg^{2+} + 2\,R\text{-}\overset{\displaystyle O}{\overset{\|}{C}}\text{-}O^{-} \longrightarrow \left(R\text{-}\overset{\displaystyle O}{\overset{\|}{C}}\text{-}O \right)_2 Mg \tag{24.2}$$

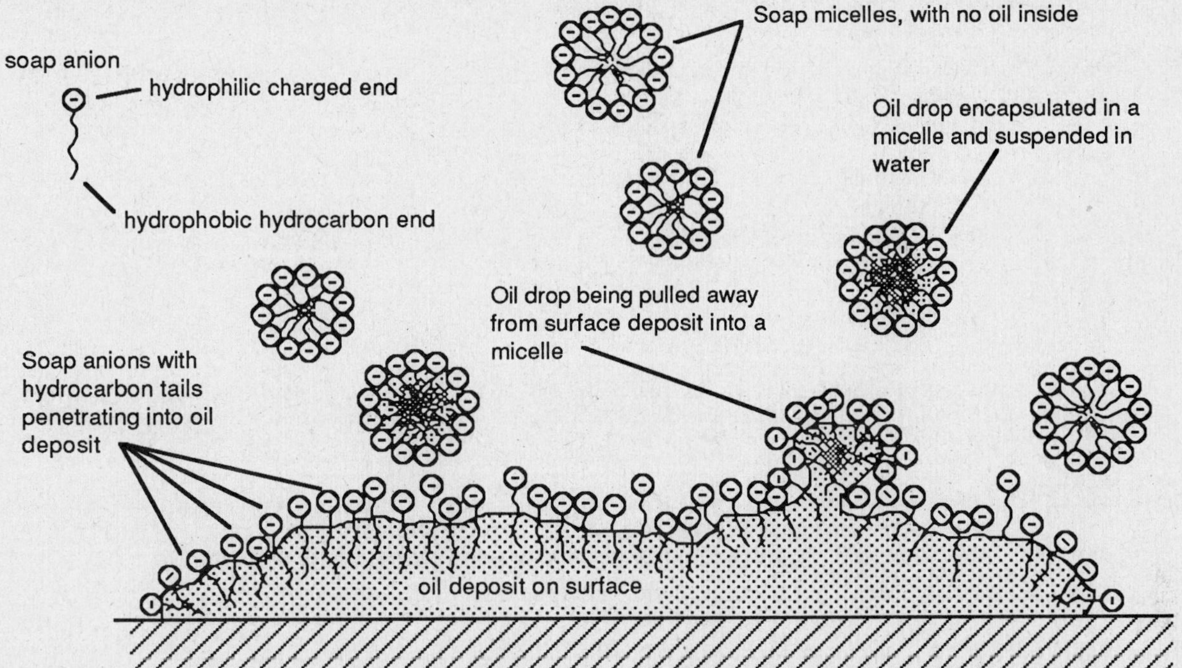

Figure 24.2.   Illustration of the cleaning action of soaps and detergents. Oil attaches to the hydrocarbon tail of soap anions and is pulled away from the surface into the water encapsulated in micelles. The micelles remain suspended in the water and can be rinsed away.

Acidic water also can destroy the cleaning ability of soaps, because the soap anions become protonated (Reaction 24.3), forming a weak fatty acid that is only slightly ionized.

$$R-\overset{\overset{\textstyle O}{\|}}{C}-O^- \ + \ H_3O^+ \ \longrightarrow \ R-\overset{\overset{\textstyle O}{\|}}{C}-OH \tag{23.4}$$

In acidic water, soap acts as a base, accepting protons from hydronium ions and neutralizing the acidity of the solution. Each soap molecule that receives a proton loses its ionic charge and becomes ineffective for cleaning. Body perspiration that accumulates on skin and clothing tends to be acidic and enough soap must be added for neutralization before cleaning action can begin. Many soaps contain additives to neutralize any acidity in the water and leave the soap anion free to combine with greases and oils.

Making Soap

Soaps are made by the chemical reaction of fats or oils with a strong base such as NaOH. The products are glycerol and a variety of soap molecules that depend on the exact length of the different hydrocarbon chains, **R, R'** and **R":**

$$
\begin{array}{ccc}
\begin{array}{l}
\text{H}\quad\text{O}\\
|\quad\ \|\\
\text{H}-\text{C}-\text{O}-\text{C}-\text{R}\\
|\quad\ \ \text{O}\\
|\quad\ \ \|\\
\text{H}-\text{C}-\text{O}-\text{C}-\text{R}'\\
|\quad\ \ \text{O}\\
|\quad\ \ \|\\
\text{H}-\text{C}-\text{O}-\text{C}-\text{R}''\\
|\\
\text{H}
\end{array}
& +\ 3\,\text{NaOH} \longrightarrow &
\begin{array}{l}
\text{H}\\
|\\
\text{H}-\text{C}-\text{OH}\\
|\\
\text{H}-\text{C}-\text{OH}\quad +\\
|\\
\text{H}-\text{C}-\text{OH}\\
|\\
\text{H}
\end{array}
\quad
\begin{array}{l}
\quad\ \ \text{O}\\
\quad\ \ \|\\
\text{Na}-\text{O}-\text{C}-\text{R}\\
\quad\ \ \text{O}\\
\quad\ \ \|\\
\text{Na}-\text{O}-\text{C}-\text{R}'\\
\quad\ \ \text{O}\\
\quad\ \ \|\\
\text{Na}-\text{O}-\text{C}-\text{R}''
\end{array}
\end{array}
$$

oil	+ strong base	$\longrightarrow$	glycerol	+ soap molecules

Synthetic detergents, often simply called "detergents" (soaps could be called "natural detergents"), were developed to overcome the limitations of soaps. Detergent molecules are similar to soap molecules in having an ionic end and a long-chain hydrocarbon end. They dissolve oils and greases and remove dirt in the same way as soaps. They have different structures, however, that do not form insoluble compounds with the metal cations in hard water, and do not become protonated in acidic water to form weak, unionized acids. Therefore, detergents do not form "bathtub ring" and wash better in acidic water. Sodium lauryl sulfate, a typical detergent, ionizes when dissolved in water as shown below:

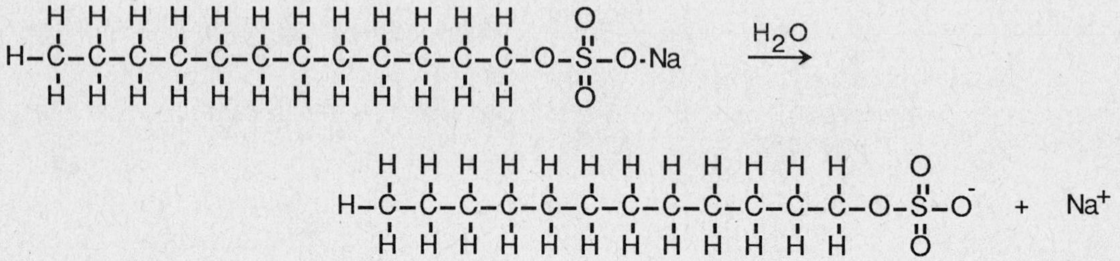

$$H-C-C-C-C-C-C-C-C-C-C-C-C-O-\overset{O}{\underset{O}{S}}-O\cdot Na \quad \xrightarrow{H_2O}$$

$$H-C-C-C-C-C-C-C-C-C-C-C-C-O-\overset{O}{\underset{O}{S}}-O^{-} \quad + \quad Na^{+}$$

PLAN OF EXPERIMENT

You first make a soap, using a vegetable oil as your starting material. Then, you compare the properties of the soap you have made with those of a commercial soap and a commercial detergent.

SAFETY

1. Wear approved eye protection.

2. Ethanol, C_2H_5OH, is very flammable. Be sure there are no open flames in the laboratory.

3. Sodium hydroxide (NaOH) solutions can cause severe burns. In addition to your safety glasses, laboratory coats or aprons are advisable. In case of NaOH contact with skin or clothing, immediately rinse the affected region copiously with water and notify your instructor.

PROCEDURE

1. Place 20 g of vegetable oil in a 200 mL Erlenmeyer flask. Add 20 mL of ethyl alcohol.
    The ethyl alcohol is added to help dissolve all the reactants and speed up the reaction. It is not used commercially in making soap.
The ethanol and oil will separate into layers. Shake the liquids up well.

2. Add 25 mL of 5 M NaOH solution. Swirl the solution well and pour it into a 250 mL beaker. Stir the mixture with a glass rod and gently heat heat it on a hotplate. Continue to heat the mixture, stirring occasionally, until it all turns pasty. This may take about 30 min. The paste is a mixture of soap and glycerol.
    When the paste begins to form, stir very carefully to prevent frothing.

3. After all the paste has formed, set the beaker on your laboratory bench top to cool.

4. Then the paste mixture is cool, add 100 mL of saturated NaCl solution to the paste mixture and stir it in thoroughly.
    As you stir, break up the chunks of paste by pressing them against the side of the beaker.

The NaCl solution provides Na+ and Cl- ions that bind to the polar water molecules, and help separate the water from the soap. This procedure is called "salting out" the soap.

5. After stirring the NaCl solution through the soap paste, filter off the soap mixture by suction filtration and wash the collected soap once with 25 mL of ice water, through the suction filter. Then, continue to suck air through the soap for about 10 min, to help dry it.

Comparing the Properties of Soaps and Detergents

Compare the properties of the soap you have made with a commercial soap and a commercial detergent.

1.   Make a solution of each soap and detergent to be tested in a 250 mL beaker by dissolving about 1 g of each in 100 mL of warm deionized water. Label the beakers to identify the solutions. These will be your soap and detergent test solutions.

2.   **Basicity:** Too much NaOH remaining in a soap can make it harsh and possibly damaging to skin and clothing.

Test the pH of each test solution by touching a clean glass rod to the solution and transferring a drop to a piece of universal indicator paper. Record your results.

3.   **Emulsifying Properties:** The ability of a soap or detergent to rinse away oil and grease depends on how well it keeps these materials in suspension as emulsions.

Label 4 clean, dry test tubes with number from 1 to 4. Transfer 10 mL portions of each of your 3 test solutions into separate test tubes. Place 10 mL of deionized water into the 4th test tube,

Indicate on the data sheet which solution is in which test tube.

Add 10 drops of kerosene or mineral oil to each test tube. Shake each test tube well and, immediately after shaking, note on the data sheet how uniformly the oil and water are distributed throughout each mixture. Set the tubes in a rack and do not disturb them for 5 min. After 5 min, record how well the oil remains in suspension for each test tube.

4.   **Behavior in Hard Water:** A solution containing about 75 mg/L each of $CaCl_2$ and $MgCl_2$ will be available in the laboratory to serve as an example of typical hard water. Place 10 mL of each of your 3 test solutions into separate clean numbered test tubes.

Indicate on the data sheet which solution is in which test tube.

Add 2 mL of the hard water solution to each test tube. After each addition, shake well and immediately observe the nature of the contents, looking for cloudiness, precipitates, films, etc. Record your observations.

Allow the test tubes to stand undisturbed for about 5 min, observe their condition again and record these observations also.

5.   **Behavior in Acidic Water:** Hydronium ions attach to the ionic end of soap molecules, destroying their cleaning ability by reducing their attraction for water molecules.

Place 10 mL of each of your test solutions in separate numbered test tubes. Add 5 drops of 3 M HCl to each test tube, shake well, and record your observations about the sudsing properties of each solution.

Then, add 10 drops of kerosene or mineral oil to each test tube and shake well. Observe the emulsifying properties of each solution immediately after shaking and after 5 min of standing undisturbed. Record your observations and compare the results with Part 3, **Emulsifying Properties.**

6.   **Effect of Phosphates:** Many detergents and a few soaps contain phosphates, which serve as bases to neutralize acidic water and, also, combine with metal cations, preventing the reaction with soap to form "bathtub ring." The amount of phosphates in different detergents varies widely.

To test for phosphates in your test solutions, place 2 mL of each into separate numbered test tubes. Add 5 drops of dilute HNO_3 (about 1 M) and 2 mL of ammonium molybdate solution (about 1% by mass) to each test tube. Warm the test tubes in a water bath, **but do not boil the solutions.** A yellow precipitate indicates phosphate. Record your results.

7.   **Cleaning Properties:** Use tap water to wash your hands with some of the soap you have made. It should lather well, unless the water is too hard. For comparison, lather your hands using deionized water.

If there was an excess of oil when the soap was made, compared to NaOH, the soap will feel greasy.

If there was an excess of NaOH when the soap was made, the soap will feel "slick," but not greasy.
Soap with too much NaOH will roughen and dry your skin. Rinse your hands well with water and dry them. Record the lathering properties and "feel" of your soap.

Rub some of the vegetable oil used to make your soap over your hands and try to wash it off with tap water alone. Then use your soap. Record the results.

Name ____ Date _______________

EXPERIMENT 24
PRELABORATORY EXERCISE

1. Why is it difficult to wash oil or grease from a surface using water alone?

2. Explain why soap anions form micelles in water.

3. What is "hard" water and why does it destroy the cleaning ability of soap?

4. Describe the advantages of synthetic detergents over soap. Can you think of any disadvantages?

Name ____ Date _______________

EXPERIMENT 24

DATA

1. Basicity: Compare your test solutions with respect to basicity.

TEST TUBE	SOLUTION	INDICATOR COLOR	pH
1.			
2.			
3.			
4.			

2. Emulsifying Properties: Describe the results of the emulsification experiment. Explain them in terms of intermolecular forces.

TEST TUBE	CONTENTS	INITIAL OBSERVATIONS	OBSERVATIONS AFTER 5 MINUTES
1.			
2.			
3.			
4.			

Explanation:

3. Reactions in Hard Water: Describe the effects of adding "hard water" to your test solutions.

TEST TUBE	CONTENTS	METAL CATION ADDED	INITIAL OBSERVATIONS	OBSERVATIONS AFTER 5 MINUTES
1.				
2.				
3.				
4.				

DATA

4. Reactions in Acidic Water: Describe the effects of adding an acid solution to your test solutions.

TEST TUBE	CONTENTS	OBSERVATIONS ABOUT SUDSING	OBSERVATIONS ABOUT OIL EMULSIFYING PROPERTIES
1.			
2.			
3.			
4.			

5. Phosphate Test:

TEST TUBE	CONTENTS	RESULTS
1.		
2.		
3.		
4.		

6. Cleaning tests: Describe the washing properties of the soap you prepared.

EXPERIMENT 25
Preparation and Spectroscopy of Transition Metal Complex Ions

PRELABORATORY PREPARATION
1. Do the Prelaboratory Exercise and turn it in at the beginning of your laboratory period.
2. Review the use of the spectrophotometer, Section XIII, Methods, page 28.
3. Review vacuum filtering technique, Section IX, Methods, page 23.
4. Review how to insert glass tubing through rubber stoppers, Section VII, Methods, page 20.

In one laboratory period, there only is time to do either Part A or Part B, along with Part C. Your instructor will assign which part each student is to do.

INTRODUCTION
The term **complex** is used to name chemical species that are made out of separate components that can exist independently. A complex has its own unique properties that are different from those of its components. Transition metal cations form complexes very readily with neutral molecules like H_2O and NH_3, and with anions like Cl^- and CN^-. Consider the transition metal **complex ion** $Co(NH_3)_4Cl_2^+$ in aqueous solution. It is made from one Co^{3+} cation, 4 ammonia molecules, and 2 Cl^- anions. The net charge of the complex ion is the sum of the charges on the cobalt and chloride ions: $(+3 - 1 - 1) = +1$. Each component of the complex ion (Co^{3+}, NH_3, and Cl^-) can exist independently in aqueous solution (although each must have a solvation sphere of water molecules around it), but the parts remain bound together in the complex ion as a single species. They differ in this respect from many other ionic species, such as Na^+ cations and Cl^- anions in the ionic compound $NaCl$, which does dissociate in solution. The complex ion also differs from ions like sulfate, SO_4^{2-}. Sulfate ions do not dissociate in solution, but that is expected, because they are not made out of components that can exist independently, under reasonably "normal" chemical conditions; neither S nor O exist as independent, stable aqueous neutral species or ions.

The behavior of complex ions is based on their structure; there is a transition metal cation at the center, surrounded by several ionic or neutral chemical groups called **ligands.** Transition metal cations are formed by losing electrons from their valence electron shell d-orbitals. Empty orbitals in the valence shell of a transition metal ion can share electron-pairs with other chemical species that have non-bonding electron pairs available to donate, forming a **coordination bond** between the metal cation and the ligand electron-donor. Ligands that group around the central cation all have at least one pair of non-bonding electrons, that coordinate with unfilled valence orbitals of the cation, to form the complex.

The electron deficient central metal cation behaves as a Lewis acid, accepting electrons from the electron-donor ligands, which serve as Lewis bases.

Due to the charge on the central cation, coordination bonding frequently is mostly ionic and only partly covalent, so that the number of ligands that can coordinate with the cation is determined more by the charge and size of the cation than by the number of valence orbitals. The number of coordinate bonds that the cation forms is called its **coordination number,** and can vary from 2 to at least 12. The most common coordination numbers are 4 and 6. Only very large, multiply charged cations form complexes with coordination numbers greater than eight.

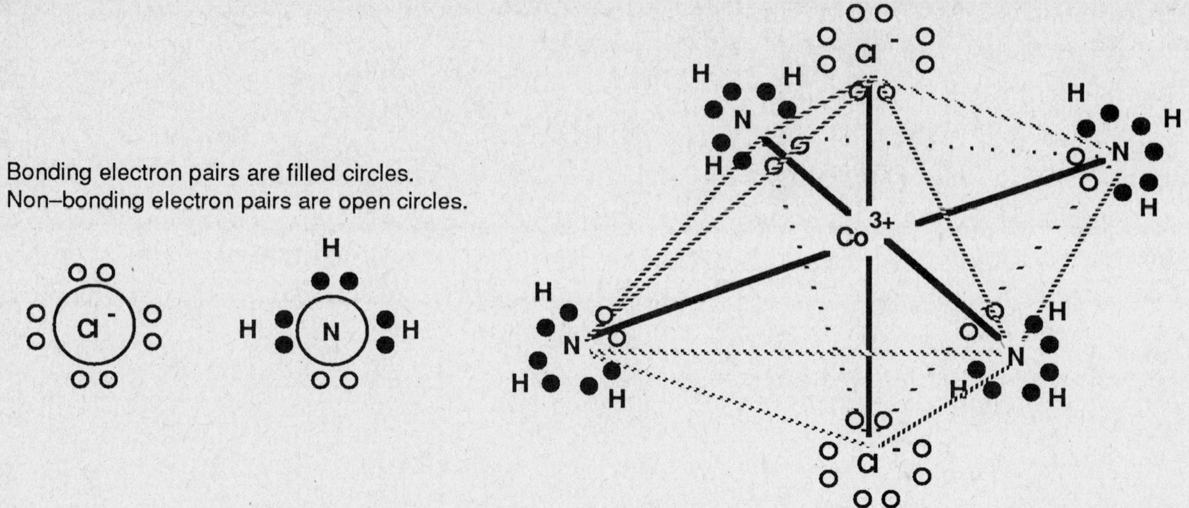

Figure 25.1.   Octahedral structure of the transition metal complex ion *trans*-$Co(NH_3)_4Cl_2^+$.

In $Co(NH_3)_4Cl_2^+$, which has 4 NH_3 and 2 Cl^- ligands, the coordination number for Co^{3+} is six. Each NH_3 group has one non-bonding electron pair and each Cl^- ion has four non-bonding pairs available for coordination (see Figure 25.1). The resulting complex ion has an octahedral symmetry, slightly distorted because the NH_3 and Cl^- ligands are not equivalent with respect to their sizes and electronegativities. The complex ion $Co(NH_3)_6^{3+}$, where all 6 ligands are completely equivalent, would form a perfect octahedron. As a "rule of thumb," the coordination number often turns out to be 2 times the charge on the cation.

The nature of the cation-ligand bond has a very strong influence on the spectral absorption spectrum of the complex ion, and spectroscopy is an important tool for studying the bonding properties of coordination complexes. The presence of the ligands around the central cation removes the energy degeneracy of the cation *d*-orbitals. With six ligands in octahedral symmetry, the geometry of the complexes studied in this experiment, two of the *d*-orbitals are raised to a higher energy than the other three (see Figure 25.2). By absorbing light energy, an electron can move from a lower to an upper level.

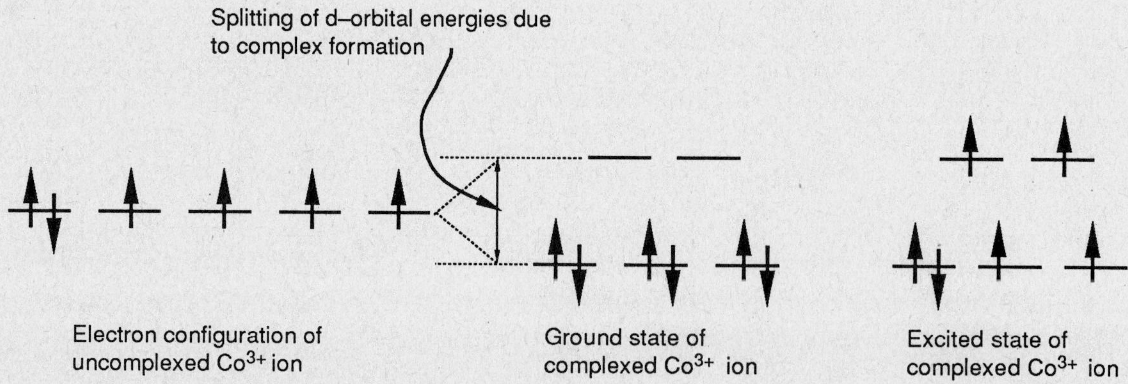

Figure 25.2.   *d*-orbital energy level diagrams for the uncomplexed Co^{3+} cation and the ground and first excited states of complexed Co^{3+} (d^6 electron structure, coordination number = 6), showing how the octahedral arrangement of six ligands splits the *d*-orbital energies into 2 higher and 3 lower levels.

The energy difference between the lower and upper levels is related to the wavelength of light absorbed, as expressed by the equation:

$$\Delta E = hc/\lambda$$

ΔE is the energy difference between lower and upper levels in joules

h = Planck's constant = 6.626×10^{-34} J s

c = speed of light = 2.998×10^{8} m/s

λ = wavelength of light absorbed in meters

Notice that, since both terms in the numerator are constants, the larger ΔE is, the smaller λ must be. In other words, when ΔE is large, the wavelength of light absorbed will be short, and if ΔE is small, a longer wavelength of light will be absorbed. The nature of the ligand determines the magnitude of ΔE. A listing of ligands, in order of how large a ΔE they produce, is called a **spectrochemical series**.

A spectrochemical series for some common ligands is:

$$CN^- > NO_2^- > en^1 > NCS^- > NH_3 > H_2O > C_2O_4{}^{2-} > OH^- > F^- > Cl^- > Br^- > I^-$$

Large ΔE   **Decreasing ΔE**   **Small ΔE**

The color of an aqueous solution of a complex ion is determined by which wavelengths of light are absorbed to excite the *d*--electrons. The color **observed** for a complex ion solution depends on what wavelengths remain after part of the light has been absorbed. The color that we see is the color of the light that is transmitted through the sample, not the color of the light absorbed by the sample. In Table 25.1, the energies absorbed and colors observed are compared for several complex ions of Co^{3+}.

Table 25.1   Relation of Wavelength Absorbed to Color Observed, for Some Complex Ions of Co^{3+}.

Complex ion	Color of ion observed	Color absorbed from Light	Wavelength of maximum absorbance (nm)
$Co(NH_3)_6{}^{3+}$	yellow	violet	430
$Co(NH_3)_5NCS^{2+}$	orange	blue	470
$Co(NH_3)_5H_2O^{3+}$	red	blue-green	500
$Co(NH_3)_5Cl^{2+}$	purple	yellow-green	530
trans-$Co(NH_3)_4Cl_2{}^+$	green	red	680

If you use the complimentary color wheel, Figure 25.3, to compare what colors are absorbed and what colors are observed for an ion, you will see that the color observed for a complex ion is complimentary to the color absorbed from the light. A compound that does not absorb at all in the visible wavelength region will appear to be transparent and colorless; one that absorbs fairly uniformly at all visible wavelengths will be gray or black. Color is the result of non-uniform absorbance in the visible region, as with $Co(NH_3)_5Cl^{2+}$ (see Table 25.1 and Figure 25.4), which has a maximum absorbance at 530 nm, the yellow-green part of the spectrum, and appears as a purple compound, the complementary color to yellow-green.

The color of a solution is the color of the light transmitted through the solution. This color is complimentary to the color of light most strongly absorbed by the solution.

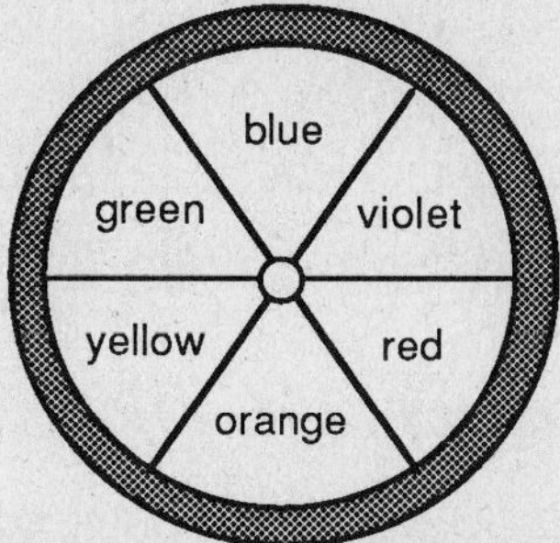

Figure 25.3.   Complimentary color wheel. Colors on opposite sides are complimentary to one another. When light is absorbed by a compound, the remaining light that is transmitted through the compound appears as the color complimentary to that absorbed; for example, if a compound absorbs yellow light, it will appear to be violet from the remaining wavelengths transmitted through it.

---

[1]en = ethylenediamine ($H_2NCH_2CH_2NH_2$)

By measuring how the absorbance spectrum of a complex ion depends on the ligands that are bonded to the central metal cation, we learn about the energy separation between the d-orbitals in the metal's valence shell and determine the order of ligands in a spectrochemical series. Figure 25.4 is the absorbance spectrum for the complex ion $Co(NH_3)_5Cl^{2+}$. The maximum absorbance occurs at 530 nm, corresponding to absorbed yellow-green color. Therefore, the observed color of the solution is the complimentary color purple, between violet and red. The energy difference between the upper and lower sets of d-orbitals is:

$$\Delta E = hc/\lambda = (6.63 \times 10^{-34} \text{ J s})(3.00 \times 10^8 \text{ m/s})/(530 \times 10^{-9} \text{ m}) = 3.75 \times 10^{-19} \text{ J}$$

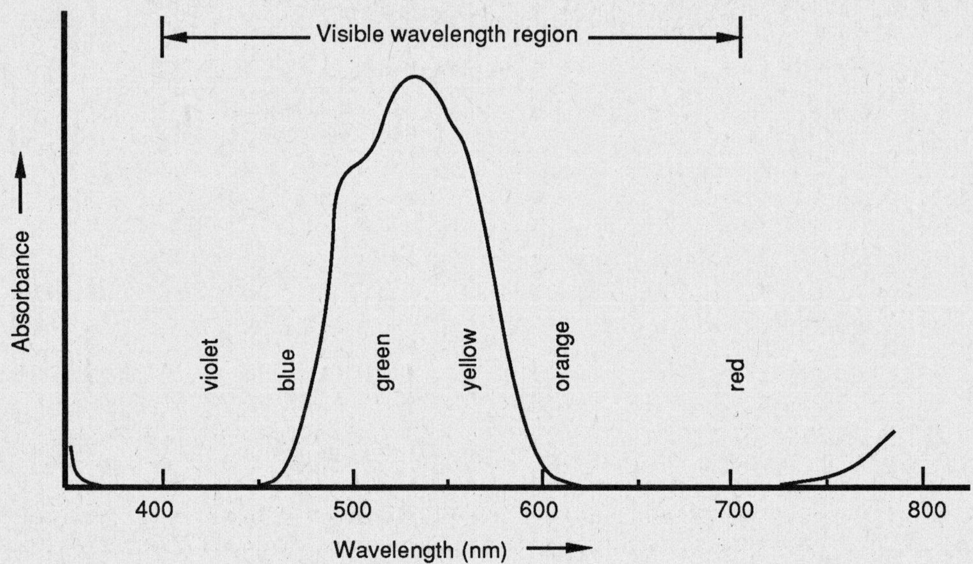

Figure 25.4. The absorption spectrum of $Co(NH_3)_5Cl^{2+}$ measured with a spectrophotometer.

PLAN OF EXPERIMENT

In Part A, the ligands around an octahedrally complexed Ni^{2+} cation are changed from H_2O to NH_3 and the effect on the absorption spectrum is measured with a spectrophotometer. In Part B, an octahedrally complexed Co^{2+} cation is oxidized to Co^{3+}, its ligands are changed from H_2O to NO_2^-, and the effect on the spectrum is measured.

You are to do only one of these parts. Your instructor will tell you whether to do Part A or Part B. In Part C, you use molecular models to study the geometry and isomers of several complex ions.

SAFETY

1.   Wear approved eye protection.

2.   Ammonia fumes are extremely irritating. Always use NH_4OH solutions in the hood.

3.   Toxic fumes of NO_2 are produced in step 6 of Part B. Be sure the water trap functions properly and perform the reaction in a hood.

PROCEDURE

You are to do **either** Part A or Part B. Your instructor will tell you which one. Part of the class might do Part A and the rest of the class Part B, to compare the two results.

A. Conversion of Hexaquonickel(II)chloride, $Ni(H_2O)_6Cl_2$, (also called *nickel chloride*) to Hexamminenickel(II)chloride, $Ni(NH_3)_6Cl_2$.

1.   Be certain that the spectrophotometer is turned on so it is warmed up for your measurements.

2.   Place 50 mL of concentrated NH_4OH solution into a small Erlenmeyer flask or bottle and cool it in an ice bath.

3.   On weighing paper, weigh about 6 g of nickel chloride, $Ni(H_2O)_6Cl_2$ to 0.1 g. Transfer to a 50 mL beaker and dissolve it in 10 mL H_2O, warming and stirring as needed.
   If the solution is not clear, centrifuge it rapidly while it is still warm.
Then, cool the solution to room temperature.

4.   Transfer about 1 mL of the clear solution to a clean graduated cylinder and dilute it to 10 mL with distilled water.

5.   Using the undiluted portion, rinse a spectrophotometer cell twice with small portions of the solution. Return the rinsing solution to the original solution. Then, fill the cell with the diluted portion of the nickel chloride solution. To remove bubbles on the cell walls, press a piece of parafilm over the top of the cell with your finger to seal it, and slowly invert and rotate the cell to bring all the bubbles to the liquid surface.
   Fingerprints on the cell in the light path can cause measuring errors. Avoid touching the lower 2/3 of the spectrometer cell.

6.   Insert the sample cell into the spectrometer and measure its absorption spectrum. Scan over the entire wavelength range, measuring at 20 nm intervals. Record the measured absorbance values on the Data sheet. Carefully follow the procedure in Section XIII, Methods, page 27.

7.   After measuring the absorption spectrum, return the nickel chloride sample *quantitatively* to the original solution.

8.   Repeat the scanning procedure with a distilled water blank. Rinse the cell 2 or 3 times with distilled water before measuring. Record the measured absorbance values on the Data sheet.

9.   **IN A HOOD**, add 12 mL of chilled, concentrated NH_4OH solution to the nickel chloride solution, while stirring. Cool the mixture by placing the beaker on ice.
   Crystals of $Ni(NH_3)_6Cl_2$ should form within about 15 minutes.
Add more NH_4OH if crystallization does not occur.

10.   Vacuum filter the crystals. If the filtered crystals are clumped together, spread them out on the filter with a spatula. With suction on, wash the crystals twice, using 2 mL of cold concentrated NH_4OH solution each time.

11.   Now, wash the crystals 3 times, using 2 mL portions of acetone each time, to remove water. Keep the suction on to draw air through the filter until the crystals are dry.

12.   Weigh your product and record the mass on the Data sheet.

13.   Dissolve the $Ni(NH_3)_6Cl_2$ crystals in 10 mL distilled water and repeat steps 4-6 to measure the absorption spectrum of the solution.
It is not necessary to run another water blank.

B. Preparation of Sodium Hexanitrocobaltate(III), $Na_3Co(NO_2)_6$

In Part B, Cobalt is oxidized from the +2 state in $Co(H_2O)_6^{2+}$ to the +3 state in $Co(NO_2)_6^{3-}$. Nitrite ion, NO_2^-, serves as both an oxidizing reagent (it is reduced to NO) and as a complexing ligand.

1.   On weighing paper, weigh about 5 g of cobalt(II)nitrate, $Co(H_2O)_6(NO_3)_2$ to 0.1 g. Record the mass on the Data sheet. Transfer the sample to a 100 mL beaker and dissolve it in 15 mL H_2O.

2.   Remove 1 mL of the solution, dilute it to 10 mL with distilled water and measure its absorbance spectrum as in steps 5-6 of Part A.

3.   After measuring the absorption spectrum, return the cobalt nitrate sample *quantitatively* to the original solution.

4.   Repeat the scanning procedure with a distilled water blank. Rinse the cell 2 or 3 times with distilled water before measuring. Record the measured absorbance values on the Data sheet.

5.   Dissolve 15 g sodium *nitrite,* $NaNO_2$, into the cobalt nitrate solution, warming and stirring as needed until dissolved.

6.   **IN A HOOD**, add to the solution, **drop by drop,** 5 mL of 50% acetic acid solution, while stirring continuously.

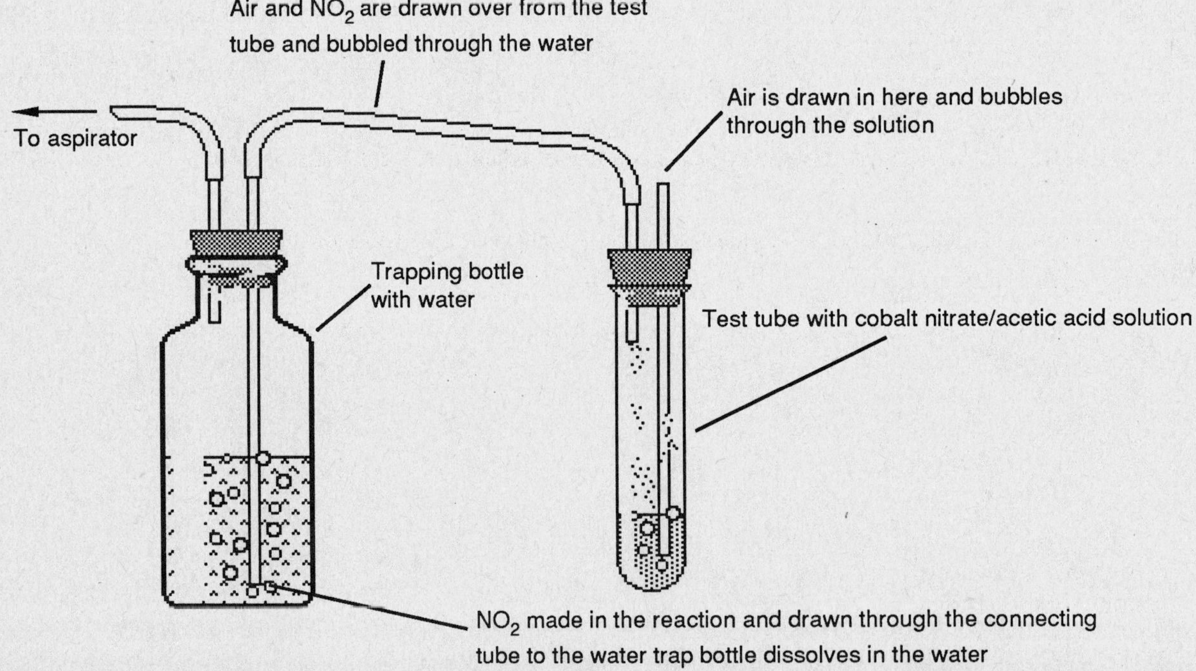

Figure 25.5.   Water trap for preventing NO_2 gas from entering the atmosphere. The aspirator creates a vacuum that draws air in through the open tube on the reaction test tube, which bubbles through the solution and effectively mixes the reagents. Then air and the product NO_2 gases are drawn into the water trap bottle where most of the NO_2 dissolves in the water. Any undissolved NO_2 is trapped by water in the aspirator flow and carried down the drain.

7.   Transfer the solution to a large test tube (25×200 mm).
     Pour it _slowly_ down the side of the test tube.
The reaction that now occurs is: $2\ NO_2^- + 2\ H_3O^+ \longrightarrow 2\ H_2O + NO(g) + NO_2(g)$
One product of this reaction is the corrosive gas NO_2, which must be removed by bubbling the gas from the test tube through a water trap made as shown in Figure 25.5. When NO_2 is dissolved in cold water, a dilute mixture of nitrous and nitric acids is produced by the reaction: $2\ NO_2 + 3\ H_2O \longrightarrow 2\ H_3O^+ + NO_2^- + NO_3^-$. This solution can safely be disposed of by pouring it down the sink drain.

8.   When the reaction in the test tube is complete, pour the solution through an ordinary filter without suction into a beaker. Cool the beaker in an ice bath.

9.   While the beaker is in the ice bath, add dropwise to the cold filtered solution (most conveniently from a buret) 25 mL ethyl alcohol, C_2H_5OH, while stirring continuously. A crystalline precipitate of sodium hexanitrocobaltate, $Na_3Co(NO_2)_6$ should form.

10.   Filter out the crystals with suction. Wash the crystals with 3 small portions of ethanol. Continue to draw air through the filter until the ethyl alcohol has evaporated. Then transfer the crystals to an evaporating dish or watch glass and dry in an oven at about 115°C for 15 min.

11.   Weigh the crystals and record their mass on the Data sheet. Calculate and record the percent yield.

12.   Dissolve about 1 g of your crystal product in 100 mL of distilled water. Transfer 10 mL of this solution to a clean spectrometer cell and measure its absorbance spectrum as in steps 5-6 of Part A.
     It is not necessary to run another water blank.

C. Models of Transition Metal Complexes
     Models of the following complexes either will be available in the laboratory or are to be constructed by you from atom models in which the sphere representing the central metal cation has holes arranged in an octahedral configuration. Make all sketches on the Data sheet.

1. $CoCl_6^{3-}$: Sketch and describe the shape of this complex ion.

2. $Co(NH_3)_4Cl_2^+$: Use a single sphere to represent NH_3. How many isomers are possible? Make a sketch of each isomer.

3. $Pt(NH_3)Cl_2$: This is a square planar complex. Show that two isomers are possible and sketch their structures.

4. $Pt(NH_3)ClBrI$: This is a hypothetical square planar complex. Show that three isomers are possible and sketch their structures.

Name ____　　Date ____________

EXPERIMENT 25
PRELABORATORY EXERCISE

1.　　**a.** Balance the half reaction for the oxidation of $Co(H_2O)_6^{2+}$ to $Co(NO_2)_6^{3-}$ and insert the correct number of electrons (e⁻) on the proper side:

$$Co(H_2O)_6^{2+} + \quad NO_2^- \longrightarrow \quad Co(NO_2)_6^{3-} + \quad H_2O$$

　　b. Balance the half reaction for the reduction of NO_2^- to NO and insert the correct number of electrons (e⁻) on the proper side:

$$NO_2^- + \quad H_3O^+ \longrightarrow \quad NO + \quad H_2O$$

　　c. Write the balanced equation for the overall redox reaction, both in ionic and in molecular form.

2.　　**a.** Use the molecular equation from **1c** to calculate the mass of $NaNO_2$ needed to react with 5 g of $Co(H_2O)_6(NO_3)_2$.

　　b. Calculate the mass of $Na_3Co(NO_2)_6$ that can be produced from 5 g of $Co(H_2O)_6(NO_3)_2$.

Name _______________________________________　　Date _____________

EXPERIMENT 25

DATA

(Observe significant figures in all calculations.)

A. Preparation of $Ni(NH_3)_6Cl_2$

1. Mass of $Ni(H_2O)_6Cl_2$: ______________

2. Equation for the reaction:

3. Theoretical yield of $Ni(NH_3)_6Cl_2$: ______________

4. Mass of $Ni(NH_3)_6Cl_2$ obtained: ______________

5. Percent yield: ______________

Spectral Data {Plot $(A_1 - A_2)$ vs. λ for the two compounds.}

$Ni(H_2O)_6Cl_2$				$Ni(NH_3)_6Cl_2$			
λ (nm)	A_1 (sample)	A_2 (blank)	$(A_1 - A_2)$	λ (nm)	A_1 (sample)	A_2 (blank)	$(A_1 - A_2)$

DATA
(Observe significant figures in all calculations.)

B. Preparation of $Na_3Co(NO_2)_6$

1. Mass of $Co(H_2O)_6(NO_3)_2$: ________________

2. Equation for the reaction (refer to the Prelaboratory Exercise):

3. Theoretical yield of $Co(H_2O)_6(NO_3)_2$: ________________

4. Mass of $Co(H_2O)_6(NO_3)_2$ obtained: ________________

5. Percent yield: ________________

Spectral Data {Plot $(A_1 - A_2)$ vs. λ for the two compounds.}

$Co(H_2O)_6(NO_3)_2$				$Na_3Co(NO_2)_6$			
λ (nm)	A_1 (sample)	A_2 (blank)	$(A_1 - A_2)$	λ (nm)	A_1 (sample)	A_2 (blank)	$(A_1 - A_2)$

Name ____   Date ____________

EXPERIMENT 25
DATA

C. Models of Transition Metal Complexes

1. $CoCl_6^{3-}$: Sketch and describe the shape of this complex ion.

2. $Co(NH_3)_4Cl_2^+$: Number of possible isomers: ____________
Make a sketch of each isomer.

3. $Pt(NH_3)Cl_2$: Sketch the two possible isomers:

4. $Pt(NH_3)ClBrI$: Sketch the three possible isomers.

Name _______________________________________   Date _____________

EXPERIMENT 25
(Additional graph paper is at the back of this book.)

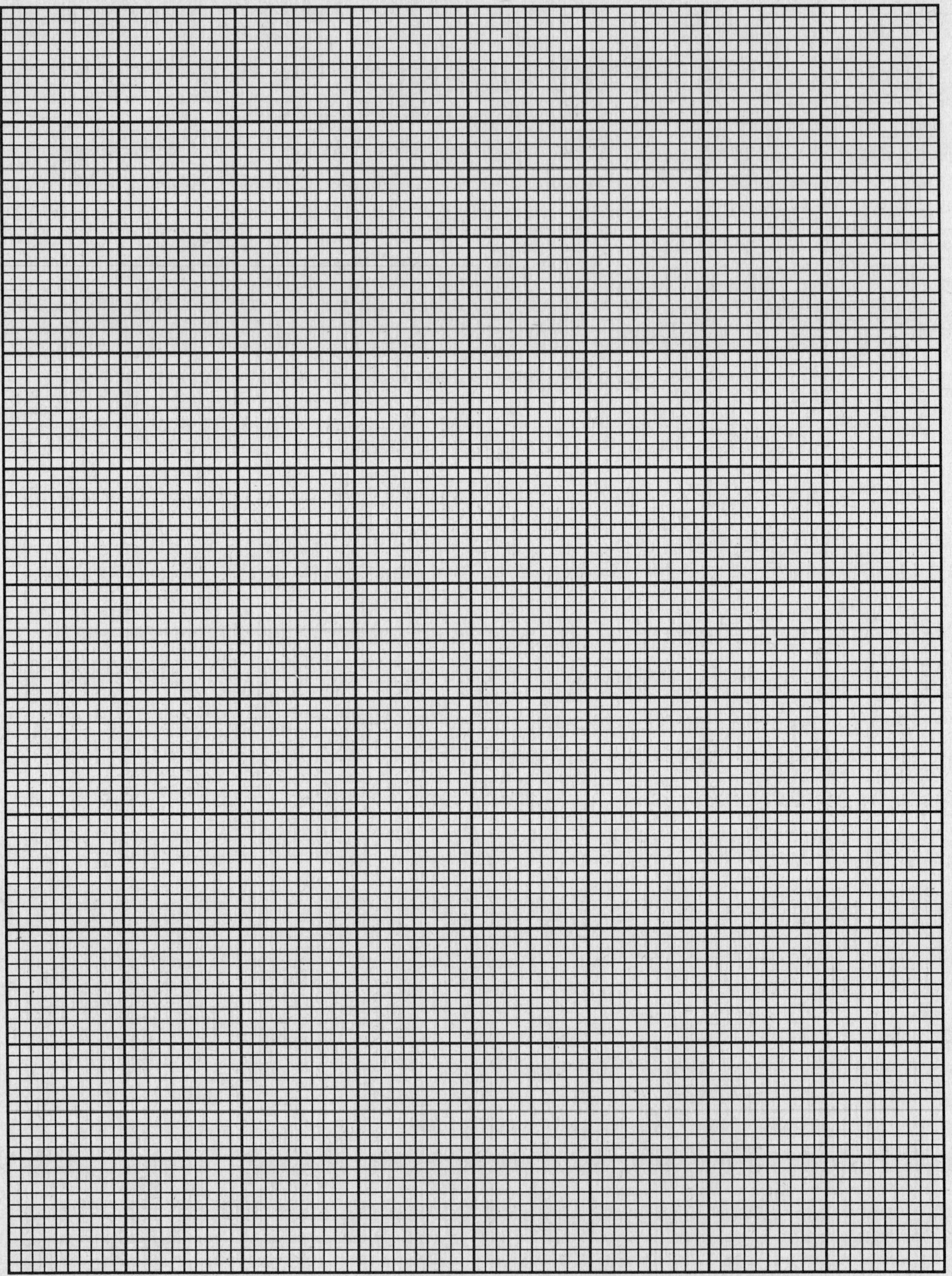

Name ____   Date _____________

EXPERIMENT 25
(Additional graph paper is at the back of this book.)

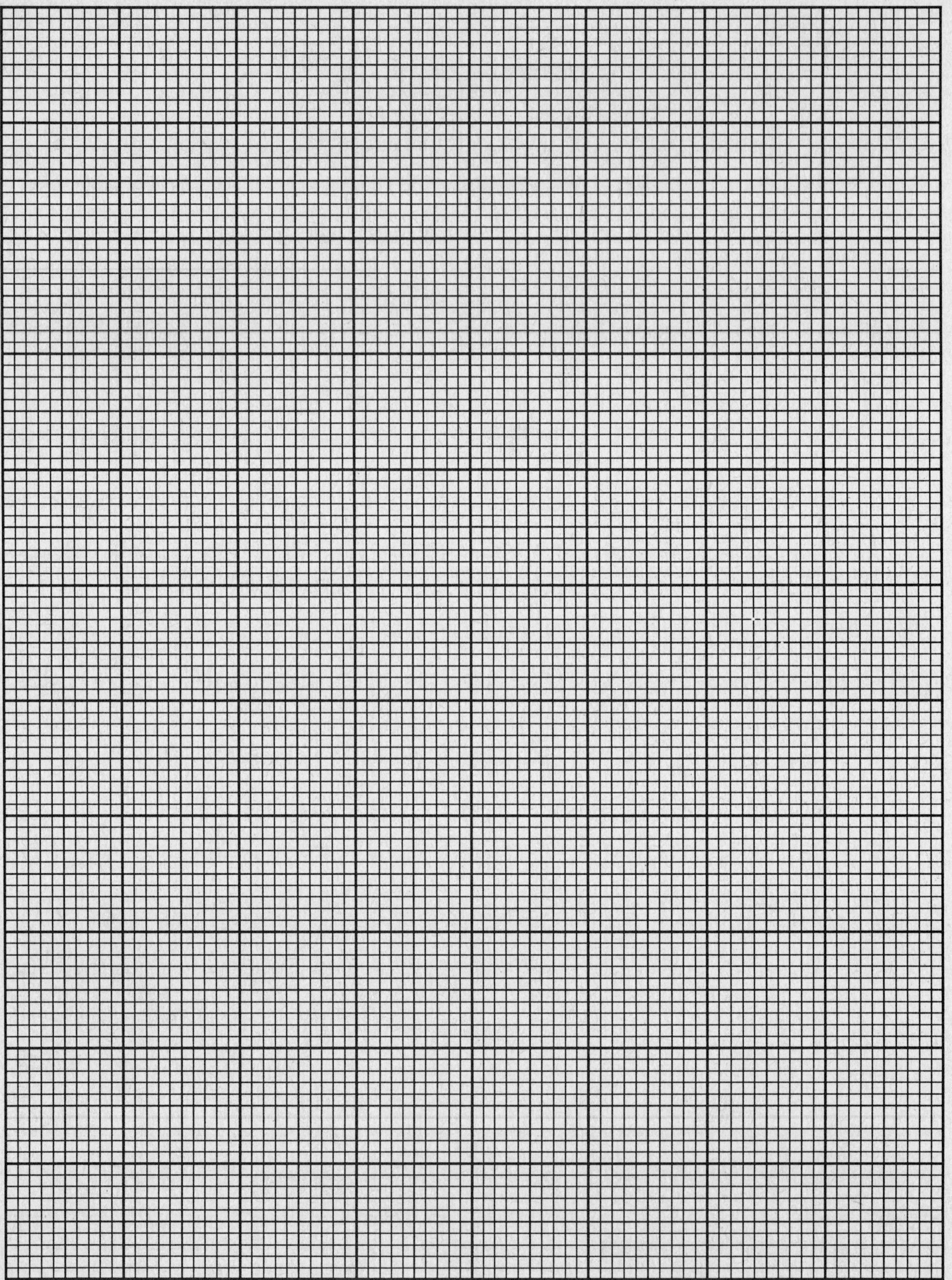

EXPERIMENT 26
Molecular Geometry: Structural Organic Chemistry and Isomerism

PRELABORATORY PREPARATION
Do the Prelaboratory Exercise and turn it in at the beginning of your laboratory period.

INTRODUCTION
Organic chemistry sometimes is called the chemistry of carbon compounds. However, there are many compounds containing one carbon atom, such as $CaCO_3$, CO_2, and $AgCN$, that are regarded as inorganic compounds. It might be better to define organic chemistry as the chemistry of compounds containing carbon-carbon bonds. This definition has the advantage of properly focussing attention on the bonding properties of carbon. Carbon-carbon bonds are strong, even when the carbon atoms are bonded at the same time to other elements. This makes possible long chains of carbon atoms in a single molecule, where many of the carbons also can have bonds to other elements.

Carbon is distinguished from all the other elements by its unique ability to form an almost unlimited number of bonds with itself and other elements. This property allows carbon to produce extended and complex molecules in linear, branched, ring, and three-dimensional structures.

It is because of carbon's unique bonding capabilities that such a wide variety of organic compounds is possible. Over seven million different organic compounds are known now and tens of thousands are added to the list every year. Carbon is only one of the 89 naturally occuring elements, but the number of known carbon compounds is far greater than the number of known compounds that contain no carbon.

In many cases, organic molecules that are different in their properties have exactly the same kinds and numbers of atoms (i.e., the same chemical formula). They differ only in the structural and geometric arrangement of their atoms.

Chemical and physical properties of molecules depend on their structure as well as their formula. Molecules with the same formula but different structures are different compounds, with different chemical behavior.

Molecules with the same formula but different structures are called **isomers.** The more atoms in a molecule, the more different ways the atoms can be arranged and, hence, the more isomers are possible. Because carbon can combine with itself almost indefinitely, the number of possible organic isomers is practically limitless. Of course, any stable arrangement of atoms must correspond to the allowable number of bonds and bond angles of the different atoms. Each different kind of atom has its own bonding limitations, determined by its electronic structure.

Carbon can have:
4 bonds in tetrahedral symmetry with bond angles of about 109°, or
3 bonds in a planar configuration with bond angles of about 120°, or
2 bonds in a linear configuration with bond angles of about 180°.

Even with these limitations, the number of possible isomers increases astronomically as the number of carbon atoms in a molecule increases. Table 26.1 shows the number of isomers theoretically possible for alkane hydrocarbons of increasing carbon content.

Alkanes contain only 2 different kinds of atoms, carbon and hydrogen, and all of the atoms are joined by single bonds. All the carbons have 4 bonds in tetrahedral symmetry.

Table 26.1. Number of isomers theoretically possible for alkanes

Number of carbon atoms	Molecular formula	Theoretical number of possible isomers
1	CH_4	1
2	C_2H_6	1
3	C_3H_8	1
4	C_4H_{10}	2
5	C_5H_{12}	3
6	C_6H_{14}	5
7	C_7H_{16}	9
8	C_8H_{18}	18
9	C_9H_{20}	35
10	$C_{10}H_{22}$	75
11	$C_{15}H_{32}$	4,347
20	$C_{20}H_{42}$	366,319
30	$C_{30}H_{62}$	4,111,846,763

Given these limitations, it is clear that the total number of alkanes must be a very small fraction of the vast number of all possible organic molecules. Even so, as the number of carbon atoms increases, the number of theoretically possible alkane isomers very rapidly grows beyond any possibility of actually trying to make or find them.

The molecules in Table 26.1 contain only 2 kinds of atoms, **C** and **H**. As more different kinds of atoms are added, the number of isomers and the range of chemical and physical properties increase correspondingly. As an example of how different isomers can have differences in physical properties, Table 26.2 shows some of the physical properties for three different isomeric compounds, with the same formula, C_3H_8O.

Table 26.2. Isomers of C_3H_8O and some of their physical properties

Isomer	Molar mass	Melting temperature (°C)	Boiling temperature (°C)	Refractive index	Flash point [a] (°C)	Solubility in water
n-Propyl alcohol	60.09	-126	97.1	1.385	15	Infinite
Isopropyl alcohol	60.09	-89.5	82.3	1.379	22	Infinite
Ethyl methyl ether	60.09	116	10.8			Slight

[a]Temperature where vapor is concentrated enough for ignition in air to occur.

It is very important to distinguish among the different possible structural forms of the same set of atoms. For this purpose, the different kinds of isomerism have been classified and a system of nomenclature for the different compounds has been developed.

There are two main kinds of isomerism—structural and geometrical.
Structural isomers have their atoms connected differently, as in the case of the two isomers of butane, shown in Figure 26.1. In structural isomers, there are different bond arrangements. Isobutane, on the right, has a **tertiary** carbon atom, one that is connected to three other carbons. Normal butane, on the left, has no tertiary carbons.

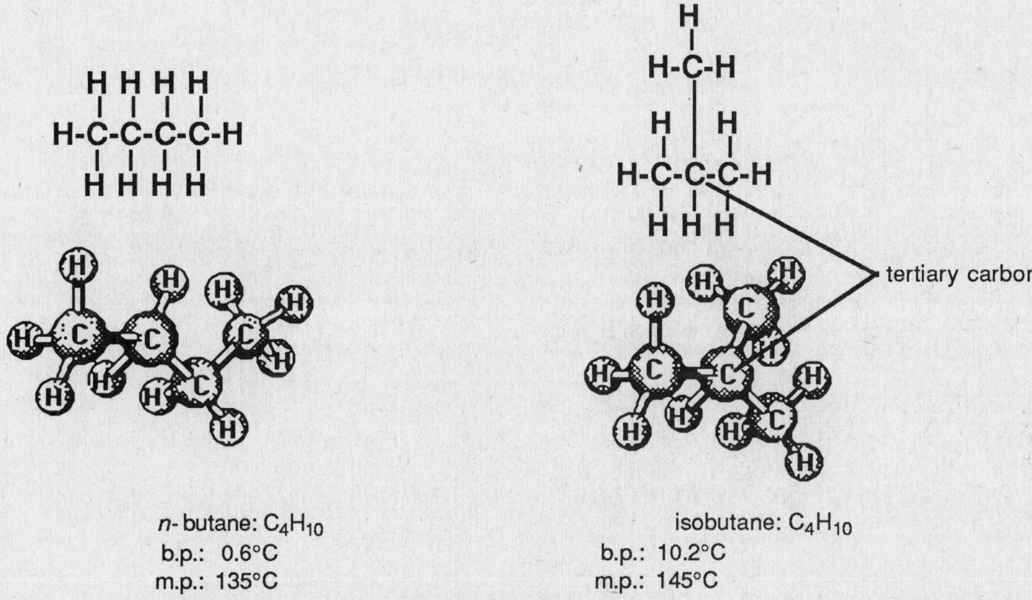

n-butane: C_4H_{10}
b.p.: 0.6°C
m.p.: 135°C

isobutane: C_4H_{10}
b.p.: 10.2°C
m.p.: 145°C

Figure 26.1. Structural isomers of butane.

Geometrical isomers have the same numbers and kinds of bonds, but the atoms have a spatial difference in their arrangement. Figure 26.2 shows the *cis* and *trans* forms of 1,2-dichloroethene. The *trans* form has its chlorine atoms on opposite sides of the double bond, while in the *cis* form, they both are on the same side.

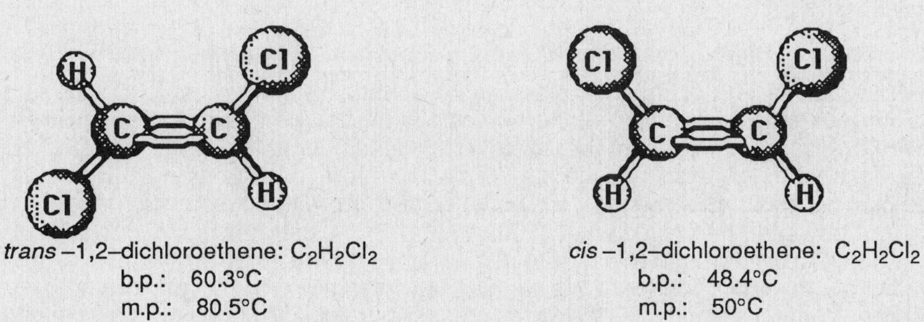

trans –1,2–dichloroethene: $C_2H_2Cl_2$
b.p.:   60.3°C
m.p.:   80.5°C

cis –1,2–dichloroethene: $C_2H_2Cl_2$
b.p.:   48.4°C
m.p.:   50°C

Figure 26.2. Geometric isomers of 1,2–dichloroethene

The double bond between the carbons prevents rotation of the left and right-hand groups relative to one another. For this, reason, the *trans* structure cannot rotate one group and become the *cis* structure.
It is because there is no rotation around the C=C bond that the *cis* and *trans* forms are truly different and behave like different compounds.
If the molecule *could* rotate around the bond between the carbons, one form would be readily converted into the other and there would be only one compound with one set of properties. Note that there is another isomer of $C_2H_2Cl_2$ that is *structurally* different (see Figure 26.3).

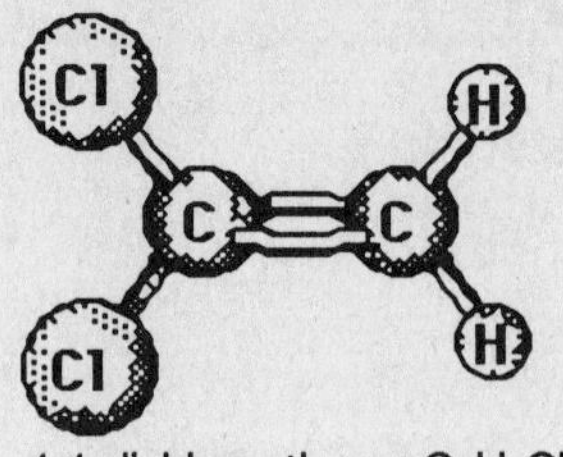

1,1-dichloroethene: $C_2H_2Cl_2$

Figure 26.3.   Structural isomer of dichloroethene

There are subclasses of structural and geometric isomers. All 3 of the isomers in Table 26.2 are structural isomers. Two of them, however, are the same kind of compound, **alcohols**, because they both contain the **-OH** functional group. Isopropyl and *n* -propyl alcohol are called **positional isomers,** because they differ only in the position of the functional group. Ethylmethyl ether has no **-OH** group and is not an alcohol. It differs from the alcohols by having the ether functional group, **-C-O-C-** .

Compounds with the same formula but different functional groups, such as the corresponding alcohols and ethers, are called *functional isomers.*

The most subtle and difficult to recognize type of isomerism is a form of geometrical isomerism where the isomers are non-superimposable mirror images of one another. In most cases, mirror image structures are superimposable and do not represent different isomers. For example, the two views of chloromethane in Figure 26.4 are mirror images, but, because one form can be easily rotated to become identical (superimposable) with the other (as you will demonstrate in this experiment), they are the same molecule and are not isomers.

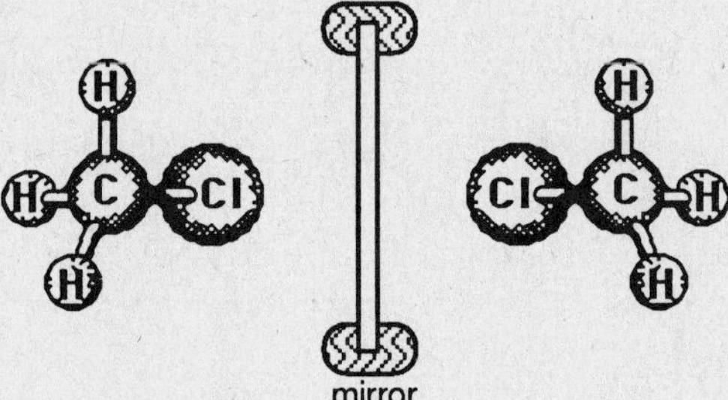

Figure 26.4.   Superimposable mirror–images of chloromethane

This example demonstrates the usefulness of molecular models, such as the "ball and stick" models in the illustrations. Structural differences and similarities among organic molecules often are difficult to perceive from 2-dimensional representations. The tetrahedral configuration of carbon bonds requires a 3-dimensional visualization that is best achieved with 3-dimensional structural models. Using such models makes it easy to see that there only is one form of chloromethane. With models, it also is easy to see that dichloro– and trichloro–methane likewise exist in only one form, with no isomers.

Consider, however, the molecule CHClBrl. Figure 26.5 illustrates this molecule and its mirror image. If you were to construct models of the forms on the right and left sides of the mirror, you would find that there is no way to rotate one form into the identical configuration of the other. The two forms are **non-superimposable.** Non-superimposable mirror images are a special kind of geometrical isomerism and such isomers are called **enantiomers.**[1] For enantiomers to exist, the molecule must be completely asymmetric, having no planes or center of symmetry.

Enantiomers are more closely alike than other kinds of isomers. In fact they can appear nearly identical. They will have the same color, melting and boiling temperatures, solubility, etc. However, they often differ in how they react with other asymmetric molecules. Biological molecules often are asymmetric, and the enantiometric distinction can be very important in biological reactions. For example, different enantiomers can have very different tastes, odors, and toxicities.

---

[1]A pair of gloves, or your two hands, can serve as examples of enantiomers. The right glove, or right hand, is a non-superimposable mirror image of the left.

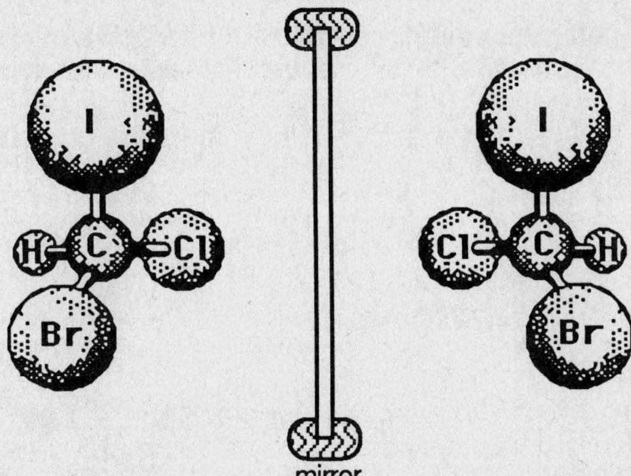

Figure 26.5.   Non–superimposable mirror–images (enantiomers) of CHClBrI.

A requirement for enantiomeric isomerism is that the molecule contain at least one tetrahedral carbon atom with all 4 bonds attached to groups that are different from each other.
It should be clear that the molecule CHClBrI meets this requirement. What about $CH_3CHClOH$? If we draw the structural diagram of the molecule,

$$\begin{array}{ccc} H & H & \\ | & | & \\ H-C-C & -OH \\ | & | & \\ H & Cl & \end{array}$$

we see that the carbon on the right has four bonds, each with a different group attached: H, OH, Cl, and CH_3. Therefore, $CH_3CHClOH$ will have enantiomeric isomers. The two enantiomers are illustrated in Figure 26.6. If you build their models, you will find that they are non–superimposable mirror images.

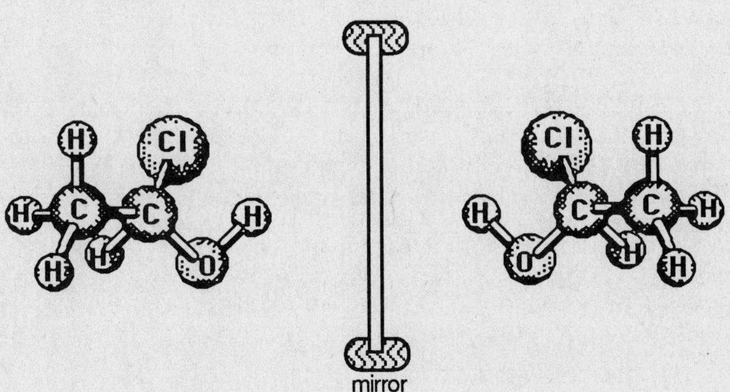

Figure 26.6.   Enantiomers of $CH_3CHClOH$

PLAN OF EXPERIMENT
Using molecular model kits, you will construct 3–dimensional models of organic molecules, to investigate molecular geometry and isomerism. You are asked to sketch **projection** and **perspective** formulas on the data sheet.

To draw *projection* formulas, simply indicate which atoms are connected to one another with solid lines to represent the bonds.

Example:

To draw *perspective* formulas as in CH_4 below:

Use a solid line (———) for bonds lying in the plane of the paper.

Use a hatched wedge-shaped line (ⅠⅠⅠ···) for bonds that project down into the paper away from the reader.

Use a solid wedge-shaped line (◀━━) for bonds that project up out of the paper toward the reader.

Example:

PROCEDURE
Answer all questions and draw all figures on the Data sheet.

A. Structural Organic Chemistry
Use a molecular model kit to perform the following operations:
1.   **a.** Construct a model of methane, CH_4.
      b. Draw a perspective formula for methane.
      c. Using a protractor, measure the H-C-H bond angles of your methane model. Record the value on the data sheet.
      d. Which hybrid orbitals does the carbon atom in methane use?

2.   **a.** From 2 models of CH_4, construct a model of ethane, CH_3CH_3.
      b. Draw a perspective formula for ethane.

3.   **a.** Replace 1 H atom in the ethane model with a Br atom, to form ethyl bromide (bromoethane), CH_3CH_2Br.
      b. How many isomers of CH_3CH_2Br can you construct?
      c. Does it matter which of the 6 H atoms is replaced by the Br atom?
      d. Are the 6 H atoms in ethane all equivalent?

4.   **a.** Replace an H atom in CH_3CH_2Br by another Br atom, to form a model of a compound with the formula $C_2H_4Br_2$.
      b. Are all of the 5 H atoms in CH_3CH_2Br equivalent? In other words, does it matter which H atom is replaced by the second Br atom to make $C_2H_4Br_2$? Explain.
      c. What type of isomerism is represented by the $C_2H_4Br_2$ models?
      d. Draw projection formulas for all possible structures of $C_2H_4Br_2$.

5.   **a.** Construct a model of ethylene (ethene), C_2H_4, from a model of ethane.
      b. Draw a projection formula for C_2H_4.
      c. Measure the H-C-C and H-C-H bond angles and record their values.
      d. Which hybrid orbitals do the carbon atoms in C_2H_4 use?

6. **a.** Construct a model of acetylene (ethyne), C_2H_2, from a model of ethylene.
   b. Draw a projection formula for C_2H_2.
   c. Measure the H-C-C bond angles and record the value.
   d. Which hybrid orbitals do the carbon atoms in C_2H_2 use?

7. **a.** From 3 models of acetylene, construct a model of the ring molecule benzene, C_6H_6. (Benzene actually can be synthesized by the trimerization of acetylene.)
   b. Draw a projection formula for benzene.
   c. Measure the H-C-C and C-C-C bond angles and record their values.
   d. Which hybrid orbitals do the carbon atoms in benzene use?

8. **a.** Replace 2 adjacent atoms of benzene with Cl atoms, to form *ortho*-dichloro-benzene, $o\text{-}C_6H_4Cl_2$.
   b. Draw all the different projection formulas for $o\text{-}C_6H_4Cl_2$.
   c. How many different isomers exist for $o\text{-}C_6H_4Cl_2$?

B. Molecular Geometry
1. **a.** To a model of benzene, add 3 model molecules of H_2 to construct a model of cyclohexane, C_6H_{12}.

  You should be able to form 2 different structures of C_6H_{12}, the so–called "boat" and "chair" forms.

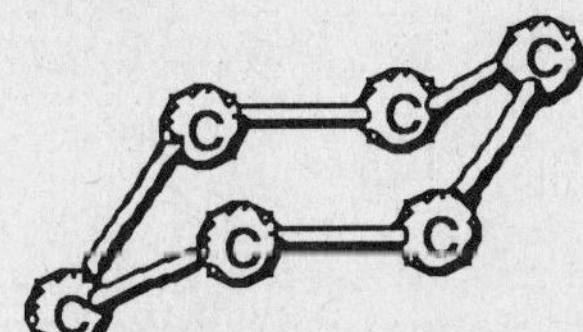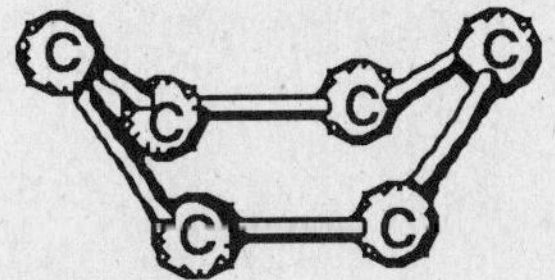

"chair" form of cyclohexane   "boat" form of cyclohexane
(Hydrogens are omitted in these figures for clarity.)

   b. Draw perspective formulas of the 2 forms of C_6H_{12} and label them.
   c. Can you convert your model from "boat" to "chair" and back, without breaking any bonds?
   d. Do these 2 forms, called **conformations**, represent isomers? Explain.

2. **a.** Make models of propane (C_3H_8), butane (C_4H_{10}), and pentane (C_5H_{12}).
   b. Draw projection formulas for all the possible isomers.

3. **a.** To a model of acetylene, add a model molecule of Br_2 to form 1,2-dibromoethene, $BrCH{=}CHBr$.
   b. Show that 2 isomers are possible for $BrCH{=}CHBr$.
   c. Draw projection formulas for the isomers of $BrCH{=}CHBr$ and label them with their correct names.
   d. What type of isomerism is represented by the $BrCH{=}CHBr$ models?

4. **a.** To a model of ethylene, add a model molecule of Br_2 to form 1,2-dibromoethane, CH_2BrCH_2Br.
   b. Draw 2 perspective formulas of CH_2BrCH_2Br, one with both Br atoms on the same side of the C-bond axis, **eclipsed form**, and one with the Br atoms on opposite sides of the C-C bond axis, **staggered form**.
C
   c. Do these 2 forms (conformations) represent different compounds? Explain.
   d. Measure the distance between the Br atoms in each model and calculate the ratio of the longer to the shorter.
   e. On the basis of this measurement, predict which conformation represents the greater population of CH_2BrCH_2Br molecules. Explain your choice.
   f. Which conformation, if any, has a dipole moment?

5. **a.** Construct a model of CHClBrF.
   b. Placing the model in front of a mirror if necessary, construct a model of the mirror image.
   c. Can the 2 models be superimposed?
   d. Draw perspective formulas of the 2 models.
   e. What type of isomerization is represented by these models?

Name ____   Date ____________

EXPERIMENT 26
PRELABORATORY EXERCISE

1. Fill in the blanks.

   a. Carbon can have 4 bonds in _________________ symmetry with bond angles of
   _________________ degrees.

   b. Carbon can have 3 bonds in _________________ configuration with bond angles of
   _________________ degrees.

   c. Carbon can have 2 bonds in _________________ configuration with bond angles of
   _________________ degrees.

2. Explain why there is such a large number and wide variety of organic compounds.

3. Define the following terms.
   isomer (the general meaning):

   structural isomer:

   geometric isomer :

   functional isomer:

   positional isomer :

   enantiometric isomer:

5. Why is a C=C double bond essential for the existence of cis and trans isomers of
1,2-dichloroethene?

Name ____   Date _____________

EXPERIMENT 26
DATA

A. Structural Organic Chemistry

1.   **b.** Draw a perspective formula for methane, CH_4.

c. Methane: H-C-H bond angle = _____________________

d. Methane: hybridization type: _____________________

2.   **b.** Draw a perspective formula for ethane, CH_3CH_3.

3.   **b.** Number of isomers of CH_3CH_2Br: _____________________

c. _____________________

d. _____________________

4.   **b.**

c. Type of isomerism: _____________________

d. Draw projection formulas for all possible structures of $C_2H_4Br_2$.

5.   **b.** Draw a projection formula for C_2H_4.

c. Ethylene: H-C-H bond angle = _____________________

d. Ethylene: H-C-C bond angle = _____________________

e. Ethylene: hybridization type: _____________________

DATA

6.   **b.** Draw a projection formula for C_2H_2.

c. Acetylene: H-C-C bond angle = _____________________

d. Acetylene: hybridization type: _____________________

7.   **b.** Draw a projection formula for benzene, C_6H_6.

c. Benzene: H-C-C bond angle = _____________________

d. Benzene: C-C-C bond angle = _____________________

e. Benzene: hybridization type: _____________________

8.   **b.** Draw all the projection formulas for $o\text{-}C_6H_4Cl_2$.

c. Number of different isomers for $o\text{-}C_6H_4Cl_2$. _____________________

B. Molecular Geometry
1.   **b.** Draw perspective formulas for 2 forms of cyclohexane, C_6H_{12}. Indicate bond angle values.

c. _____________

d. _____________

2.   **b.** Draw projection formulas for all the possible isomers of C_3H_8, C_4H_{10}, and C_5H_{12}.

Name ____   Date ____________

EXPERIMENT 26
DATA

3.   **c.** Draw projection formulas for the isomers of BrCH=CHBr.

4.   **b.** Draw perspective formulas for the eclipsed and staggered forms of CH_2BrCH_2Br.

c. _____________

d. Distance between Br atoms: _______________;   ratio: _______________

e.

f.

5.   **c.** _____________

d. Draw perspective formulas of the 2 models.

e. Type of isomerization: _______________

EXPERIMENT 27
Synthesis of Aspirin

PRELABORATORY PREPARATION
1. Do the Prelaboratory Exercise and turn it in at the beginning of your laboratory period.
2. Review suction filtering technique, Section IX, Methods, page 23.

INTRODUCTION
This experiment can be done in one laboratory period if Part D, the preparation of copper aspirinate, is excluded. If Part D is included, do it at the beginning of the next laboratory period.

Without question, aspirin is one of the most generally useful medicinal agents ever discovered. Annual consumption in the United States alone is several million kilograms. Since 1899, when the German scientist Dreser introduced it as a mild analgesic and antipyretic, aspirin has become the layman's first line of defense against minor discomforts such as colds and headaches. It also is used extensively in the treatment of arthritis and rheumatism and for reducing fever accompanying infections. The manufacture of aspirin utilizes a reaction between salicylic acid and acetic anhydride.

Salicylic acid can be synthesized from phenol, which in turn can be prepared from benzene:

$$C_6H_6 \longrightarrow C_6H_5OH \longrightarrow C_6H_4(OH)(COOH) \qquad (27.1)$$

$$\text{benzene} \qquad\quad \text{phenol} \qquad\qquad \text{salicylic acid}$$

The conversion of salicylic acid to aspirin (acetylsalicylic acid) occurs as follows:

$$\text{(27.2)}$$

salicylic acid   acetic anhydride   acetylsalicylic acid (aspirin)   acetic acid

Aspirin is an effective **chelating agent**, and the asterisks mark oxygen atoms that can form chelate bonds.

A chelating agent is a compound with a molecular structure that allows it to use two or more different atoms in its molecule to form bonds to a single metal ion.

The term **chelate** comes from the Greek work *chele,* which means *a claw.* You can see from Figure 27.1 how the aspirin molecule resembles a claw gripping a copper ion. The aspirin molecule can assume the clawlike position because there is free rotation about many of its single bonds.

Chelate bonding occurs when non-bonding electron pairs, of which the chelating oxygen atoms each have two sets, can enter into empty or partially empty orbitals on the metal ion. It can be a very strong bond, and chelating agents such as EDTA (ethylenediamine tetraacetic acid) are sometimes used in cases of metal poisoning to "complex" with the metals, removing them from the blood and body tissues and allowing them to be excreted from the body as relatively inert substances.

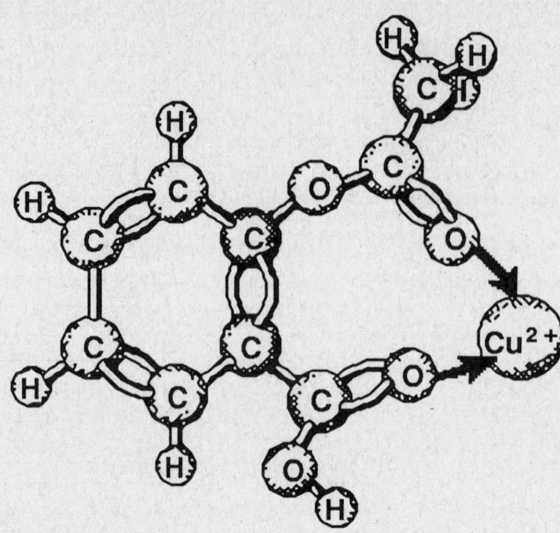

Figure 27.1.   Aspirin chelating a Cu^{2+} ion, as visualized with a ball-and-stick model. The arrows indicate that electron pairs on the oxygen atoms are entering into an interaction with orbitals on the Cu^{2+} ion.

It is thought that the pain-killing properties of aspirin are due, at least partly, to its chelating ability. With many body body stresses that cause pain, such as arthritis, infections and muscular soreness, the concentration of Cu^{2+} ions in the blood rises to several times higher than normal. These copper ions normally are in the tissue cells, and the depletion of Cu^{2+} in the cells can evoke a pain response. Aspirin may chelate the copper ions in the blood and carry them back to the cells, where enzymes can metabolize the aspirin and release the copper back into the cell structure, relieving the pain.

PLAN OF EXPERIMENT
You first synthesize aspirin using Reaction 27.2. Recrystallizing your product will improve its purity. You estimate the purity of your recrystallized product by comparing its melting point with the handbook value. Finally, you make copper aspirinate, a chelate compound of aspirin and copper.

SAFETY

1.   Wear approved eye protection.

2.   Ethanol, which is flammable, is used in the recrystallization and copper aspirinate synthesis procedures.

Be certain that there are no open flames in the laboratory during these operations.

3.   Acetic anhydride and sulfuric acid are very corrosive materials and can cause severe skin burns. Handle

them with care. If they should contact your skin, wash the affected area immediately with soap and water.

PROCEDURE
A. First Crystallization of Crude Product
1.   Weigh to ±0.1 g on tared weighing paper, about 2 g of salicylic acid and transfer the sample to a 250 mL Erlenmeyer flask.

2.   Carefully add 5 mL of acetic anhydride and 10 drops of concentrated sulfuric acid (to catalyze the reaction).
   THE FIRST FEW DROPS OF ACID CAN CAUSE A SUDDEN ERUPTION FROM THE FLASK.
   Add the reagents very slowly and carefully to avoid a rapid reaction that might cause spattering.
Gently swirl the flask more or less continuously for 10 min.

3.   Set the flask aside for another 10-15 min to allow complete reaction.

4.   During this waiting time, set up a Büchner funnel suction filtering apparatus, with an aspirator trap between the filter and the sink aspirator. (See Section IX, Methods, page 22, for the set-up and filtering procedure.)

5.   The reaction in the flask generates heat. After it has stood for 10-15 min, cool it under running water. Then add just 50 mL cold distilled water to the flask and swirl to mix. Record the exact amount of water used on the Data sheet. A precipitate should appear. If necessary, cool again to get a good precipitate.
To calculate your percent yield, it is necessary to account for product that dissolves in the water you add to the solution and in the washing water. Keep an accurate record of the amount of water used.

6.   Filter the product and wash twice with 10 mL portions of cold distilled water.
Be sure the aspirator faucet is turned on full force during all filtering and washing operations, to prevent water backing up into your apparatus. To turn the suction off and on, open and close the pinch clamp on the aspirator trap.

When adding sample or wash water to the filter, turn the suction *off* by opening the pinch clamp. When all the liquid has been added, turn the suction *on* by closing the pinch clamp to draw the liquid through the filter.
Record the amount of wash water used.

7.   Continue to draw air through the sample to remove excess water and speed drying

8.   Place the damp filter paper containing the precipitate on top of a dry piece of filter paper. Place the double layer of filter paper and precipitate on a watch glass to dry.

B. Recrystallization
Recrystallization is a good method for purifying crystalline compounds. In the first crystallization, some impurities inevitably become trapped in the solid material, both within the crystal lattice and at the crystal boundaries between crystals. Each time a compound is dissolved and recrystallized, more of these impurities are removed. One or two recrystallization steps, carefully done, can yield a much purer product than the first crystallized sample. You will make one recrystallization and then qualitatively assess the purity of your product by comparing its melting point with the handbook value for pure aspirin crystals.
Be certain there are no open flames in the laboratory.

1.   Prepare a water bath, using a 600 mL beaker about 1/3 full of water on a hotplate.

2.   Set aside about 1-2 mg of your first product. You will not recrystallize this portion but will compare its melting point with that of your recrystallized product.

3.   Dissolve the remaining aspirin crystals in 20 mL of ethyl alcohol in a 100 mL beaker and warm the solution to about 50°C in the water bath.

4.   While the solution is warming, add to it 50 mL of distilled water pre-warmed to about 50°C. Hold the solution at 50°C until all the crystals are dissolved. Record the amount of water used.

5.   When all the crystals are dissolved, hold the solution at 50°C another 2 or 3 min. Then remove the solution from the bath, cover the beaker with a watch glass and allow it to air cool slowly for about 15 min. You should be able to watch the crystals grow. If you have a good crop of crystals, proceed to step 6.
If no product has formed, try the following, in the order given, to promote crystal growth :
a.   Using a stirring rod to scratch the inside of the beaker.
b.   Adding a tiny amount of your product from Part A to the beaker.
c.   Cooling the flask in ice.
Once crystallization begins, leave the beaker undisturbed for about 5 min. Then proceed to step 6.

6.   Collect the crystals by suction filtration, washing them twice with 10 mL portions of cold distilled water as in step 6, Part A. Record the amount of wash water used.

7.   Continue to draw air through the sample to remove excess water and speed drying

8.   Place the damp filter paper containing the precipitate on top of a dry piece of filter paper. Place the double layer of filter paper and precipitate on a watch glass to dry.

9.   Dry your product in an electric oven at 75°C for about 30 min. Then transfer the dry sample to a tared weighing bottle. If preferred, the product may be air dried on the filter assembly by suction in about 20 min.
Also dry and weigh the small portion of your first crude product that you set aside earlier. Record the masses on the Data sheet.

10.   On the Data sheet, compare the colors and crystal forms of the products obtained in Parts A and B.

11.   Calculate the maximum theoretical yield of aspirin from Equation 27.2 and your starting mass of salicylic acid.

12.   Some of your product is lost by dissolving in the water added to the flask and used to wash the precipitate. Calculate the maximum mass of aspirin that could have dissolved in the amount of water used to dilute and to wash the sample.
    Take the solubility of aspirin to be 0.25 g per 100 mL of water.

13.   Calculate your actual percent yield of aspirin, corrected for the amount dissolved in water used in the synthesis.

C. Measuring the Melting Point
    The melting point of a pure crystalline material is a characteristic property that is affected very little by changing environmental conditions, such as atmospheric pressure. Impurities cause defects in the crystal lattice which affect the melting point. The presence of impurities almost always lowers the melting point by some amount that depends on the quantity and nature of the impurity. The melting point of a pure crystalline substance usually is quite sharp; the substance melts completely over a narrow temperature range. Impurities often broaden the temperature range over which melting occurs. A measured melting point that occurs over a narrow temperature range and agrees closely with the handbook value is a good indicator of pure product.
    The handbook value[1] for the melting point of aspirin is 135°C. The melt resolidifies at 118°C.
    Your instructor will demonstrate the use of the melting point apparatus. Because the melting point of aspirin is higher than the boiling point of water, you must use a liquid such as mineral oil in the heating bath.
    Measure the melting points of your first crude product and your recrystallized product. Record the temperature where melting is complete and the temperature range over which melting occurs.

If you do not do Part D, put your product into a small capped vial labelled with your name and turn it into your laboratory instructor.

D. Formation of Copper Aspirinate, a Chelate Compound
There may not be time to include Part D in one laboratory period.
1.   Weigh out 0.3 g of cupric acetate into a clean, dry, preweighed watchglass or onto a piece of pre-weighed weighing paper.

2.   Transfer the sample to a 400 mL beaker containing 150 mL of cold distilled water.

3.   Stir until the solid is completely dissolved. Record the color of the solution.

4.   Weigh out 0.5 g of your dry recrystallized aspirin (or obtain pure aspirin from your instructor) and place it in a 100 mL beaker. Add 20 mL ethyl alcohol and stir until the aspirin dissolves.

5.   Add the aspirin solution to the cupric acetate solution. Stir the combined solution thoroughly and heat on a hotplate to 50-60°C.

6.   Remove the solution from the hotplate and place it in an ice bath. Stir occasionally and watch the dark blue crystals form. This may take up to 15 min. The crystals are copper aspirinate, a chelate compound.

7.   Suction filter the crystals and discard the filtrate.

8.   Place the crystals on a watch glass to dry.

9.   How could you distinguish between the copper acetate and copper aspirinate crystals? Both crystals are blue, so how can you be sure that the aspirin did not cause the copper acetate to recrystallize from the solution? Propose a test method on the Question sheet.
    If you have time, test your method for distinguishing between the two kinds of crystals.

10.   Put your remaining product into a small capped vial labelled with your name and turn it in to your laboratory instructor.

---

[1] *CRC Handbook,* 59th ed., 1978, Chemical Rubber Company Press, West Palm Beach, Florida.

Name _______________________________________ Date ______________

EXPERIMENT 27
PRELABORATORY EXERCISE

1.   What is the theoretical yield in grams of aspirin if 2.173 g of salicylic acid reacts with 5.1 mL of acetic anhydride (density = 1.08 g/mL) according to Reaction 27.2? Molecular masses are: salicylic acid = 138.1 g/mol; acetic anhydride = 102.1 g/mol; acetylsalicylic acid = 180.2 g/mol.

Theoretical yield = ____________ g

2.   The total quantity of water used to dilute the solutions and wash the precipitates during the first and second crystalizations was 145 mL. Given the solubility of aspirin to be 0.25 g/100 mL of water, calculate the grams of aspirin that could have dissolved in this volume of water.

Grams of aspirin dissolved = ____________g

3.   The total mass of dry aspirin collected in both crystallizations was 2.314 g. What was the percent actual yield, corrected for the amount that dissolved?

Actual percent yield, corrected for amount dissolved = ____________%

Name ____   Date ______________

EXPERIMENT 27

DATA
(Observe significant figures in all calculations.)

1. Mass of salicylic acid: ______________

2. Volume of acetic anhydride: ______________

3. Theoretical yield of aspirin, uncorrected for solubility: ______________

4. Volume of water used for dilution in Part A: ______________

5. Volume of water used for washing in Part A: ______________

6. Volume of water used for dilution in Part B: ______________

7. Volume of water used for washing in Part B: ______________

8. Total volume of water used in Parts A and B: ______________

9. Mass of aspirin that can dissolve; Solubility = 0.25 g/100 mL H_2O: ______________

10. Theoretical yield of aspirin, corrected for solubility: ______________

11. Mass of dry aspirin obtained in the two crystallizations: ______________

12. Percent yield: ______________

13. Compare the colors and crystal forms of the two products obtained in Parts A and B:

Calculations:

13. Melting point of first crude product: ______________

14. Melting point of recrystallized product; Handbook value = 135°C: ______________

15. Observations about melting point temperature range:

16. Conclusions about purity of both products:

17. What test can tell the difference between copper acetate and copper aspirinate?

18. Results of applying your test to your crystals:

EXPERIMENT 28

Preparation of Ethyl Alcohol: Reactions of Alcohols

PRELABORATORY PREPARATION

1. The fermentation for this experiment must be started during the prior laboratory period.
2. Do the Prelaboratory Exercise and turn it in at the beginning of your laboratory period.
3. Review distillation technique, Section XI, Methods, page 24.
4. Review how to fit glass tubing into rubber stoppers, Section VII, Methods, page 20.

INTRODUCTION

Alcohols are made by putting the hydroxyl functional group, **–OH,** on one or more carbon atoms in a hydrocarbon structure. For example, ethanol (ethyl alcohol) is made by replacing a hydrogen atom in ethane with a hydroxyl group.

ethane — ethanol (ethyl alcohol)

Figure 28.1. Structures of ethane and ethanol.

There are two ways of naming alcohols:

1. Substitute the ending *ol* for the final *e* in the name of the hydrocarbon from which the alcohol is derived.
2. Add the word *alcohol* after the name of the alkyl radical formed by removing one hydrogen atom from the hydrocarbon.

EXAMPLE:

The simplest alkane is **methane**, CH_4. Remove one hydrogen atom to form the radical CH_3, called the **methyl** radical. Add an –OH group to form CH_3OH, an alcohol that may be called either **methanol** or **methyl alcohol**. Similarly, with ethane, drawn structurally above, removing a hydrogen atom forms the ethyl radical, CH_3CH_2. Add a hydroxyl group to form CH_3CH_2OH, called either **ethanol** or **ethyl alcohol.**

If there are three or more carbons in the hydrocarbon starting compound, it is always possible to put the hydroxyl group on different carbons that are not equivalent, making different alcohols that are **structural isomers.** Structural isomers have the same formula but different structural arrangements of the atoms (see the discussion of isomers in Experiment 26). For example, by replacing one hydrogen atom with an –OH group in propane, an alkane with three carbons, two isomeric alcohols can be made:

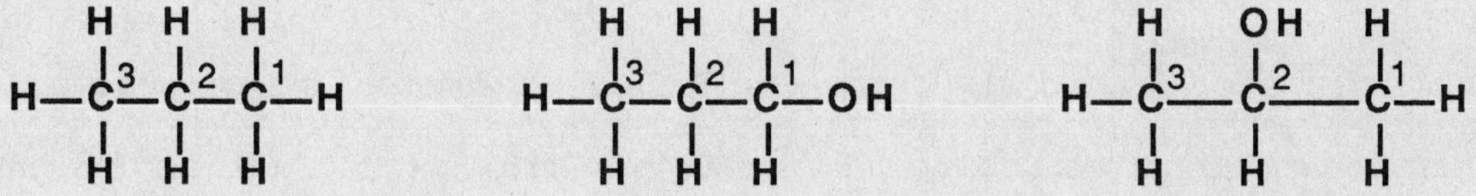

propane — 1-propanol (propyl alcohol) — 2-propanol (isopropyl alcohol)

Figure 28.2. Structures of propane and isomers of propanol.

The number beside each carbon is used for showing where the hydroxyl groups are located. In complicated structures, there is a standard system for numbering the carbons.

More than one hydroxyl group can be added, to make a wide variety of **polyalcohols,** as shown below for propanediol and propanetriol:

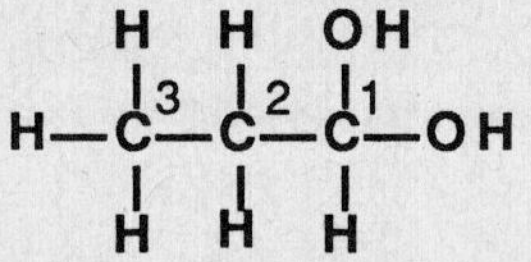

1,1-propanediol (a)   1,2-propanediol (b)   1,3-propanediol (c)   1,2,3-propanetriol (d)

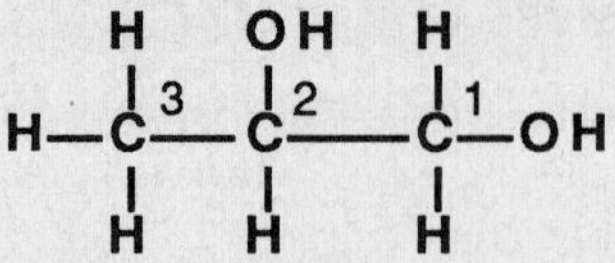

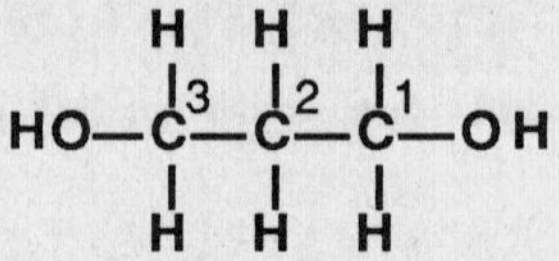

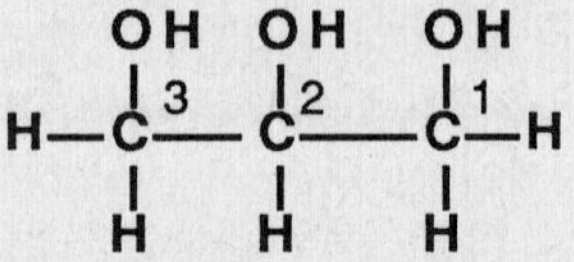

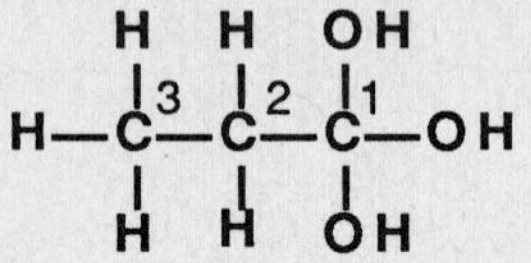

1,1,1-propanetriol (e)   1,1,2-propanetriol (f)   1,1,3-propanetriol (g)   1,2,2-propanetriol (h)

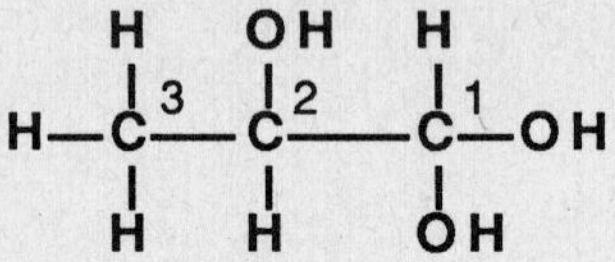

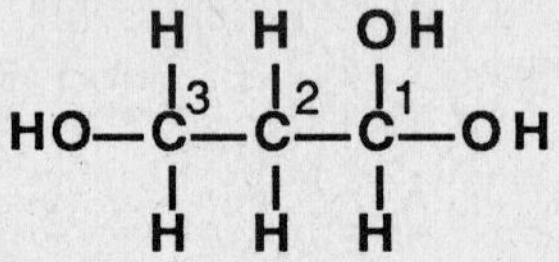

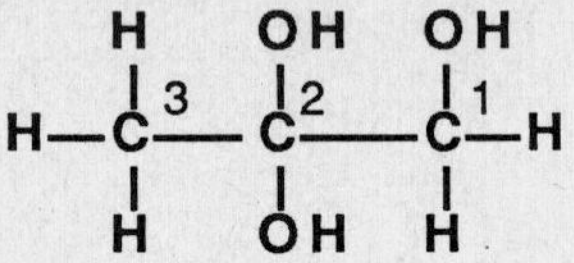

Figure 28.3.   Structural isomers of propanediol and propanetriol.

Molecules a–c represent the 3 possible structural isomers of propanediol and molecules d–h are the 5 possible structural isomers of propanetriol. If you visualize the process of flipping the molecules over, you will see that these are the maximum number of unique structural isomers for these alcohols, because, for example, 1,1,1-propanetriol is equivalent to 3,3,3-propanetriol, and 1,2,2-propanetriol is equivalent to 3,2,2-propanetriol.

Another structural classification of alcohols having only one hydroxyl group is based on the number of carbon atoms directly attached to the hydroxyl-bearing carbon.

- In a *primary alcohol,* the hydroxyl-bearing carbon is attached to one other carbon atom.
- In a *secondary alcohol,* the hydroxyl-bearing carbon is attached to two other carbon atoms.
- In a *tertiary alcohol,* the hydroxyl-bearing carbon is attached to three other carbon atoms.

In Figure 28.2, 1-propanol is a **primary** alcohol because the hydroxyl-bearing carbon is attached directly to only one other carbon, and 2-propanol is a **secondary** alcohol. A **tertiary** alcohol must have at least one hydrocarbon side-chain. The location of the side-chain also is identified by the carbon number, as in the examples of tertiary alcohols shown in Figure 28.4.

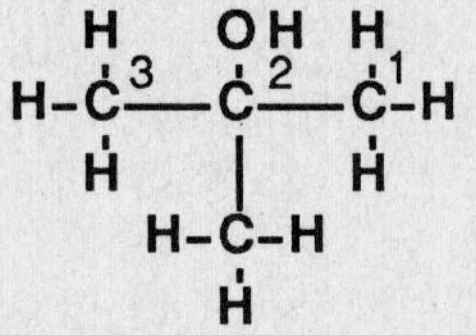

2-methyl-2-propanol 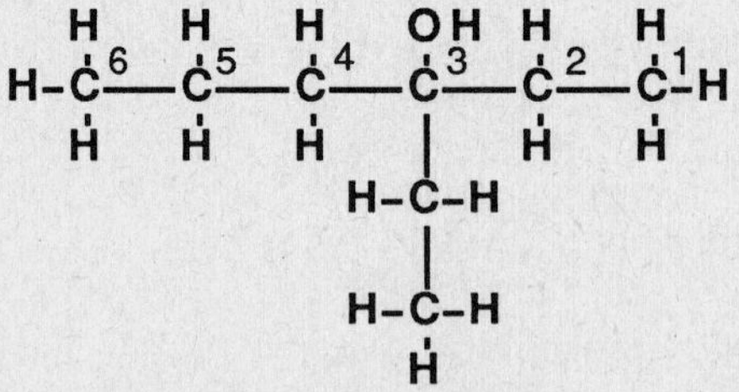   3-ethyl-3-hexanol

Figure 28.4.   Structures of two tertiary alcohols.

Chemical Reactions of Alcohols

The chemical reactivity of alcohols is mainly determined by reactions of the hydroxyl group. This fact illustrates one of the most important principles of organic chemistry:

The majority of organic chemical reactions involve only the functional groups, and the remainder of the carbon skeleton generally is unchanged.

In this experiment, you investigate oxidation and halogenation reactions of alcohols. Oxidation of an organic molecule generally involves the gain of oxygen or the loss of hydrogen. Halogenation involves replacing the hydroxyl group by a halogen atom.

The products from alcohol oxidation depend on the alcohol structure.

Oxidation of *primary alcohols* first gives aldehydes, which are oxidized further to carboxylic acids.
(In the structural formulas below, **R** represents a general alkyl chain.)

$$R\overset{\overset{\displaystyle O\ H}{|}}{\underset{\underset{\displaystyle H}{|}}{C}}H \xrightarrow{\text{oxidation}} R\overset{\overset{\displaystyle O}{\|}}{C}H \xrightarrow{\text{oxidation}} R\overset{\overset{\displaystyle O}{\|}}{C}OH$$

primary alcohol   aldehyde   carboxylic acid

Oxidation of *secondary alcohols* yields ketones.

$$R\overset{\overset{\displaystyle O\ H}{|}}{\underset{\underset{\displaystyle H}{|}}{C}}R' \xrightarrow{\text{oxidation}} R\overset{\overset{\displaystyle O}{\|}}{C}R'$$

secondary alcohol   ketone

***Tertiary alcohols* are very difficult to oxidize, and then it usually results in fragmentation of the carbon skeleton.**

$$R\overset{\overset{\displaystyle O\ H}{|}}{\underset{\underset{\displaystyle R''}{|}}{C}}R' \xrightarrow{\text{oxidation}} \text{no reaction}$$

tertiary alcohol

Laboratory oxidation of alcohols is readily accomplished with sodium dichromate ($Na_2Cr_2O_7$) in acid solution. Reactions 28.1-2 show the dichromate oxidation of ethanol to acetaldehyde and then to acetic acid.

$$3\,CH_3CH_2OH + Cr_2O_7^{2-} + 8\,H^+ \longrightarrow 3\,CH_3CHO + 2\,Cr^{3+} + 7\,H_2O \qquad (28.1)$$
$$\text{ethanol} \qquad \text{dichromate} \qquad\qquad \text{acetaldehyde}$$

$$3\,CH_3CHO + Cr_2O_7^{2-} + 8\,H^+ \longrightarrow 3\,CH_3COOH + 2\,Cr^{3+} + 4\,H_2O \qquad (28.2)$$
$$\text{acetaldehyde} \qquad \text{dichromate} \qquad\qquad \text{acetic acid}$$

When Reactions 28.1-2 occur, the aqueous solution changes from orange, the color of $Cr_2O_7^{2-}$, to dark green, the color of Cr^{3+}.

Oxidation reactions of alcohols are of interest in considering the relative toxicities of methanol, a severe poison, and ethanol, the alcohol used in beverages. In the body, oxidation of alcohols occurs mainly in the liver. Methanol is first oxidized to formaldehyde, which then is oxidized further to formic acid, both of which are strong poisons and very disruptive to metabolic processes. On the other hand, ethanol is oxidized first to acetaldehyde, which is 1/60 as toxic as formaldehyde, and then to acetic acid, which is produced in small amounts by normal body metabolism and is involved in the formation of proteins. Eventually, the acetic acid is oxidized further to CO_2 and H_2O. The acceptability of ethanol as a beverage depends on the fact that the oxidation products are metabolically tolerable in moderate amounts.

Substitution Reactions of Alcohols

Reactions of alcohols with hydrogen halides are typical substitution reactions in which the hydroxyl group is replaced by a halogen atom:

$$ROH + HX \longrightarrow RX + HOH \qquad (X = I, Br, Cl, \text{ in order of decreasing reactivity.})$$

Alcohol reactivity with hydrogen halides also depends on the alcohol structure; tertiary alcohols react faster than secondary alcohols, which react faster than primary alcohols. With HCl, tertiary alcohols react readily, but the less reactive secondary and primary alcohols require a catalyst, such as anhydrous $ZnCl_2$:

$$(CH_3)_3COH + HCl \xrightarrow{25°C} (CH_3)_3CCl + HOH \qquad \text{(fast)}$$
tertiary alcohol

$$(CH_3)_2CHOH + HCl \xrightarrow{ZnCl_2} (CH_3)_2CHCl + HOH \qquad \text{(slower)}$$
secondary alcohol

$$CH_3CH_2OH + HCl \xrightarrow{ZnCl_2} CH_3CH_2Cl + HOH \qquad \text{(still slower)}$$
primary alcohol

Oxidation and halogen substitution reactions can be used as tests to identify primary, secondary, and tertiary alcohols by their different reactivities.

Preparation of Ethanol

Commercial production of ethanol is mainly by the hydration of ethylene,

$$H_2C=CH_2 + H_2O \xrightarrow[\text{catalyst}]{\substack{300°C \\ \text{high pressure}}} H-CH_2-CH_2-OH$$

and by the enzyme-catalyzed fermentation of carbohydrates (sugars and starches). The overall reaction for fermentation of the sugar sucrose is

$$\underset{\text{sucrose}}{C_{12}H_{22}O_{11}} + H_2O \xrightarrow{\text{enzymes}} 4\,\underset{\text{ethanol}}{C_2H_5OH} + 4\,CO_2 \qquad (28.4)$$

Fermentation processes were used in making bread, beer, and wine long before the beginning of recorded history. Efforts to improve these processes were important to establishing the science of chemistry. Ethyl alcohol was the first pure organic compound known. Acetic acid, made by souring (oxidizing) wine, was the first acid to be made.

Oxygen must be excluded from the fermentation vessel, or the enzymes will produce very little ethanol. In the absence of air, Reaction 28.4 continues until the ethanol concentration reaches 14 to 16 percent, the yeast being unable to grow in solutions of higher ethanol content. To obtain higher concentrations of ethanol, the solution must be concentrated by distillation. This is why wines aged in the bottle are limited in their alcoholic content to a maximum of about 16 percent. Wines and other spirits aged in casks can lose some water through the porous wood, allowing an increase in ethanol concentration. Stronger alcoholic beverages such as whiskey, rum, vodka, brandy and liqueur are concentrated by distillation prior to being aged in casks, to obtain their high alcoholic content.

In the United States, alcoholic content of beverages is expressed in **degrees proof.** One degree proof is equal to 0.5% by volume. Therefore, a beverage containing 40% ethanol is said to be 80 proof. This practice may have begun in frontier days, when a pioneer felt it necessary to test his drink purchased in a bar by ignition, to see if it had been diluted. The story is that reliable ignition was regarded as 100% proof that the drink contained at least 50% ethanol, so a beverage with 50% ethanol was said to be 100 proof.

Alcoholic beverages are taxed by most governments, but ethanol used for laboratory or industrial purposes is not. To prevent untaxed laboratory ethanol from being diverted into illegal beverages, it is **denatured** by adding small amounts of toxic impurities that are very difficult to remove, even by distillation.

PLAN OF EXPERIMENT

In this experiment, ethanol is produced by a fermentation process very similar to the commercial methods used to make alcoholic beverages. The complex fermentation reactions require the presence of several different enzymes, each of which catalyzes a particular step in the overall process of about 12 separate reaction steps. The enzymes for this experiment are provided by growing yeast cells. They begin the fermentation process by converting sucrose to the simpler sugars glucose and fructose, both of which are monosaccharides with the same formula: $C_6H_{12}O_6$. Glucose and fructose then yield ethanol by many individual enzymatic reactions, the last of which is the reduction of acetaldehyde to ethanol.

Several conditions are necessary for efficient fermentation:
- **The essential nutrient phosphate must be provided.**
- **pH must be controlled so it does not fall too low.**
- **Air must be excluded from the fermentation reaction.**

A mixture of phosphate and tartrate salts, known as Pasteur's salts[1], provide phosphate and pH buffering. Abundant oxygen causes the yeast cells to grow rapidly by incorporating most of the sugar into their structure, producing very little ethanol in the process. With limited oxygen, not much new yeast material is built and most of the sugar is excreted as ethanol, a waste product of yeast metabolism. Air is excluded by fastening a deflated rubber balloon over the reaction vessel to seal it. As Reaction 28.4 proceeds, the initial oxygen is consumed by the yeast and expansion of the balloon contains the CO_2 that is produced, without a significant buildup of pressure.

When fermentation is completed, at least 2 days later, you concentrate your product by distillation and test it by ignition. Then you compare its oxidation and halogenation reactions with those of an ethanol standard and with secondary and tertiary alcohols.

---

[1]Louis Pasteur first observed that fermentation was markedly improved by adding certain salts.

SAFETY

1.   Wear approved eye protection.

2.   Be sure there are no open flames in the laboratory during the second laboratory period.

3.   Be careful when igniting your product that no other inflammable materials are nearby and that no product has spilled on the bench top.

PROCEDURE
First Laboratory Period

1.   Put 25 g of table sugar (sucrose) and 250 mL of water into a 1 L Erlenmeyer flask.

2.   Stir until the sugar is dissolved. Only if necessary, warm the flask very gently to aid dissolution.

3.   Add to the sugar solution about 25 mL of "Pasteur's salt" solution[2].

4.   Make a paste of about 10 g of Brewer's yeast with water and add it to the solution. With a wash bottle, wash down any yeast adhering to the sides of the flask. Stopper the flask and shake the mixture well.

5.   Obtain a one-hole stopper that fits your flask and carefully insert a snugly-fitting glass tube through it. The tube should extend about 3 cm above the stopper and not more than 2–3 cm below it.
         Follow the procedure in Section VII, Methods, page 19, with care when inserting the glass tube. The tube diameter must be large enough so the neck of the balloon stretches over it with a snug fit.

6.   Use this stopper and tube to snugly close the reaction flask and attach a rubber balloon, as shown in Figure 28.5. The balloon will exclude air from the fermentation process and will allow the expansion of the CO_2 gas that is produced.

7.   Stand the flask in a warm location (20-25°C) until the next laboratory period (at least 2 days).
         When you return to complete the experiment, expansion of the balloon with CO_2 gas will indicate that the fermentation was successful.

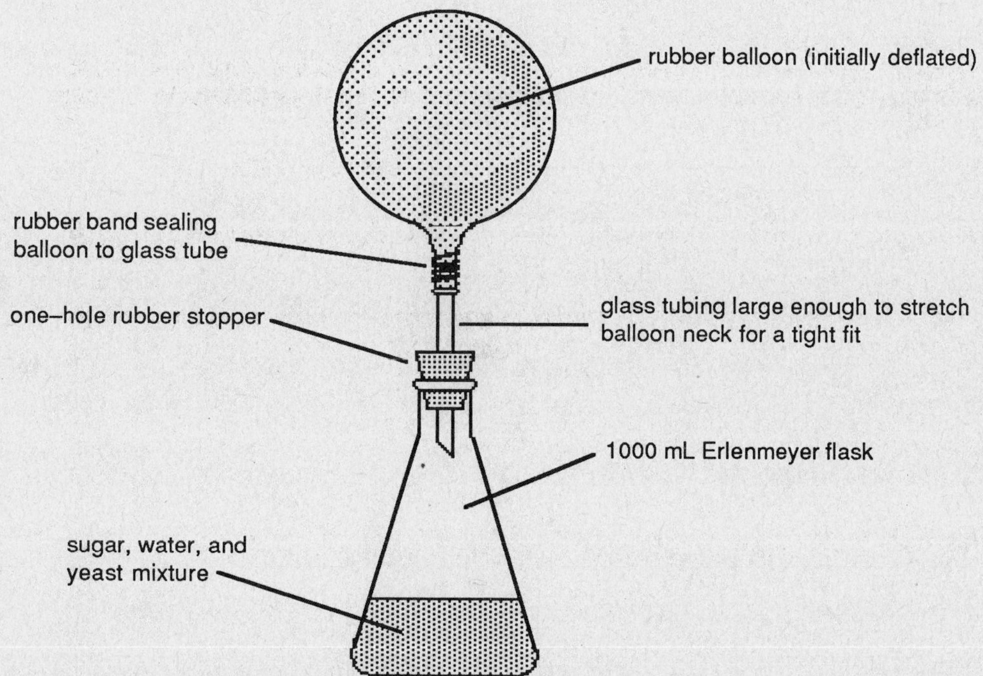

Figure 28.5.   Balloon-sealed reaction flask for fermentation of sucrose to ethanol.

---

[2]If not provided, make the solution by dissolving 0.25 g ammonium tartrate, 0.005 g calcium phosphate, and 0.05 g potassium phosphate in 25 mL of water.

Second Laboratory Period
Part A: Distillation
1. Collect the parts for the distillation apparatus as shown in Figure 0.14, Methods.

2. Pour about 50 mL of the liquid from your sugar-yeast mixture into the distilling flask. Add a few boiling chips to the flask.
Be careful to pour only the liquid, leaving the yeast in the fermentation flask.

3. Assemble the distillation apparatus as shown in Figure 0.14. Use a heating mantle. The thermometer bulb should be located just below the the outlet to the condenser, to insure accurate readings of the vapor stream. Connect the heating mantle to a voltage controller (Variac) and begin heating at a dial setting of about 80.

4. When boiling starts, reduce the voltage so that the liquid distills over into the receiving flask at about one drop per second.
Do not let the temperature rise above 85°C.

5. Stop the distillation when you have collected about 15 mL of distillate.

6. Clean and dry the distilling flask and use it to redistill your first 15 mL of product.

7. Stop the second distillation when you have collected about 10 mL of distillate.
Save the second distillate for use in Parts B, C, and D.

Part B: Combustion Test
The lighter alcohols are very flammable. Ethanol burns readily in air with a non-luminous flame, producing carbon dioxide and water. If your product burns in this fashion, it indicates fairly pure ethanol.

1. Pour 4 mL of your second distillate into a watch glass.
Wipe up any spilled liquid and be sure no flammable materials are nearby. Do not let your hair hang near the watchglass.

2. Carefully ignite the liquid with a match. Record the results.

C. Substitution Reactions of Alcohols: Hydroxyl Replacement by Chlorine
The substitution reaction is:

$$\text{R-CH}_2\text{-OH} + \text{HCl} \xrightarrow{\substack{\text{ZnCl}_2 \\ \text{catalyst}}} \text{R-CH}_2\text{-Cl} + \text{HOH}$$

alcohol   halogen   alkyl halide
acid

The alkyl halide formed is not soluble in alcohol or water and separates out, forming a cloudy, heterogeneous mixture. When an alcohol is added to the acid–catalyst solution:

- **Tertiary alcohols react almost instantly, forming a very cloudy mixture (two layers may form).**

- **Secondary alcohols dissolve at first to give a clear solution. Then, as the reaction proceeds, the solution turns cloudy after about 10 min.**

- **Primary alcohols react only with heating or not at all.**

1. Number 4 clean, dry test tubes and put 5 mL of the HCl-ZnCl$_2$ solution (available from your instructor) in each one. Swirl the solution in each tube to be sure the solutions are well mixed.

2. **To tube No. 1:** add 1 mL of laboratory ethanol.
To tube No. 2: add 1 mL of 2-propanol.
To tube No. 3: add 1 mL of 2-methyl-2-propanol (*tert*- butyl alcohol).
(It melts at 25.1°C and will have to be warmed slightly.)
To tube No. 4: add 1 mL of your alcohol distillate.
Swirl each solution well.

3.　　Prepare a warm water bath by heating a 500 mL beaker half full of water to 75-85°C on a hotplate. Place the 4 test tubes in the water bath and observe each tube closely for 10 min. Record your observations.

Leave all solutions in the water bath for another hour, making occasional observations of any changes. *During this time, begin Part D.*

Part D: Oxidation of Alcohols

The oxidation reactions are 28.12. Only primary and secondary alcohols will be oxidized by sodium dichromate. If oxidation occurs the solution color will change from orange to green.

1.　　Number 4 clean, dry test tubes.

2.　　**To tube No. 1:**　　add 1 mL of laboratory ethanol.
　　　　To tube No. 2:　　add 1 mL of 2-propanol.
　　　　To tube No. 3:　　add 1 mL of 2-methyl-2-propanol (*tert*- butyl alcohol).
　　　　　　　　　　　　(It melts at 25.1°C and will have to be warmed slightly.)
　　　　To tube No. 4:　　add 1 mL of your alcohol distillate.

2.　　Add 2 mL of 5% $Na_2Cr_2O_7$ to each test tube. Record the color of each solution.

3.　　Add 10 drops of 6 M H_2SO_4 to each test tube.

4.　　Prepare a boiling water bath with a 500 mL beaker half full of water. Place the test tubes in this bath and turn off the hotplate.

5.　　After 5 min, remove the test tubes from the bath and record any color changes that have occurred.

Name ____   Date ____________

EXPERIMENT 28
PRELABORATORY EXERCISE

1.   Write the structural formulas of:
   a.   2-butanol

   b.   1,3-pentanediol

   c.   1,2,2-hexanetriol

2.   What class of compound is produced by oxidation of a:

   a.   primary alcohol?   _______________________

   b.   secondary alcohol?   _______________________

3.   Draw structural formulas for all the possible isomers of an alcohol with 2 hydroxyl groups and four carbon atoms. Name the different atoms. (Refer to the discussion of isomers in Experiment 26 for guidance if necessary.)

4.   Which should react the fastest with HCl? (circle the correct answer):
   a. 1-propanol   **b.** 2-propanol   **c.** 2-methyl-2-propanol

5.   Which should react the fastest with dichromate ion? (circle the correct answer):
   a. 1-propanol   **b.** 2-propanol   **c.** 2-methyl-2-propanol

Name ____ Date ____________

EXPERIMENT 28
DATA

Part B: Combustion Test

1.　　Results of distillate combustion test:

Part C: Substitution Reactions

Test tube no.	Contents	Appearance of solution			
		Right after mixing	After 5 min	After 10 min	After 1 hr
1					
2					
3					
4					

From the results above, what type of alcohol (primary, etc.) was produced in your distillation? Explain your reasoning:

D. Oxidation reactions

Test tube no.	Contents	Initial color	Final color
1			
2			
3			
4			

From the results above, what type of alcohol (primary, etc.) was produced in your fermentation? Explain your reasoning:

Were the results of Parts C and D consistent with respect to the type of alcohol produced in your fermentation?

Make any other significant observations from any part of the experiment:

EXPERIMENT 29
Determining the Equilibrium Constant of an Esterification Reaction

PRELABORATORY PREPARATION
1. Do the Prelaboratory Exercise and turn it in at the beginning of your laboratory period.
2. Review the definitions of normality and equivalents in "Chemical Units," page xi.

INTRODUCTION
The formation of an ester by the reaction of a carboxylic acid (general formula: RCOOH) with an alcohol (general formula: ROH) is a good example of a reversible reaction. Although all reactions are reversible in principle, the term **reversible reaction** ordinarily is used to mean that appreciable quantities of the initial reactants still remain when equilibrium is reached. Under such conditions, it is easy to observe a shift in the equilibrium position that is caused by altering the concentrations of the equilibrium species (Le Chatelier's Principle). The general reaction between a carboxylic acid and an alcohol may be written:

$$R-\overset{\overset{\textstyle O}{\|}}{C}-OH \ + \ R'-OH \ \rightleftharpoons \ R-\overset{\overset{\textstyle O}{\|}}{C}-O-R' \ + \ H_2O \qquad (29.1)$$

$$\text{organic acid} \qquad\qquad \text{alcohol} \qquad\qquad\qquad \text{ester}$$

This is a very useful reaction for studying the basic principles of reversible reactions because no side products are formed, the rates of the forward and reverse reactions are convenient for simple experiments, and the analytical procedures are simple and accurate.

The forward reaction of 29.1 is called *esterification* **and the reverse reaction is called** *hydrolysis.*
The equilibrium constant for Reaction 29.1 is written:

$$K = \frac{[RCOOR'][H_2O]}{[RCOOH][R'OH]} \qquad (29.2)$$

For a given reaction, the value of the equilibrium constant depends only on the temperature. Esterifications generally reach equilibrium when between 50% and 75% of the initial reactants have been consumed.
This leads to equilibrium constants that tend to be between 1 and 9.
The role of the acid catalyst is to speed up both the forward and reverse reactions, so that the equilibrium is reached more quickly; the catalyst has no effect on the final concentrations of reagents at equilibrium.

Esters
Esters are usually noncorrosive, nontoxic, volatile liquids with pleasant, fruity odors. They are common products of natural processes. Fats and vegetable oils are esters of long-chain fatty acids and glycerol. In fruits and flowers, esters contribute characteristic odors and tastes. They are easy to synthesize and have many commercial uses in perfumes, synthetic flavorings, cosmetics, and solvents for laquers. Some esters can be polymerized, forming the polyester plastics used in fabrics, of which Dacron and Mylar are examples.

Naming Esters
To name an ester, substitute the ending *-ate* for *-ic* in the name of the reactant acid, and put the name of the reactant alcohol at the beginning.
- The ester made from acetic acid and methyl alcohol is called **methyl acetate**.
- The ester made from propionic acid and isobutyl alcohol is called **isobutyl propionate**.

PLAN OF EXPERIMENT
You prepare 3 different esters to note their fragrances, and to measure the equilibrium constant of one of the reactions. The refluxing procedure you use here is similar to the method used for commercial preparation of esters. In commercial production, the yield is increased by removing the ester as it is formed, thus preventing the attainment of equilibrium. This experiment shows one method for removing the product from the reaction mixture.

The principle for determining the equilibrium constant is as follows:

1. Measure the moles of carboxylic acid that remain after equilibrium is reached by titrating with a standardized base. It is necessary to correct for the presence of a known amount of H_2SO_4 catalyst that also is in the equilibrium mixture.

2. Subtract the equilibrium moles of carboxylic acid from the initial number of moles to determine the number of carboxylic acid moles that have reacted.

3. Knowing how many moles of acid have reacted, calculate the moles of alcohol remaining and the moles of products formed from the reaction stoichiometry.

4. With this data, calculate the equilibrium constant using Equation 29.2.

EXAMPLE:

The following reagents were placed into the boiling flask of a reflux apparatus:

22.2 mL (0.300 moles) of propionic acid
10.1 mL (0.251 moles) of methyl alcohol
2.1 mL concentrated H_2SO_4

The mixture was refluxed until equilibrium was reached, forming the ester product **methyl propionate** by the reaction:

$$C_3H_5COOH + CH_3OH \underset{}{\overset{\text{mineral acid catalyst}}{\rightleftharpoons}} C_6H_5COOCH_3 + H_2O \qquad (29.3)$$

propionic acid   methyl alcohol   methyl propionate

The moles of propionic acid remaining at equilibrium was determined as follows:

a. 2.1 mL of concentrated H_2SO_4 was pipetted into an Erlenmeyer flask that already contained 20 mL of distilled water. Phenolphthalein indicator was added and this solution was titrated to the end point with 41.2 mL of 2.02 N NaOH.

This titration showed that the 2.1 mL of H_2SO_4 used as a catalyst contained: (41.2 mL)(2.02 meq/mL) = 83.2 meq of acid.

b. 5.00 mL of the equilibrium mixture was pipetted into a flask and titrated to the endpoint with 15.9 mL of 2.02 N NaOH.

This titration shows that the 5.00 mL of equilibrium mixture contained:
(15.9 mL)(2.02 meq/mL) = 32.1 meq of total acid (propionic + sulfuric).

c. The total volume of the refluxed equilibrium mixture was determined to be:

22.2 mL propionic acid + 10.1 mL methyl alcohol + 2.1 mL sulfuric acid = 34.4 mL total

Since a 5.00 mL sample of this mixture contained 32.1 meq of acid, the total 34.4 mL of mixture must have contained:

$$\left(\frac{32.1 \text{ meq}}{5 \text{ mL}} \right) \times 34.4 \text{ mL} = 221 \text{ meq of acid total}$$

Of the 221 meq of acid in the equilibrium mixture, 83.2 meq was from the sulfuric acid catalyst (see step a). Therefore, the meq of propionic acid in the equilibrium mixture was:

221 - 83.2 = 138 meq propionic acid at equilibrium

Propionic acid has just one ionizable proton. Therefore, moles of propionic acid equals equivalents, so that there were 138 millimoles of propionic acid left at equilibrium.

Thus, the moles of propionic acid consumed in the reaction was:
0.300 initial moles - 0.138 final moles = 0.162 moles consumed

From Reaction 29.3, we see that the moles of methyl propionate ester formed, the moles of water formed, and the moles of methyl alcohol consumed, are each equal to the moles of propionic acid consumed. Now you can list the moles of all the species present in the 34.4 mL of equilibrium mixture:

Equilibrium moles of C_3H_5COOH = 0.138
Equilibrium moles of CH_3OH = 0.251 - 0.162 = 0.089
Equilibrium moles of $C_6H_5COOCH_3$ = 0.162
Equilibrium moles of H_2O = 0.162

The moles per liter are:

$$[C_3H_5COOH] = \frac{0.138\ mol}{34.4\ mL} \times \frac{1000\ mL}{1\ L} = 4.01\ mol/L$$

$$[CH_3OH] = \frac{0.089\ mol}{34.4\ mL} \times \frac{1000\ mL}{1\ L} = 2.59\ mol/L$$

$$[C_6H_5COOCH_3] = \frac{0.162\ mol}{34.4\ mL} \times \frac{1000\ mL}{1\ L} = 4.71\ mol/L$$

$$[H_2O] = \frac{0.162\ mol}{34.4\ mL} \times \frac{1000\ mL}{1\ L} = 4.71\ mol/L$$

Then, the equilibrium constant, Equation 29.2, is:

$$K = \frac{(4.71)(4.71)}{(4.01)(2.59)} = 2.14$$

You probably will not have time to do the equilibrium calculations for your experiment during the laboratory period. Therefore, you are to make carbon copies or photocopies of your data sheets for turning in at the end of this laboratory period. Take your original data sheets home, complete your calculations, and turn the completed experiment in at your next laboratory period.

SAFETY

1. Wear approved eye protection.

2. Be sure there are no open flames in the laboratory.

3. Acetic acid and sulfuric acid are very corrosive. Handle them with great care. If they contact skin or clothing, rinse immediately with plenty of water.

4. Acetic acid vapors are very irritating. Dispense acetic acid only in a hood and avoid inhalation.

5. Avoid inhaling the alcohol vapors. Be cautious when testing the ester odors by following the proper procedure for testing odors as described in the Procedure section.

6. Do not taste any of the esters, even though they may smell good and some are the same as used in commercial flavoring agents.

7. Butyric acid has a very unpleasant odor that is hard to remove from skin and clothing. Handle it carefully, avoiding contact.

8. Use a pipet bulb for all pipetting. Never pipet by mouth.

PROCEDURE

The reflux apparatus, Figure 29.1, is used to bring one reaction to complete equilibrium for measuring the equilibrium constant. If you have enough time, you can partially run several other reactions (Procedure steps 9–13) to note the odors of the different esters produced and to observe the effect of the catalyst.

1.   Collect the parts of the reflux apparatus, Figure 29.1, and pipet 14.6 mL (0.250 mol) of ethanol and 14.3 mL (0.250 mol) of acetic acid into the 250 mL boiling flask. Record the volume of each reagent added.

2.   Carefully pipet 1.10 mL concentrated sulfuric acid catalyst into the boiling flask. Record the volume of acid added.

3.   Add several new boiling chips to the flask.
Fresh boiling chips must be added *before* heating begins. If boiling chips are added to a hot liquid, they can initiate sudden boiling with spattering or even explosion.

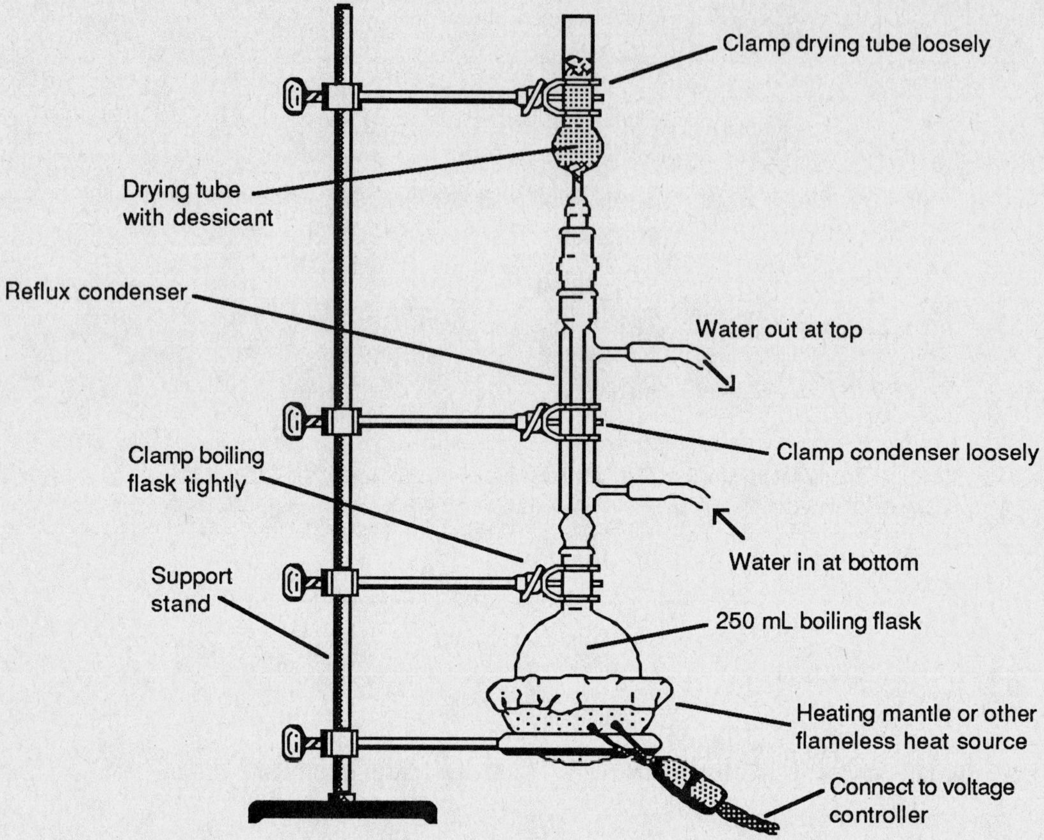

Figure 29.1.   Assembled reflux apparatus.

4.   Clamp the boiling flask firmly to the support stand. Attach the condenser, drying tube, and water hoses as shown in Figure 29.1.

> **The boiling flask should be clamped tightly, but the condenser and drying tube should be clamped loosely, for support only. This avoids strains on the glass joints. Be sure water enters at the bottom and exits at the top of the condenser.**

Have your instructor check the apparatus before continuing.

5.   Begin heating the flask by turning up the voltage regulator to a setting given by your instructor. Record the starting time of refluxing, when the solution starts to boil.

> **Reflux the solution for 15 min.**

6.   While refluxing progresses, fill a 250 mL beaker half full with tap water, add several fresh boiling chips, and begin heating to a boil on a hotplate. This will be a water bath for step 12.

7.   While refluxing progresses and after starting the hot water bath:
    a.   Place about 20 mL distilled water in a 125 mL Erlenmeyer flask.
    b.   Then carefully pipet 1.10 mL concentrated sulfuric acid into the flask.
    c.   Add 4 drops of phenolphthalein indicator.
    d.   Titrate to the equivalence point with 1 N standardized NaOH. Record the
       exact normality and milliliters used of the NaOH.

> **This data allows you to calculate the equivalents of sulfuric acid added as a catalyst, and is needed as a correction factor in the calculation of the equilibrium constant.**

8.   When refluxing is completed, remove or turn off the heating mantle and allow the apparatus to cool for 10 min before removing the flask.

STEPS 9–13, IN WHICH YOU MAKE SEVERAL DIFFERENT ESTERS AND NOTE THEIR ODORS, ARE NOT RELATED TO THE EQUILIBRIUM EXPERIMENT. YOUR INSTRUCTOR WILL DECIDE IF YOU HAVE TIME TO DO THEM.

After finishing the titration in step 7, you should have about 15 minutes free while the reflux ends and the apparatus cools. Use this time to make a variety of esters and observe their odors. **While you are doing this, keep an eye on your refluxing apparatus and the time remaining for refluxing.**

9.   Number 5 test tubes.
>   **In tubes nos. 1 and 2,** put 2 mL methyl alcohol and 2 g salicylic acid.
>   **In tube no. 3,** put 2 mL isoamyl alcohol and 3 mL acetic acid.
>   **In tube no. 4,** put 2 mL ethyl alcohol and 3 mL butyric acid.
>   **In tube no. 5,** put 2 mL octyl alcohol and 3 mL acetic acid.

10.   *Carefully* note the odors from each tube. Use the following procedure:
First, fill your lungs with air.
Then, hold the tube about 5 cm in front of your nose and, with your free hand, wave the fumes at the top of the tube toward your nose.
Write your observations on the Data sheet.

11.   Add 10-15 drops of concentrated sulfuric acid to each tube, ***except for no. 1.***

12.   Place all 5 test tubes in the 250 mL boiling water bath that you prepared in step 6. Leave them in the bath for 5 min, then turn off the heat.
If your reflux apparatus has cooled by this time, go to step 14 and finish the equilibrium part of the experiment. Then return to step 13.

13.   Carefully test the odors from each test tube in the same manner as step 10.
If no odor is detectable, pour the contents of the test tube into a beaker containing 10 mL of hot water, to volatilize the ester.
Look for odors such as wintergreen, pear, banana, orange, and pineapple. Write your observations on the Data sheet.

14.   When your reflux apparatus has cooled, remove the flask.
Remember that the condenser and drying tube are clamped loosely. Do not let them fall.
Pipet a 5.00 mL portion of the solution into a 125 mL Erlenmeyer flask. Replace the flask on the condenser and resume heating. Record on the Data sheet the time when boiling resumes. The mixture should reflux another 10 min.

15.   Add 20 mL distilled water and 4 drops of phenolphthalein to the 5.00 mL portion removed in step 14. Titrate the solution with 1 N standardized NaOH. Record the amount of NaOH required.
The purpose of this titration is to determine how much carboxylic acid remains after 15 min of refluxing.

16.   After the solution has refluxed another 10 min, allow it to cool and titrate another 5.00 mL portion, as in step 15. Record the amount of standard NaOH used.
If the amount of NaOH used to titrate the 2 portions was the same to within 1 mL, then equilibrium was reached after the first 15 min of refluxing. If equilibrium has not been reached, continue refluxing for 10 min periods until 2 successive titrations of 5.00 mL portions agree to within 1 mL of NaOH used.

17.   **Demonstration of Le Chatelier's Principle**
After equilibrium has been reached and the titration data recorded, we can observe the effect of adding an excess of one reactant.
Add 12.2 mL of ethanol to the equilibrium mixture and reflux for 15 min. After cooling, add 4 drops of phenolphthalein indicator and titrate a 5.00 mL portion as before. Record the amount of standard NaOH used.

18.   To separate the ester product, mix the solution remaining in the boiling flask with an approximately equal volume of water. Stir well and the allow 2 layers to separate.
     If 2 distinct layers do not form, add 5 g of sodium chloride to the mixture and stir again.
The ethyl acetate product you have made is only slightly soluble in water and even less soluble in salt water. Adding salt to force the separation is called "salting out" the ester.

19.   Pipet a few mL of each layer into separate test tubes and carefully test their odors. Add 1 drop of water to each test tube and record on the Data sheet whether or not it dissolves.

20.   If time permits, test the solubility, **in a hood,** of your ethyl acetate product in ethanol and in ether. Record the results.

21.   Turn in a carbon copy or photocopy of your Data sheets to your instructor. Complete your calculations at home and turn in the completed sheets at your next laboratory period.

Name ____   Date _____________

EXPERIMENT 29
PRELABORATORY EXERCISE

1.   After performing an esterification reaction, you have an equilibrium sample containing your product and an unknown amount of benzoic acid (a carboxylic acid with one ionizable proton). At the beginning of the reaction, 13.0 mL of concentrated H_2SO_4 was added as a catalyst. To measure the benzoic acid in the equilibrium mixture:

a. 13.0 mL of pure concentrated H_2SO_4 was titrated with standardized 1.19 N NaOH. This titration required 12.2 mL of the NaOH standard.

b. The entire equilibrium mixture, which contains the benzoic acid plus 13.0 mL concentrated H_2SO_4, was titrated with the same standard NaOH. This titration required 21.3 mL of NaOH.

Calculate the moles of benzoic acid in the equilibrium mixture.

2.   The ester **methyl benzoate** is formed by the reaction:

$$C_6H_5COOH + CH_3OH \underset{}{\overset{\text{mineral acid catalyst}}{\rightleftharpoons}} C_6H_5COOCH_3 + H_2O$$

benzoic acid   methyl alcohol   methyl benzoate

When 96.5 mL (1.00 mole) of benzoic acid and 40.4 mL (1.00 mole) of methyl alcohol react, with 13.1 mL of H_2SO_4 present as a catalyst, it is found (by a titration analysis similar to that of question 1) that at equilibrium, there are 0.433 moles of benzoic acid remaining. The total volume of the solution is the sum of the volumes of reactants and catalyst or 150 mL.

a. Calculate the concentrations of all the reactants and products present at equilibrium.

b. Calculate the equilibrium product.

Name ____   Date ____________

EXPERIMENT 29

DATA
(Observe significant figures in all calculations.)
Make photocopies or carbon copies of the data recorded on these data pages.

Initial Quantities

1. Ethanol (C_2H_5OH),   initial moles: _____________ ; initial volume: ___________

2. Acetic acid (CH_3COOH),   initial moles: _____________ ; initial volume: ___________

3. Total of reactants,   initial moles: _____________ ; initial volume: ___________

4. Sulfuric acid (H_2SO_4),   initial volume: ___________

5. Instructor's check of reflux apparatus: _____________

6. Time first reflux began: _____________

7. Time first reflux ended: _____________

Titration of H_2SO_4 Catalyst

8. Normality of NaOH standard: _____________

9. Volume of concentrated H_2SO_4:   _____________

10. Final buret reading: _____________

11. Initial buret reading: _____________

13. Volume of NaOH titrant required: _____________

Titration of First Reflux Portion

14. Volume of first reflux portion: _____________

15. Final buret reading: _____________

16. Initial buret reading: _____________

17. Volume of NaOH titrant required: _____________

Titration of Additional Reflux Portions

18. Time second reflux began: _____________

19. Time second reflux ended: _____________

20. Volume of second reflux portion: _____________

21. Final buret reading: _____________

22. Initial buret reading: _____________

23. Volume of NaOH titrant required: _____________

> **Use Steps 24–29 only if necessary.**

24. Time third reflux began: _____________

25. Time third reflux ended: _____________

26. Volume of third reflux portion: _____________

27. Final buret reading: _____________

28. Initial buret reading: _____________

29. Volume of NaOH titrant required: _____________

EXPERIMENT 29
DATA

Demonstration of Le Chatelier's Principle

30. Volume of ethanol added to equilibrium mixture: _______________

31. Time reflux began: _______________

32. Time reflux ended: _______________

33. Volume of sample portion from new equilibrium: _______________

34. Final buret reading: _______________

35. Initial buret reading: _______________

36. Volume of NaOH titrant required: _______________

Separation of Ester Product

37. Did two layers form? _______________ ;   With or without added salt? _______________

38. In which layer does water dissolve? _______________

39. What is the composition of the upper layer? _______________

40. What is the composition of the lower layer? _______________

41. Is ethyl acetate soluble in ethanol? _______________

42. Is ethyl acetate soluble in ether? _______________

43. Describe the odor of ethyl acetate: ____

Odor Tests (Procedure Steps 9–13)

44. Odors of test tube mixtures before adding H_2SO_4 (Step 10):

       Test tube 1: ____

       Test tube 2: ____

       Test tube 3: ____

       Test tube 4: ____

       Test tube 5: ____

45. Odors of test tube mixtures after reaction with H_2SO_4 (Step 11):

       Test tube 1 (no H_2SO_4): ____

       Test tube 2: ____

       Test tube 3: ____

       Test tube 4: ____

       Test tube 5: ____

Describe the effect of adding H_2SO_4: ____

Name ____   Date __________

EXPERIMENT 29
DATA

Calculations:

46. Use the volume of standard NaOH used to titrate 1.1 mL of H_2SO_4 catalyst to calculate the equivalents of H_2SO_4 added to the reflux mixture.

47. Use data from the titrations of the 2 reflux portions that agreed within 1 mL, to calculate the equivalents of total acid in each 5 mL portion. Average these 2 values.

48. Assume that the volume of the equilibrium mixture was the same as the volume of the initial reactants, plus catalyst. Calculate the number of equivalents in the total sample at equilibrium, using the result from Step 47. Then subtract the equivalents of H_2SO_4 catalyst present to obtain the equivalents and moles of acetic acid remaining at equilibrium.

49. Write the balanced equation for the reaction between ethanol and acetic acid. Name all the compounds. Use the stoichiometry to calculate the moles of all species present at equilibrium. Express their concentrations in moles per liter.
     Remember that your volume was about 30 mL.

50. Calculate the equilibrium constant for this reaction.

EXPERIMENT 30

Determination of the Rate Law and the Effect of Temperature for the Oxidation of Iodide Ion by Hydrogen Peroxide: A Clock Reaction

PRELABORATORY PREPARATION
1. Do the Prelaboratory Exercise and turn it in at the beginning of your laboratory period.
2. Review use of the buret, Section IV, Methods, page 17.

INTRODUCTION
When studying chemical reactions, three characteristics are of primary interest: the position of equilibrium, the rate of reaction, and the particular sequence of reaction steps by which the reactants are converted into the products. In this experiment, you measure the **reaction rate** of a particular reaction and determine how the rate depends on the concentrations of the individual reactants.

The term *rate of a reaction* **refers to how fast a product is formed or a reactant is consumed. It is found by measuring the change in concentration, ΔC, of a product or reactant that occurs in a given time, Δt.**

$$\text{rate} = \Delta C/\Delta t$$

Reaction rates vary widely. Some reactions are explosively fast, going to completion in microseconds or faster, while others may be so slow, requiring years or longer for completion, that it appears nothing at all is happening. For example, magnesium metal burns (reacts with oxygen) very rapidly in air and is used for signal flares, but another metal, silver, reacts extremely slowly with oxygen, even when placed in a flame. As an example of contrasting ionic reactions, silver ions react rapidly with chloride ions, to form solid silver chloride: $Ag^+(aq) + Cl^-(aq) \longrightarrow AgCl(c)$; however, magnesium ions react slowly with oxalate ions to form solid magnesium oxalate: $Mg^{2+}(aq) + C_2O_4^{2-}(aq) \longrightarrow MgC_2O_4(c)$. These examples illustrate that the nature of the reactants is one factor that governs the rate of a reaction; some reactions are inherently fast and some are inherently slow. Certain experimental variables also can influence the rate of a reaction.

The rate of any particular homogeneous reaction[1] depends first of all on the nature of the reactants. Experimental variables that also influence the reaction rate are the reactant concentrations, the temperature, and the possible presence of a catalyst.

Effect of Reactant Concentration
The effect of reactant concentration cannot be predicted from the reaction equation; it must be determined by experiment.
For many reactions, the experimentally determined relation between reaction rate and reactant concentration can be written in a simple mathematical form. For example, in the generalized reaction

$$aA + bB + cC \longrightarrow \text{products}$$

the relation between the reaction rate and the reactant concentration can often be written:

$$\text{rate} = k[A]^x [B]^y [C]^z \tag{30.1}$$

Equation 30.1 is the **rate equation** for the generalized reaction, the proportionality constant **k** is called the **rate constant**, the exponents **x, y,** and **z** are called the **orders of the reaction** with respect to reactants **A, B,** and **C.**

Since reactant concentrations change as the reactants are consumed, the rate of reaction will change as a reaction proceeds, generally becoming slower. Equilibrium is reached when a reaction rate slows to zero.
For this reason, it is common to compare *initial* **reaction rates,** based on the starting reactant concentrations. For most reactions with simple rate equations, the individual reaction orders (**x, y,** and **z**) lie between -2 and +2, and can be integral (e.g. -2, 0, +1) or fractional (e.g., -3/2, +1/2).

The reaction orders for each reactant can be determined by changing the reactant concentrations one at a time and measuring the corresponding reaction rate changes.

---

[1]For heterogeneous reactions, the rate also depends on the particle size or surface area of solid or liquid reactants.

EXAMPLE 1: Finding the Reaction Orders

In this example, the method used for finding the reaction order of each reactant is the same as that used in the procedure for this experiment.

Bromate ion and bromide ion react in acidic solution by:

$$BrO_3^- + 5\,Br^- + 6\,H^+ \longrightarrow 3\,Br_2 + 3\,H_2O \tag{30.2}$$

The rate of the reaction can be determined by measuring the increase in concentration of Br_2 per second. Such measurements show that the reaction rate depends on the reactant concentrations according to the rate equation:

$$\text{rate} = R = k[BrO_3^-]^x[Br^-]^y[H^+]^z \tag{30.3}$$

To find the reaction orders, the reaction rates were measured for the series of experiments in Table 1 at 27°C . In each experiment, the amounts of the reactants were varied and sufficient water was added to make the total volume of the reaction mixture the same for all the experiments.

Experiment No.	0.0020 M KBr (mL)	0.0020 M KBrO$_3$ (mL)	0.020 M HCl (mL)	H$_2$O (mL)	Total Vol. (mL)	Reaction rate at 27°C (mol/L-s)
1	10	10	10	70	100	1.44×10^{-12}
2	20	10	10	60	100	2.88×10^{-12}
3	10	20	10	60	100	2.88×10^{-12}
4	10	10	20	60	100	5.76×10^{-12}

Compare experiments 1 and 2:

The volume of Br^- added from the 0.0020 M stock solution was doubled while the volumes of BrO_3^- and H^+ were not changed. As a result, the reaction rate doubled. Because extra water was added so that the total volume of solution remained constant, doubling the volume of Br^- also doubled its concentration. Using the rate equation 30.3 and noting that only $[Br^-]$ was changed, we can write expressions for the rates of experiments 1 and 2.

For experiment 1: $R_1 = k[BrO_3^-]^x[Br^-]_1^y[H^+]^z$ $\qquad$ (30.4)

For experiment 2: $R_2 = k[BrO_3^-]^x[Br^-]_2^y[H^+]^z$ $\qquad$ (30.5)

Divide Equation 30.4 by Equation 30.5 to get Equation 30.6 which shows how the ratio of rates depends on the ratio of changed concentrations. Since **k**, **$[BrO_3^-]^x$**, and **$[H^+]^z$** are the same in both Equations 30.4 and 30.5, they cancel out of the ratio.

$$\frac{R_1}{R_2} = \left(\frac{[Br^-]_1}{[Br^-]_2} \right)^y \tag{30.6}$$

The table shows that $R_2 = 2\,R_1$, and $[Br^-]_2 = 2\,[Br^-]_1$, so that Equation 30.6 becomes:

$$\frac{R_1}{2\,R_1} = \left(\frac{[Br^-]_1}{2\,[Br^-]_1} \right)^y \quad \text{or} \quad \frac{1}{2} = \left(\frac{1}{2} \right)^y$$

This result shows that the reaction order for Br^- is **y = 1.**

Notice that it is not necessary to know the numerical value for *k* in order to find the reaction orders. In this case, it was easy to determine the value of **y** by inspection. If **y** were not a simple integer, it could be found by taking the logarithm of both sides of Equation 30.6.

$$\log \frac{R_1}{R_2} = y\,\log \frac{[Br^-]_1}{[Br^-]_2} \quad ; \quad y = \frac{\log\,(R_1/R_2)}{\log\,([Br^-]_1/[Br^-]_2)}$$

Compare experiments 1 and 3:

In these experiments, the concentration of BrO_3^- was doubled while the concentrations of Br^- and H^+ were not changed. As a result, the reaction rate doubled. Analyzing experiments 1 and 3 in the same way as we did experiments 1 and 2, we find that the reaction order for BrO_3^- is **x = 1.**

Compare experiments 1 and 4:

In these experiments, the concentration of H^+ was doubled while the concentrations of Br^- and BrO_3^- were not changed. As a result, the reaction rate quadrupled. Analyzing experiments 1 and 3 in the same way as we did experiments 1 and 2, we find that the reaction order for H^+ is **z = 2.**

Using the measured reaction orders, the rate Equation 30.3 may be written:

$$R = k[BrO_3^-][Br^-][H^+]^2 \tag{30.7}$$

We still need to find the value of the rate constant **k.**

EXAMPLE 2: Finding the Rate Constant

The rate constant is found as follows:
   a. Select any one of the experiments from the table of experiments.
   b. Substitute the experimental values for the reactant concentrations and reaction rate into the rate equation 30.7.
   c. Solve the equation for **k.**

Notice, however, that the table of experiments gives the amounts of reactants in terms of the volume of stock solution used, while Equation 30.7 requires that the amounts of reactants be expressed in terms of solution concentration, moles/L.

Amounts of reactants must be converted to moles/L in the reaction mixture before calculating the rate constant.

To find the concentrations of each reactant in the reaction mixture:
   a. Calculate the moles of each reactant from the stock solution concentration and volume used.
   b. Divide the moles of reactant by the *total* volume of the reaction mixture to get the reactant concentration.

Problem:

Calculate the rate constant for Reaction 30.2, using data from experiment 1.

Solution:

First find the molar concentrations in experiment 1 for all the reactants.

BrO_3^-:

$$\text{moles } BrO_3^- = Vol_{KBrO_3\text{ stock}} \times M_{KBrO_3\text{ stock}} = Vol_{mixture} \times M_{KBrO_3\text{ mixture}}$$

$$= (10 \times 10^{-3}\text{ L})(0.0020\text{ M}) = 2.0 \times 10^{-5}\text{ moles}$$

$$M_{BrO_3^-} = M_{KBrO_3\text{ mixture}} = \frac{\text{moles } BrO_3^-}{Vol_{mixture}} = \frac{2.0 \times 10^{-5}\text{ moles}}{100 \times 10^{-3}\text{ L}} = 2.0 \times 10^{-4}\text{ mol L}^{-1}$$

Br^-:

$$\text{moles } Br^- = Vol_{KBr\text{ stock}} \times M_{KBr\text{ stock}} = Vol_{mixture} \times M_{KBr\text{ mixture}}$$

$$= (10 \times 10^{-3}\text{ L})(0.0020\text{ M}) = 2.0 \times 10^{-5}\text{ moles}$$

$$M_{Br^-} = M_{KBr\text{ mixture}} = \frac{\text{moles } Br^-}{Vol_{mixture}} = \frac{2.0 \times 10^{-5}\text{ moles}}{100 \times 10^{-3}\text{ L}} = 2.0 \times 10^{-4}\text{ mol L}^{-1}$$

H^+:

$$\text{moles } H^+ = Vol_{HCl\text{ stock}} \times M_{HCl\text{ stock}} = Vol_{mixture} \times M_{HCl\text{ mixture}}$$

$$= (10 \times 10^{-3}\text{ L})(0.020\text{ M}) = 2.0 \times 10^{-4}\text{ moles}$$

$$M_{H^+} = M_{HCl\text{ mixture}} = \frac{\text{moles } H^+}{Vol_{mixture}} = \frac{2.0 \times 10^{-4}\text{ moles}}{100 \times 10^{-3}\text{ L}} = 2.0 \times 10^{-3}\text{ mol L}^{-1}$$

Substitute these concentrations into Equation 30.7, along with the reaction rate for experiment 1, and solve for **k.**

$$R = k[BrO_3^-][Br^-][H^+]^2 \tag{30.7}$$

$$1.44 \times 10^{-12}\text{ mol L}^{-1}\text{ s}^{-1} = k\,(2.0 \times 10^{-4}\text{ mol L}^{-1})(2.0 \times 10^{-4}\text{ mol L}^{-1})(2.0 \times 10^{-3}\text{ mol L}^{-1})^2$$

$$k = \frac{1.44 \times 10^{-12}\text{ mol L}^{-1}\text{ s}^{-1}}{(2.0 \times 10^{-4}\text{ mol L}^{-1})(2.0 \times 10^{-4}\text{ mol L}^{-1})(2.0 \times 10^{-3}\text{ mol L}^{-1})^2} = 9.00\text{ L}^3\text{ mol}^{-3}\text{ s}^{-1}$$

The units for **k** may seem peculiar but remember that **k** is a proportionality constant and its units are appropriate to make the units for R come out correctly as *mol L^{-1} s^{-1}*. Normally, one would calculate values of **k** for each of the experiments and average them to obtain the best value.

Using experiment 1 only, the final result for the general rate equation is:

$$R = (9.00 \text{ L}^3 \text{ mol}^{-3} \text{ s}^{-1})[BrO_3^-][Br^-][H^+]^2$$

Influence of Temperature

Increasing the temperature of the reaction mixture generally increases the reaction rate. **The reaction rate constant depends on the temperature and is different for experiments done at different temperatures. However, the reactant orders, x and y, do not change with temperature.**

If you measure the different rate constants at two different temperatures, you can calculate the **activation energy** of the reaction from:

$$\log \frac{k_2}{k_1} = \frac{E_a}{2.3 \text{ R}}\left(\frac{T_2 - T_1}{T_2 T_1}\right) \tag{30.8}$$

In Equation 30.8, k_2 is the rate constant at temperature T_2, k_1 is the rate constant at temperature T_1 (all temperatures, of course, are in degrees Kelvin), and E_a is the activation energy of the reaction. A chemical reaction generally requires that some reactant bonds be broken and new bonds formed. The activation energy is an indication of how much kinetic energy there must be in a collision between two reactant molecules in order for the necessary bond breaking and reforming to take place.

EXAMPLE 3: Finding the Activation Energy

From the experiment table, the rate of experiment 1 at 27°C was:

$$R_{1, \, 27°C} = 1.44 \times 10^{-12} \text{ mol L}^{-1} \text{ s}^{-1}$$

When experiment 1 in the experiment table was repeated at 43°C, the rate of the reaction was:

$$R_{1, \, 43°C} = 4.63 \times 10^{-12} \text{ mol L}^{-1} \text{ s}^{-1}$$

In Example 2, the reaction rate constant at 27°C was found to be: $k_{27°C} = 9.00 \text{ L}^3 \text{ mol}^{-3} \text{ s}^{-1}$
At 43°C, the rate constant is:

$$k_{43°C} = \frac{4.63 \times 10^{-12} \text{ mol L}^{-1} \text{ s}^{-1}}{(2.0 \times 10^{-4} \text{ mol L}^{-1})(2.0 \times 10^{-4} \text{ mol L}^{-1})(2.0 \times 10^{-3} \text{ mol L}^{-1})^2} = 28.9 \text{ L}^3 \text{ mol}^{-3} \text{ s}^{-1}$$

Substituting these values into Equation 30.8 gives:

$$\log \frac{28.9}{9.00} = \frac{E_a}{2.3(8.31)}\left\{\frac{(316 - 300)K}{(316 \text{ K})(300 \text{ K})}\right\}$$

$$E_a = \frac{1.81 \times 10^6}{16} \log 3.21 = (1.13 \times 10^5)(0.507) = \mathbf{5.74 \times 10^4 \text{ Joules} = 57.4 \text{ kJ}}$$

PLAN OF EXPERIMENT

In the reaction of this experiment, iodide ion is oxidized to tri-iodide ion by hydrogen peroxide in acidic solution:

$$H_2O_2(aq) + 3 \, I^-(aq) + 2 \, H^+ \longrightarrow I_3^-(aq) + 2 \, H_2O(aq) \tag{30.9}$$

The rate equation for Reaction 30.4 may be written as:

$$\text{rate} = k[H_2O_2]^x[I^-]^y[H^+]^z \tag{30.10}$$

By changing the concentrations of H_2O_2, I^-, and H^+ one at a time, as in Example 1 above, and measuring how the rate of H_2O_2 decomposition changes, you will determine the orders x, y, and z in Equation 30.10. **In this experiment, the exponents x, y, and z are integers. Therefore, you should round your measured values to the nearest integral value.**

If you double the concentration of one reactant while the concentrations of the other reactants remain constant, and the reaction rate:
- **does not change**, the reaction order for that reactant is **0**.
- **doubles**, the reaction order for that reactant is **1**.
- **increases four times**, the reaction order for that reactant is **2**.
- **increases eight times**, the reaction order for that reactant is **3**.

After values for the reaction orders **x**, **y**, and **z** have been found, the rate constant **k** in Equation 30.10 can be calculated. Insert your measured values for the reaction rate, reaction orders and the concentrations used, and solve Equation 30.10 for **k**. The reaction rate for oxidation of I^- is equal to the rate of consumption of $S_2O_3^-$, measured as **moles L^{-1} s^{-1}**.

"Clock" Method for Timing the Reaction Rate

You time the reaction by using a built-in chemical "alarm-clock" that causes the reaction solution to suddenly change color after a given amount of I^- has been oxidized. The alarm-clock is made by adding a measured amount of sodium thiosulfate and a small amount of starch indicator to the solution.

Thiosulfate ion does not react with the reactants in Equation 30.9, but does react very rapidly with tri-iodide ion, according to Equation 30.11:

$$2\,S_2O_3^{2-}(aq) + I_3^-(aq) \longrightarrow 3\,I^-(aq) + S_4O_6^{2-}(aq) \qquad \text{(very fast)} \qquad (30.11)$$

Therefore, as long as any thiosulfate ion is present in the solution, all the I_3^- formed by Reaction 30.9 is immediately converted back to I^-. The starch indicator reacts only with I_3^-, forming a complex that turns the solution a deep blue color. Only when all the thiosulfate ion is used up by Reaction 30.11 can the concentration of I_3^- increase beyond a trace. When this happens, the solution suddenly turns blue because of the starch-tri-iodide complex that is formed. To summarize how the chemical alarm-clock works:

- The same fixed amount of $S_2O_3^-$ is added to each test solution.

- The $S_2O_3^-$ reacts with I_3^- as rapidly as it is formed in Reaction 30.9.

  Therefore, the rate of $S_2O_3^-$ consumption is exactly equal to the rate of Reaction 30.9.

- When all the $S_2O_3^-$ is gone, the I_3^- concentration increases and reacts with the starch indicator to turn the solution blue.

- The time interval between the start of the experiment and the moment the reaction solution turns blue is **inversely** proportional to the rate of Reaction 30.9.

Influence of Temperature (Optional)

The effect of temperature on reaction rate is observed by warming one of the the reaction solutions to around 40°C and measuring the new reaction rate. You then calculate a value for the rate constant at the higher temperature and calculate the **activation energy** of the reaction from Equation 30.8.

SAFETY	
1.	Wear approved eye protection.

PROCEDURE

Work with a partner.

  It is very important that all glassware be clean and that only distilled or deionized water be used in this experiment.

This is because traces of impurities, particularly metal ions such as iron or copper, can catalyze the decomposition of hydrogen peroxide.[2] The starch and hydrogen peroxide solutions must be prepared not more than 24 hours before use.[3] All reagents are dispensed from burets that have come to room temperature.

---

[2]If necessary, one drop of EDTA solution (0.1 M EDTA: 37.2 g ethylenediaminetetraacetic acid disodium salt per liter of H_2O) may be added to each reaction mixture to sequester trace metal ions and minimize their effects.

[3] 1% boiled starch stock solution: Mix 10 g soluble starch in a little cold water to just dissolve it and pour into 1 L of boiling water. Then stir the solution, allow to cool, and store in a refrigerator. Make the starch solution no more than 2-3 days before the experiment. The 3% hydrogen peroxide stock solution should be standardized the day of the experiment for best results.

A. Experiments with [H_2O_2] Constant

1.   Prepare 4 reaction mixtures containing all the reagents *except H_2O_2* in clean, **labelled**, 400 mL beakers according to the table below. Measure 10 mL of H_2O_2 into a separate clean small beaker.
Measure the reagent volumes very carefully.

Beaker	H_2O (mL)	0.05 M KI (mL)	0.05 M $Na_2S_2O_3$ (mL)	0.1% Starch (mL)	0.05 M Buffer[4] (mL)	Total Vol. (mL)	0.8 M H_2O_2 (mL)
1	90	60	5	5	30	200	10
2	100	50	5	5	30	200	10
3	120	30	5	5	30	200	10
4	125	25	5	5	30	200	10

Measure the temperature of each solution. They should be within 1°C (±0.5 °C) of each other. If they are not, place the beakers close to one another and stir them gently until their temperatures are within 1°C.[5]

2. *Beaker 1:* Have a clock or timer ready to use. Check and record the temperature of mixture 1 on the Data sheet. Watch the sweep-second of the clock carefully and just as it passes the 12, pour the H_2O_2 quickly into the reaction mixture in beaker 1. (Or start the timer at the same time that you add the H_2O_2 to the reaction mixture.) Stir continuously and watch the solution for the first appearance of color. Stop the timer or note the sweep-second hand position when the color appears.
          Watch carefully because the color appears quite suddenly.
Record the elapsed time on the Data sheet.

3. *Beakers 2-4:* Before each measurement, carefully measure 10 mL of H_2O_2 into a separate clean small beaker. Carry out the measurement as in step 2. Record the elapsed time on the Data sheet.

4. Measure the pH of each beaker with pH paper and record the value on the Data sheet.

B. Experiments with [KI] Constant

1.   Prepare 4 reaction mixtures containing all the reagents *except H_2O_2* in clean, **labelled**, 400 mL beakers according to the table below. Notice that the reaction mixture in beaker 8 is the same as that in beaker 4, which will yield a repeat measurement.
          Measure the reagent volumes very carefully.

Beaker	H_2O (mL)	0.05 M KI (mL)	0.05 M $Na_2S_2O_3$ (mL)	0.1% Starch (mL)	0.05 M Buffer[6] (mL)	Total Vol. (mL)	0.8 M H_2O_2 (mL)
5	105	25	5	5	30	200	30
6	115	25	5	5	30	200	20
7	120	25	5	5	30	200	15
8	125	25	5	5	30	200	10

Measure the temperature of each solution. They should be within 1°C (±0.5 °C) of each other. If they are not, place the beakers close to one another and stir them gently until their temperatures are within 1°C.

2. *Beakers 5-8:* Before each measurement, carefully measure the correct amount of H_2O_2 into a separate clean small beaker. Carry out the measurement as in step 2, Part A. Record the elapsed time on the Data sheet.

3. Measure the pH of each beaker with pH paper and record the value on the Data sheet.

---

[4]Use either a commercial acetate buffer, pH = 4.7, or make one by combining equal volumes of 0.05 M sodium acetate and 0.05 M acetic acid.

[5] Alternatively, you can have a large beaker of cold water and a large beaker of warm water ready for cooling or warming the solutions to within 1°C. Each reaction beaker should be adjusted to the temperature of the first experiment just before adding the H_2O_2 solution. Do not bother to adjust the temperature of the H_2O_2 solution for it will cause a negligible change.

[6]Use either a commercial acetate buffer, pH = 4.7, or make one by combining equal volumes of 0.05 M sodium acetate and 0.05 M acetic acid.

C. Experiment at a Different pH

In Parts A and B, all the reaction solutions were buffered to about pH = 4.7, so the hydrogen ion concentrations all were the same. According to reaction 30.4, however, H^+ is a reactant and its concentration should be varied to determine the reaction order with respect to H^+. In experiments 1-8, the pH was 4.7, so that $[H^+] = 1.9 \times 10^{-5}$ mol/L. In experiment 9 we use the same concentrations of H_2O_2 and KI as in experiments 4 and 8, but add 25 mL of 0.30 M acetic acid (CH_3COOH) which increases $[H^+]$ a factor of ten, to $[H^+] = 1.9 \times 10^{-4}$ and pH = 3.7.

1.　Prepare the reaction mixture containing all the reagents *except H_2O_2* in a clean, **labelled**, 400 mL beaker according to the table below. The concentrations of H_2O_2 and KI are the same as in experiments 4 and 8, but the concentration of H^+ is 10 times greater.
Measure the reagent volumes very carefully.

Beaker	H_2O (mL)	0.05 M KI (mL)	0.05 M $Na_2S_2O_3$ (mL)	0.1% Starch (mL)	0.05 M Buffer (mL)	Total Vol. (mL)	0.3 M acetic acid (mL)		0.8 M H_2O_2 (mL)	
9	100	25	5	5	30	200	25		10	

2. *Beaker 9 :* 　Before the measurement, carefully measure 10 mL of H_2O_2 into a separate clean small beaker. Carry out the measurement as in step 2, Part A. Record the elapsed time on the Data sheet.

3. Measure the pH of beaker 9 with pH paper and record the value on the Data sheet.

D. Experiment at a Different Temperature (Optional)

In Parts A, B, and C, all the reaction solutions were at the same temperature. To determine the effect of temperature and measure the activation energy, we repeat one measurement at a higher temperature.

1.　Prepare the reaction mixture containing all the reagents *except H_2O_2* in a clean, **labelled**, 400 mL beaker according to the table below. The concentrations of H_2O_2 and KI are the same as in experiments 4 and 8, but the reaction temperature will be higher.
Measure the reagent volumes very carefully.

Beaker	H_2O (mL)	0.05 M KI (mL)	0.05 M $Na_2S_2O_3$ (mL)	0.1% Starch (mL)	0.05 M Buffer[7] (mL)	Total Vol. (mL)		0.8 M H_2O_2 (mL)	
10	125	25	5	5	30	200		10	

2.　Heat about 400 mL of water over a burner or on a hot plate to a little below boiling.

3.　Place 200 mL of cold water in an 800 mL beaker. Add hot water with stirring to this beaker until the water temperature is about 20°C higher than the room temperature of rate experiments 1-9.

4.　Put the beaker containing the reaction mixture into the water bath. Stir and measure the temperature until it stabilizes. Record the final temperature on the Data sheet.

5. *Beaker 10:* 　Before the measurement, carefully measure 10 mL of room temperature H_2O_2 into a separate clean small beaker. Carry out the measurement as in step 2, Part A, adding room temperature H_2O_2 to the warm reaction mixture. Record the elapsed time on the Data sheet. You will assume that the addition of 10 mL of room temperature H_2O_2 solution to 190 mL of warmer solution does change the reaction temperature significantly.

6. Measure the pH of beaker 10 with pH paper and record the value on the Data sheet.

---

[7]Use either a commercial acetate buffer, pH = 4.7, or make one by combining equal volumes of 0.05 M sodium acetate and 0.05 M acetic acid.

Name ____   Date ____________

EXPERIMENT 30
PRELABORATORY EXERCISE

1.   A series of measurements on the aqueous clock reaction
$$6\,I^- + BrO_3^- + 6\,H^+ \longrightarrow 3\,I_2 + Br^- + 3\,H_2O$$
produced the following experimental data:

Experiment No.	0.020 M KI (mL)	0.0020 M Na$_2$S$_2$O$_3$ (mL)	0.080 M KBrO$_3$ (mL)	0.20 M HCl (mL)	H$_2$O (mL)	Total Vol. (mL)	Elapsed time (s)	Temp °C
1	10	10	10	10	60	100	90	27
2	20	10	10	10	50	100	42	27
3	10	10	20	10	50	100	45	27
4	10	10	10	20	50	100	21	27
5	10	10	10	10	60	100	38	40

The elapsed time is the time for the solution to change color, which indicates the time required for a fixed number of moles of iodide ion to react with $S_2O_3^{2-}$. (See *"Clock" Method for Timing the Reaction Rate* in the Introduction.)

a.   Find the concentration in the reaction mixture for each reactant and then calculate the reaction rate for each experiment, i.e., the moles of I^- reacting per liter per second (see Example 2). Enter the values in the table below.

Experiment No.	[KI] (M)	[Na$_2$S$_2$O$_3$] (M)	[KBrO$_3$] (M)	[HCl] (M)	H$_2$O (mL)	Total Vol. (mL)	Reaction rate (mol/L-s)	Temp °C
1					60	100		27
2					50	100		27
3					50	100		27
4					50	100		27
5					60	100		40

b.   The rate equation for the reaction has the form:
$$R = k[I^-]^x[BrO_3^-]^y[H^+]^z$$
Use data from the tables to calculate the reaction order for each reactant (see Example 1). All reaction orders are integers.

$$x = \underline{\hspace{2cm}} \; ; \; y = \underline{\hspace{2cm}} \; ; \; z = \underline{\hspace{2cm}}$$

c.   Calculate the rate constant **k** and write the general rate equation using numerical values for **k**, **x**, and **y** (see Example 2).

$$k = \underline{\hspace{3cm}} \; ; \; R = \underline{\hspace{6cm}}$$

d.   Calculate the activation energy for the reaction (see Example 3). $E_a = \underline{\hspace{4cm}}$

Calculations:

Name ____   Date _____________

EXPERIMENT 30

DATA

(Observe significant figures in all calculations.)

A. Experiments with [H_2O_2] Constant

Beaker	H_2O (mL)	0.05 M KI (mL)	0.05 M $Na_2S_2O_3$ (mL)	0.1% Starch (mL)	0.05 M Buffer (mL)	Total Vol. (mL)	0.8 M H_2O_2 (mL)	pH	Temp °C	Elapsed time (s)
1	90	60	5	5	30	200	10			
2	100	50	5	5	30	200	10			
3	120	30	5	5	30	200	10			
4	125	25	5	5	30	200	10			

Reaction order of I^-: $y =$ _____

B. Experiments with [KI] Constant

Beaker	H_2O (mL)	0.05 M KI (mL)	0.05 M $Na_2S_2O_3$ (mL)	0.1% Starch (mL)	0.05 M Buffer (mL)	Total Vol. (mL)	0.8 M H_2O_2 (mL)	pH	Temp °C	Elapsed time (s)
5	105	25	5	5	30	200	30			
6	115	25	5	5	30	200	20			
7	120	25	5	5	30	200	15			
8	125	25	5	5	30	200	10			

Reaction order of H_2O_2: $x =$ _____

C. Experiment at a Different pH

Beaker	H_2O (mL)	KI (mL)	$Na_2S_2O_3$ (mL)	Starch (mL)	Buffer (mL)	Total Vol (mL)	0.3 M acetic acid (mL)	pH	H_2O_2 (mL)	Temp °C	Elapsed time (s)
9	125	25	5	5	30	200	25		10		

Reaction order of H^+: $z =$ _____

D. Experiment at a Different Temperature (Optional)

Beaker	H_2O (mL)	0.05 M KI (mL)	0.05 M $Na_2S_2O_3$ (mL)	0.1% Starch (mL)	0.05 M Buffer (mL)	Total Vol. (mL)	0.8 M H_2O_2 (mL)	pH	Temp °C	Elapsed time (s)
10	125	25	5	5	30	200	10			

Activation energy: $E_a =$ ________________________

Calculations:

Name ____   Date ______________

EXPERIMENT 30
QUESTIONS

1. For each experiment, calculate the reactant concentrations, reaction rates and the rate constant. Enter the values in the table below. Determine $[H^+]$ from the pH.

Beaker	$[KI]$ (M)	$[Na_2S_2O_3]$ (M)	$[H_2O_2]$ (M)	$[H^+]$ (M)	R (mol L^{-1} s^{-1})	k (L^3 mol^{-3} s^{-1})
1						
2						
3						
4						
5						
6						
7						
8						

Average value of k = _______________________

2. **a.** Write the general rate equation for Reaction 30.9, using your average value for the rate constant and the experimental reaction orders for each of the reactants.

b. What is the *overall* reaction order?   **Overall order** = _________

3. How can you set the "alarm" of the clock reaction so that it will run for a longer time and consume more of the reactants?

Calculations:

EXPERIMENT 31
Electrochemistry:
Reduction Potential Series and Electrolysis

PRELABORATORY PREPARATION
1. Do the Prelaboratory Exercise and turn it in at the beginning of your laboratory period.

INTRODUCTION
The atoms in every metal have a certain tendency to dissolve into a solution as positively charged cations. If a piece of metal is immersed in water, some of the metal atoms will leave the solid and enter the water as positive metal ions, leaving electrons behind in the solid:

$$\text{metal atom} \longrightarrow \text{cation} + \text{electrons} \qquad (31.1)$$
$$\text{(in solid metal)} \qquad \text{(in solution)} \quad \text{(in solid metal)}$$

The excess electrons left in the metal cause the solid metal to build up a negative charge. The positively charged cations entering the solution cause the solution to build up a positive charge. As this process continues, metal atoms trying to dissolve as cations have a more and more difficult task, because the metal and solution have acquired charges that resist further reaction. Because of the repulsive force between like charges, as the metal becomes more and more negative it resists acquiring still more electrons from dissolving metal atoms. Similarly, the positively charged solution resists the entry of newly dissolved cations. At the same time, the negatively charged piece of metal attracts the cations back to its surface, speeding up the reverse reaction:

$$\text{cation} + \text{electrons} \longrightarrow \text{metal atom} \qquad (31.2)$$
$$\text{(in solution)} \quad \text{(in solid metal)} \qquad \text{(in solid metal)}$$

Reaction (31.1) can proceed, continuing to dissolve metal, only if the solution and the metal do not build up progressively larger charges. There are two ways to accomplish this:
1. Let cations entering the solution replace cations already in the solution.
2. Construct an electrochemical cell.

Replacement of Cations in Solution
Suppose metal **A** dissolves into a solution that already contains cations of a different metal **B**. The increase of positive charge in the solution and negative charge on the metal can attract cations of metal **B** to the solid metal where they accept electrons and change to the solid metal form. An example shown in Figure 31.1 is the dissolving of zinc in a solution of copper sulfate. The zinc goes into solution as cations (forward Reaction 31.3) and copper cations plate out on the zinc metal as a layer of copper metal (reverse Reaction 31.4). In this way, neither the metal nor the solution builds up an electrical charge, both remaining neutral.

Of course, once metallic copper has formed, it too tries to dissolve (forward Reaction 31.4). To maintain electrical neutrality, only one metal can dissolve; the other must return from solution to the solid form. Thus, zinc and copper solid metals are in competition to dissolve as cations. For reaction 31.3 to proceed in the forward direction, Zn^{2+} must continue to replace Cu^{2+} in the solution. In other words, Reaction 31.3 must go more easily to the right than Reaction 31.4 goes to the right; then Reaction 31.4 is forced to go to the left.

$$Zn \rightleftharpoons Zn^{2+} + 2e^- \qquad (31.3)$$
$$Cu \rightleftharpoons Cu^{2+} + 2e^- \qquad (31.4)$$

This does, in fact, happen, and we say that Zn replaces Cu in solution. Because the Zn metal atoms lose electrons to become cations, the Zn is **oxidized**. The copper cations *gain* electrons to become copper metal atoms, and so the Cu^{2+} ions are **reduced**. Such reactions are called **oxidation-reduction** reactions, or sometimes **redox** reactions. The overall reaction is:

$$Zn + Cu^{2+} \longrightarrow Zn^{2+} + Cu \qquad (31.5)$$
$$\text{oxidation number:} \quad 0 \quad\quad +2 \quad\quad\quad +2 \quad\quad 0$$

The oxidation number of Zn becomes more positive, and so the Zn is oxidized. Conversely, the oxidation number of Cu^{2+} becomes more negative, and therefore the Cu^{2+} is reduced.

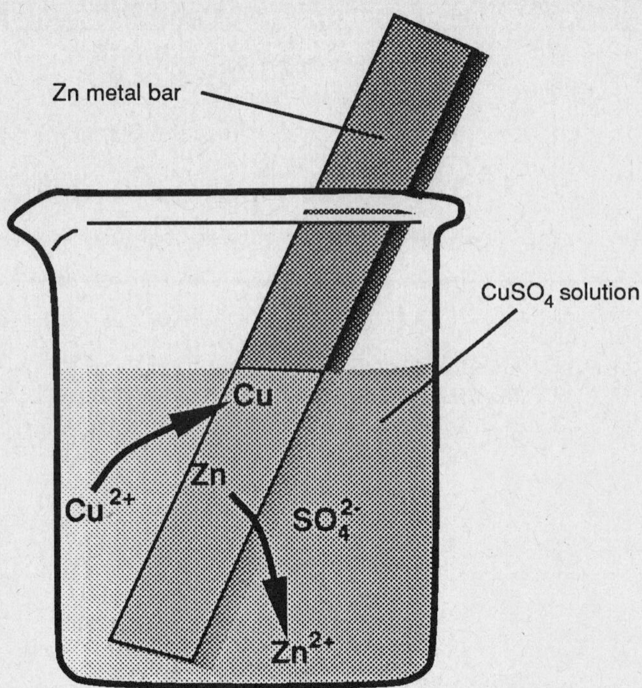

Figure 31.1. Zn^{2+} replacing Cu^{2+} in a $CuSO_4$ solution by direct reaction.

In order for one element to be oxidized and lose electrons, another must be reduced and gain those electrons. Oxidation and reduction always accompany one another. The driving force for such reactions is called the **electromotive force**. Zn possesses a greater tendency to go into solution as cations than does Cu. Your textbook has a table of **standard electrode reduction potentials** that expresses this tendency numerically in volts.

The more *positive* the standard reduction potential, the greater is the tendency for the *reduction* reaction to occur.

The more *negative* the standard reduction potential, the greater is the tendency for the *oxidation* reaction to occur.

EXAMPLE 1:

Compare the reduction potentials of Zn and Cu to determine which metal will replace the other in solution. From a table of standard reduction potentials, one finds:

$$Zn^{2+} + 2e^- \rightleftharpoons Zn \qquad E° = -0.763 \text{ volts} \tag{31.6}$$
$$Cu^{2+} + 2e^- \rightleftharpoons Cu \qquad E° = +0.340 \text{ volts} \tag{31.7}$$

Since the copper reduction potential is more positive, copper will be reduced, precipitating out of solution (forward Reaction 31.7), and zinc will be oxidized, dissolving into solution (reverse Reaction 31.6).

The standard electrode potential depends on the concentration of the cation in the solution in contact with the electrode. The temperature also affects the standard electrode potential. Standard electrode potentials are determined at 25°C with the electrode in contact with a solution having an activity of 1.0 for the cation corresponding to the electrode metal.

Electrochemical Cells

Another way to retain the electrical neutrality of the metal and solution that is necessary in order to dissolve the Zn is to construct an **electrochemical cell**, see Figure 31.2 where a Zn electrode is immersed in a solution of $ZnSO_4$ and a Cu electrode is immersed in a solution of $CuSO_4$. Picture the situation of Figure 31.2 with the wire connection between the Zn and Cu electrodes removed. Then both Zn and Cu metal atoms would tend to dissolve as cations, charging their respective metal bars negative and their solutions positive. The charge accumulation would soon stop the reactions because there are no different metal cations in their respective solutions to replace. However, the Zn electrode would attain a *greater* negative charge than would the Cu electrode, because the driving force for Reaction (31.3) is greater than that for Reaction (31.4). If you now connect the wire conductor between these electrodes, the greater negative charge on the Zn electrode will drive electrons over to the Cu electrode. The increase in negative charge on the Cu metal will then attract Cu^{2+} ions out of solution to deposit on the electrode as Cu metal.

The porous plug slows down the mixing of the two solutions while still allowing ions to move between the solutions and maintain electrical neutrality. With the porous barrier, Cu metal is prevented from plating on the Zn electrode and affecting the nature of the surface. Likewise, the Cu electrode surface remains constant in composition during the reaction. The reaction will proceed as in Equation (31.5), with Zn^{2+} accumulating in the solution on the left-hand side of the cell and Cu^{2+} being depleted from the solution on the right-hand side. The wire, which allows electrons to transfer from left to right in Figure 31.2, keeps the metal bars electrically neutral, but the accumulation of Zn^{2+} and the depletion of Cu^{2+} in the solutions would soon make the solution on the left positive and that on the right negative if the connecting passage were not present. A charge imbalance in the solution is avoided because SO_4^{2-} ions can pass through the porous plug, being attracted into the positive region near the Zn bar and repelled from the negative zone near the Cu bar.

The assembly shown in Figure 31.2 is called an *electrochemical cell*. It allows the oxidation and reduction reactions to be separated spatially, using a connecting wire to transfer the electrons. Because Reaction (31.3) has a greater tendency to go forward than does (31.4), the Zn electrode will remain more negative throughout the reaction than the Cu electrode.

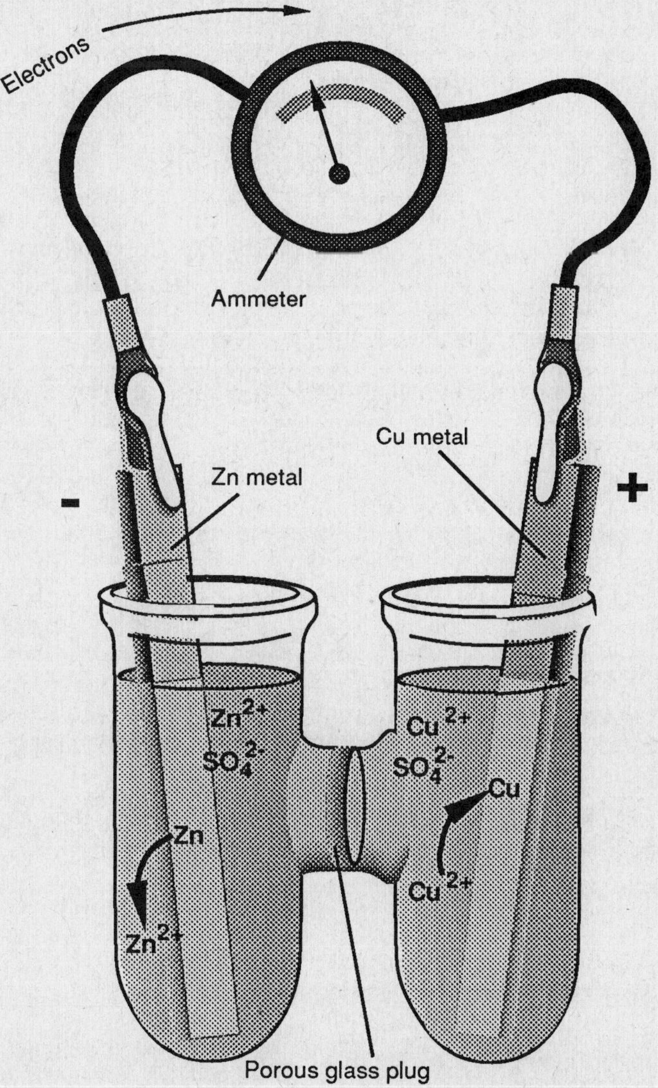

Figure 31.2. Electrochemical cell. Zn and Cu electrodes are immersed in solutions of $ZnSO_4$ and $CuSO_4$ respectively. The two solutions are prevented from freely mixing by a porous plug barrier which permits a potential difference between the solutions to drive charged ions through it but greatly restricts random diffusion. The Zn electrode where oxidation occurs is called the **anode**, and the Cu electrode where reduction occurs is called the **cathode**. The potential difference between the two electrodes is measured with a voltmeter.

Electrolysis Cells

If an external voltage is applied to an electrochemical cell, the voltage difference between the electrodes can be controlled experimentally and electrons can be made to flow through the connecting wire in either direction. In the copper-zinc cell of Figure 31.2, zinc dissolves and copper precipitates in spontaneous reactions determined by their respective reduction potentials. But if an external voltage is applied so as to make the copper electrode more negative than the zinc, electrons will flow the other way from Cu to Zn. Then the reactions will be reversed and Cu will dissolve and Zn precipitate. Forcing an oxidation-reduction reaction to occur by an external voltage is called **electrolysis.**

PLAN OF EXPERIMENT

RANKING OF METALS IN THE REDUCTION POTENTIAL SERIES

You will construct electrochemical cells using a variety of different metal electrodes in order to determine the ranking of these metals with regard to their tendency to undergo the reduction Reaction 31.1. For each metal pair, you can tell which way the electrons move by observing the direction that the needle on a voltmeter deflects or the sign (+ or -) of the voltage on a digital meter. Because a voltmeter measures the voltage *difference* between its two terminals, the magnitude of the voltage reading indicates the *difference* in reduction potential between the two electrodes of the cell.[1]

The metal that gains electrons (the metal *toward* which electrons move) has a more *positive* reduction potential than its partner in the cell.

If metal **A** has a more positive reduction potential than metal **B**, and metal **B** has a more positive reduction potential than metal **C**, then the reduction potential of metal **A** is more positive than that of metal **C**. You can use this principle to make predictions about the direction of electron flow and the magnitude of the voltage generated for metal pairs that have been measured with other metals but not together. To make the voltmeter indicate a positive voltage (needle deflection toward the right or a plus sign on a digital meter), electrons must enter the meter's *positive* connection and exit at the meter's *negative* connection.

When the voltmeter indicates a positive voltage, its positive terminal is connected to the electrode undergoing an oxidation reaction and its negative terminal is connected to the electrode undergoing a reduction reaction.

The sign convention on voltmeters is *opposite* to that in electrochemistry. The voltmeter indicates a positive voltage when the positive meter terminal is connected to the cell electrode that has the more negative charge. This is the cell **anode** where oxidation is occurring and is the source of electrons. Reduction occurs at the cell **cathode.**

Because a voltmeter registers only a voltage *difference*, you cannot directly measure the individual reduction potentials for each electrode of a cell. To circumvent this problem, we will use a copper electrode in contact with 0.1 M $Cu(NO_3)_2$ as a reference standard having a reduction potential of +0.34 volts. The reduction potentials of other metals can be calculated from:

(measured positive cell voltage) = (reduction potential of cathode) - (reduction potential of anode)

$$(31.6)$$

EXAMPLE 1:

The voltage measured for a cell made with copper and zinc electrodes was +1.07 volts. The copper electrode was in contact with 0.1 M $Cu(NO_3)_2$ and the zinc electrode with 0.1 M $Zn(NO_3)_2$. To make the meter read a positive voltage, it was necessary to connect the copper electrode to the negative terminal of the meter and the zinc electrode to the positive terminal.

    a. Which electrode is the cell cathode and which is the anode?

    b. Which metal dissolves and which plates out as solid metal?

    c. What is the measured reduction potential for zinc?

SOLUTION:

    a. The copper electrode was connected to the negative terminal and the zinc electrode was connected to the positive terminal. Therefore, copper is the cathode and zinc is the anode of the cell.

    b. Reduction occurs at the cathode, so copper plates out in the reduction reaction:

$$Cu^{2+} + 2\,e^- \longrightarrow Cu$$

---

[1]Different types of voltmeters differ in their ability to accurately measure the voltage of electrochemical cells. The most accurate measurements are made with a *null* type of voltmeter that balances an external reference voltage against the cell voltage. Normal electronic voltmeters usually will give a fairly accurate reading ($\pm$ 10%). Mechanical magnetic coil voltmeters always give a low reading. The purpose of this experiment is not to measure accurate reduction potentials but to find the ranking of different metals in the reduction potential series, and any meter type is suitable for this.

Oxidation occurs at the anode, so zinc dissolves in the oxidation reaction:
$$Zn \longrightarrow Zn^{2+} + 2\,e^-$$
The overall cell reaction is Reaction 31.5.

c. Cu is our reference electrode with a known reduction potential of +0.34 volts. Cu is the cathode and Zn is the anode. The Zn reduction potential may be determined using Equation 31.6:
$$+1.07\ V = +0.34\ V - (Zn\ reduction\ potential)$$
$$Zn\ reduction\ potential = E°_{Zn} = (+0.34 - 1.07)\ V = \textbf{-0.73 V}$$

ELECTROLYSIS

Using an external voltage source, you will electrolyze two cells in series. Neither cell would undergo a spontaneous oxidation-reduction reaction without the external voltage. One cell consists of two graphite electrodes immersed in KI solutions and the other contains two copper electrodes immersed in a single $CuSO_4$ solution (see Figure 31.3).

The reactions in the graphite-KI cell are:

$$2\,I^- \longrightarrow I_2(aq) + 2\,e^- \qquad \text{oxidation}$$
$$2\,H_2O + 2\,e^- \longrightarrow 2\,OH^-\ (aq) + H_2(g) \qquad \text{reduction}$$

The reactions in the Cu-$CuSO_4$ cell are:

$$Cu \longrightarrow Cu^{2+} + 2\,e^- \qquad \text{oxidation}$$
$$Cu^{2+} + 2\,e^- \longrightarrow Cu \qquad \text{reduction}$$

In the Cu-$CuSO_4$ cell, the overall chemical change is the transfer of copper metal from one electrode to the other.

Data from this experiment allows you to calculate the molar mass of copper as follows:

1. Measure the pH of the KI solution before and after electrolysis with a pH meter to determine $[OH^-]$. Then measure the volume of the KI solution and calculate the moles of OH^- formed in the graphite-KI cell. The moles of OH^- formed are equal to the moles of electrons tranferred in the electrolysis.
This same number of electrons will have passed through the Cu-$CuSO_4$ cell also.

2. Determine the mass of copper transferred in the Cu-$CuSO_4$ cell.
One copper atom is transferred for every two electrons that pass through the cell.
The number of moles in the mass of copper transferred is equal to one-half the moles of electrons passing through the cell. The moles of electrons in turn is equal to the moles of OH^- measured in the titration.
The molar mass of copper is equal to the mass of Cu transferred divided by 2 times the moles of OH^- measured in the titration.

SAFETY

1. Wear approved eye protection.

2. Many of the solutions used are toxic. If skin or clothing contact occurs, wash immediately with soap and large amounts of running water.

At the end of the laboratory period, wash your hands carefully with soap and water.

3. Electrolysis apparatus may operate at dangerous voltages and currents. Follow the instructor's directions carefully to avoid serious shock hazards. Make all connections securely before turning the power supply on.

Be especially careful not to short the leads to the power supply.

PROCEDURE
Work with a partner.

A variety of metal electrodes will be available in the laboratory (Cu, Zn, Al, Fe, Pb, etc.). Carbon electrodes, which are inert to most oxidation-reduction reactions also will be available. Use them in various combinations in order to determine the ranking of the standard reduction potentials of the different metals.

A. RANKING OF METALS IN THE REDUCTION POTENTIAL SERIES
1. Obtain a glass jar, porous cup, electrode supports, wires, and one of each kind of electrode available.
Clean the electrodes well with steel wool before using them.

2.   Assemble the apparatus as shown in Figure 31.3, using Cu and any other of your electrodes.
The first several electrode pairs you measure should include Cu as one of the electrodes.
Using the appropriate matching solutions [e.g., $Pb(NO_3)_2$ with the Pb electrode] place one solution and its corresponding electrode in the glass jar and the other electrode with its solution in the porous cup. All solutions are 0.1 M.
   Use electrode holders if they are available. Try to repeat the electrode positions and spacing as closely as possible in each cell. [2]

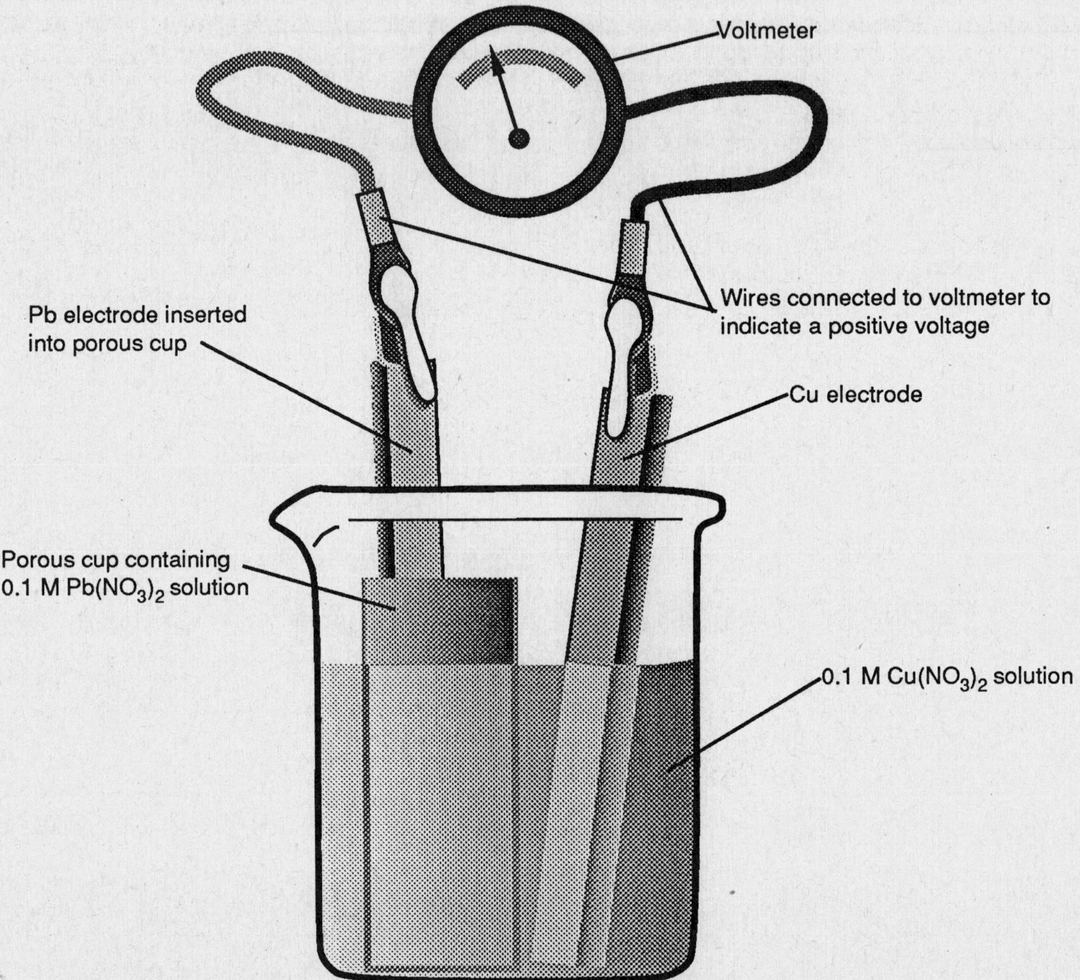

Figure 31.3.   Lead-copper cell illustrating a typical cell configuration for comparing the reduction potentials of different metal electrodes. The Pb electrode is immersed in O.1 M $Pb(NO_3)_2$ contained in a porous cup to prevent mixing of the solutions. Other cell configurations for separating the solutions might be used in your laboratory.

3.   There will be voltmeters available in the lab. Use one to measure the voltage of your cell, noticing in particular which cell electrode is the anode and which the cathode. Record this information and the cell voltage in the first table of your data sheet. Write the electrode reactions in the second table.
   When the voltmeter indicates a positive voltage, its positive terminal is connected to the cell anode (where oxidation is taking place) and its negative terminal is connected to the cell cathode (where reduction is taking place).

---

[2]This is especially important if voltmeters with low input resistance are used. Electrochemical cells have high internal resistance and an effort should be made to keep the cell voltage drop consistent from cell to cell.

4.   Make similar measurements using all your other electrodes, each one paired with Cu.
Some metals, when paired with copper, will not produce an easily measured cell voltage.
To rank these metals more accurately, measure several metal pairs that do not include copper, each pair including one metal that did not produce an easily measured voltage when paired with Cu. For the other electrode, any other metal whose electrode potential is known from comparison with copper can be used as a reference electrode. Measure enough electrode pairs to allow you to determine an electrode potential for each metal available.

5.   You will not have a silver electrode available but you can use the carbon electrode in $AgNO_3$ solution to determine the reduction potential for $Ag^+ + e^- \longrightarrow Ag$. Pair the carbon electrode with metals that have more negative reduction potentials than Ag. Use a metal in its appropriate solution as one half of the cell and carbon immersed in $AgNO_3$ as the other half. If the cell causes the reduction of Ag^+ to Ag metal plated onto the carbon electrode, then the other electrode must be the cell anode and be undergoing an oxidation reaction. If you know the reduction potential for the other metal, the reduction potential for silver can be found from Equation 31.6.

6.   Tabulate the reduction reactions and your measured reduction potentials in order, from most positive to most negative, on the Data sheet. Put the accepted standard electrode potentials from the reduction potential table in your text into your table also.

7.   When you are finished, clean your electrodes and return all solutions to the marked bottles.

B. ELECTROLYSIS
1.   Set up the apparatus in Figure 31.4 using 250 mL beakers containing about 150 mL of solution. The KI and $CuSO_4$ solutions will be available in the laboratory. With a pH meter, measure and record the pH of the KI solution.

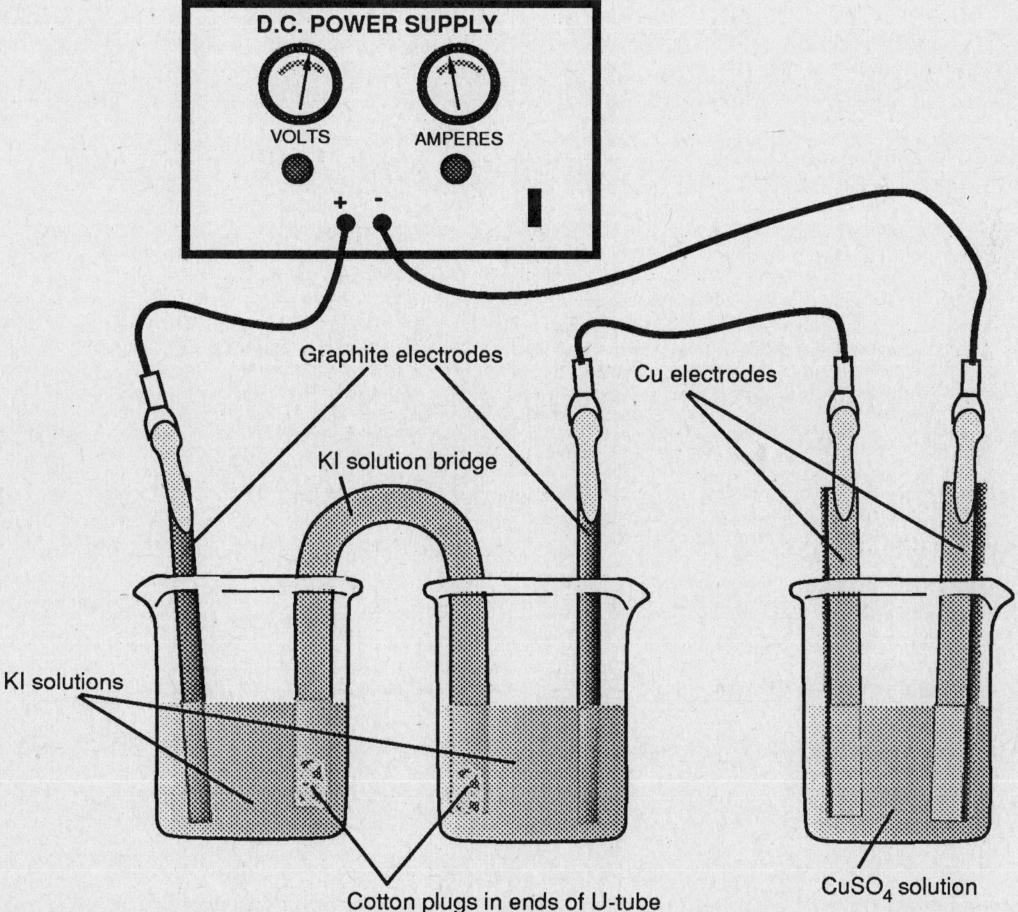

Figure 31.4   Electrolysis apparatus.

2.   Use steel wool or sandpaper to clean two copper strips. Scratch identifying numbers **1** and **2** on them. Wash them in water and dry them carefully on a watch glass placed on a *warm* hot plate or by holding them high over a flame.
     The Cu strips must be heat-dried very gently to avoid any oxide discoloration.

3.   Weigh the copper strips carefully to 0.1 mg and record their masses on the Data sheet.

4.   Assemble the copper strips and graphite rods into your electrolysis apparatus. Run the electrolysis at about 6V for 20 minutes. Record the exact starting and stopping times.
     Be careful not to move the electrodes during the electrolysis. Record the current (amperes) every 4 or 5 minutes.

5.   When electrolysis is finished, disconnect the voltage source. Remove the copper electrodes and dry them as before. Reweigh the electrodes and record their new masses.

6.   The pH of one of the KI solutions will be higher than at the start of electrolysis. Measure and record the higher pH[3]. Then measure the volume of this solution with a graduated cylinder. Assume the final and initial volumes are equal.

7.   Calculate the moles of electrons that have passed through the cells during electrolysis and the molar mass of Cu.

8.   Use the average current through the apparatus and the total electrolysis time to calculate the moles of electrons that have been transported. Compare this value with that obtained from the pH measurement.

---

[3]Alternatively, the pH of the KI solutions can be determined by titration with standardized 0.1 M HCl solution. Use phenolphthalein indicator.

Name ____   Date ____________

EXPERIMENT 31
PRELABORATORY EXERCISE

1. The voltage measured from a cell constructed with Cu and Mg electrodes immersed in their corresponding nitrate solutions was +2.72 V. The Mg electrode was connected to the positive voltmeter terminal and the copper electrode to the negative.

    a. Which electrode is the cell cathode and which the anode?   **anode: _______; cathode:_______**

    b. Which metal dissolves and which plates out as solid metal? **dissolves:_______; plates out:______**

    c. What is the measured reduction potential for zinc? $E° =$ ___________________

2. The electrolysis apparatus of Figure 31.4 was operated for 18.0 minutes using nickel electrodes (in $NiSO_4$ solution) instead of copper. At the end of electrolysis, it was found that the average value of the mass change of the two nickel electrodes was 0.369g. In the KI solution which increased in pH, the volume was 164 mL and the final pH = 12.1. Assume that the nickel cation was Ni^{2+}. What was the measured molar mass of nickel?

Measured molar mass of Ni = _________________

Calculations:

Name ____   Date ____________

EXPERIMENT 31

DATA
(Observe significant figures in all calculations.)

A. RANKING OF METALS IN THE REDUCTION POTENTIAL SERIES

1. Experimental Results

Electrode pair	Measured cell voltage	Anode metal	Cathode metal	Oxidation reaction (anode reaction)	Reduction reaction (cathode reaction)

2. Reduction Potentials for Each Metal Used
(Arrange in order from most positive to most negative.)

Metal	Reduction half-reaction (cathode reaction)	Measured E (volts)	Reference table value for $E°$ (volts)

DATA
(Observe significant figures in all calculations.)

B. ELECTROLYSIS

1. Initial pH of KI solution: _______ ; Hydroxyl ion concentration in initial KI solution: _______________

2. Mass of Cu strips after cleaning and marking,   No. 1: _______________ ; No. 2: _______________

3. Electrolysis times: Starting time: _____________ ; Ending time: _____________ ; Total time: _________

4. Current readings:

Amperes						
Time						

Average current: _____________________

5. Mass of Cu strips after electrolysis,   No. 1: _______________ ; No. 2: _______________

   Mass change for each strip (absolute value),   No. 1: _______________ ; No. 2: _______________

   Average mass change of Cu strips: _______________

6. Final pH of KI solution which increased in pH: _______________

7. Hydroxyl ion concentration in final KI solution: _______________

8. Volume of final KI solution: _______________

9. Moles of OH⁻ produced during electrolysis: _______________

10. Moles of electrons transported (from OH⁻ measurement):   _______________

11. Moles of electrons transported (from average current):   _______________

12. Molar mass of Cu (from Cu^{2+} cation):   _______________

Calculations:

Name ____   Date ____________

EXPERIMENT 31
QUESTIONS

1. How would the following experimental errors affect the value of the molar mass of Cu (+, -, or 0)?

    a. The final pH measured for the KI solution is too low. ________

    b. The ammeter indicates too high a current.   ________

    c. The initial mass measured for just *one* of the Cu
    electrodes was too high.   ________

2. Why is it important to keep the electrode positions and spacing as alike as possible from cell to cell when comparing reduction potentials of different metals?

3. Explain why it was necessary to know the reduction potential for the reaction:
$$Cu^{2+} + 2\,e^- \longrightarrow Cu \qquad E° = 0.34 \text{ V}$$
and use it as a reference standard for these experiments.

4. Was there a significant difference in the number of moles of electrons transported during the electrolysis as determined by the pH measurement and by the average current measurement? If so, suggest some reasons why the two measurements might give different results. Tell which measurement you think should be the more accurate and explain your reasons.

Introduction to Semimicro Qualitative Analysis
(Background material for Experiments 32-34)

INTRODUCTION

The next group of experiments are a series of studies in qualitative analysis, in which you learn how to identify ions that are present in unknown solutions and solids. The method of identification is based on the behavior of different ions when they are made to react with certain reagents. As you perform and come to understand these analyses, you will make use of much of the solution and equilibrium chemistry that you have learned in the classroom. In this respect, the qualitative analysis sequence serves an integrating role, presenting a series of problems to be solved by applying your overall chemical knowledge. In addition, these experiments can be a lot of fun. Each experiment presents a puzzle which is solved "detective fashion" by assembling a collection of chemical clues into an airtight case for the correct identifications. As a bonus, the clues often take the form of colorful solutions and precipitates.

OVERALL PLAN

Only cations of Groups 1, 2, and 3 will be analyzed in this brief introduction to qualitative analysis.
The general idea is to prove the presence or absence of a given cation in an unknown sample by observing chemical reactions of the unknown with a particular set of reagents. The process of identification is simplified by chemically separating smaller groups from the large class of all cations. Identification tests for the ions in these smaller groups are less ambiguous and easier to perform than would be the case if the entire class of cations were analyzed as a whole.

OVERVIEW OF CATION ANALYSIS SCHEME

(Although we will only do experiments on Groups 1, 2, and 3, all 5 cation groups are discussed below.)

Cations are classified into five different groups. By mixing a series of reagents with the unknown, in a carefully structured sequence, the different cation groups can be separated from one another.

Cation Group 1: Cations that are precipitated as chlorides from cold, dilute acidic solution. These are Ag^+, Hg_2^{2+}, and Pb^{2+}.

Cation Group 2: Cations that are not precipitated as chlorides, but precipitated as sulfides from acidic solutions. These are Cu^{2+}, Cd^{2+}, Hg^{2+}, Pb^{2+}, Bi^{3+}, Sn^{2+}, Sn^{4+}, As^{3+}, As^{5+}, Sb^{3+}, and Sb^{5+}.

Cation Group 3: Cations that are not precipitated as chlorides or sulfides from acidic solutions, but precipitated as hydroxides or sulfides from ammoniacal solutions. These are Al^{3+}, Cr^{3+}, Fe^{3+}, Mn^{2+}, Co^{2+}, Ni^{2+}, and Zn^{2+}.

Cation Group 4: Cations that are not precipitated under any of the above conditions, but precipitated as carbonates from ammoniacal solutions. These are Ca^{2+}, Sr^{2+}, and Ba^{2+}.

Cation Group 5: Cations not precipitated at all under the conditions above, remaining in solution after Groups 1-4 have been separated. These are Na^+, K^+, Mg^{2+}, and NH_4^+.

The reagent for a particular group will precipitate not only the cations in that group, but cations in all preceding groups as well. It will not precipitate cations in groups that follow. The system of analysis presented here requires that each group of cations be removed in numerical order.

SPECIAL TECHNIQUES AND PRECAUTIONS FOR QUALITATIVE ANALYSIS
Organization and Cleanliness

It is important to be orderly and systematic in your laboratory procedure and in the recording of your experimental results. Your recorded observations should include:

1. Color of solutions and precipitates.
2. Color changes that occur during mixing or heating.
3. Time needed for precipitation or other reaction.
4. Character of precipitate: crystalline, flocculant (fluffy), colloidal, etc.
5. Evolution of gases.
6. Odors generated by reactions.

Recorded observations can be arranged conveniently into a flow chart. Figure 31.1 is an example of a flow chart for a sample that is to be analyzed for Group 1 cations.

Figure 31.1. Sample flow chart for cation analysis, designed for Group 1 cations

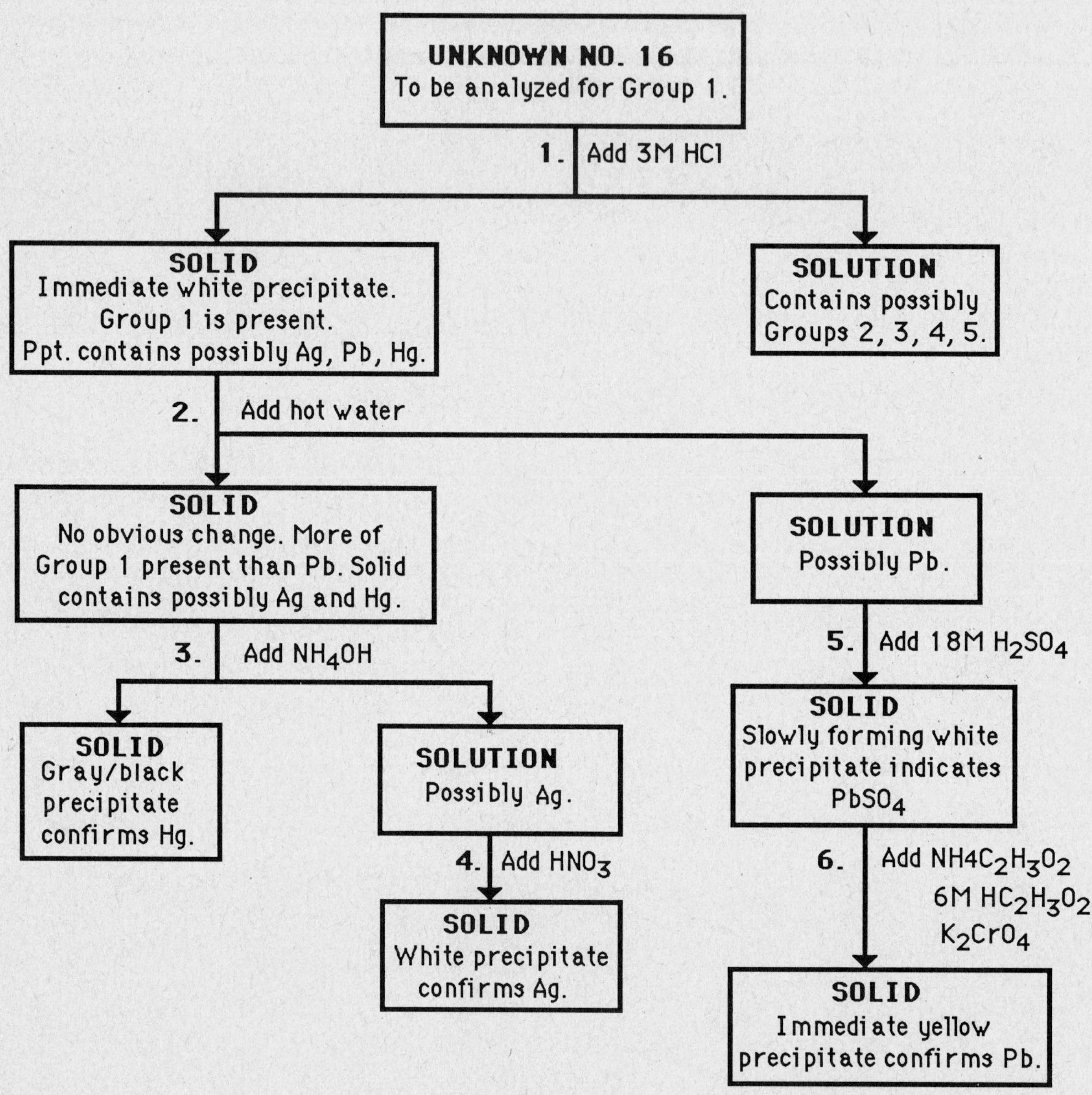

TIPS FOR SUCCESS

Organize your working area carefully. Put all unnecessary equipment away. Arrange clean stirring rods, dropping pipets, pH paper, wash bottle, and reagent bottles conveniently on spread-out clean paper towels. Do not rush through an experiment. Be sure to read the procedure beforehand.

Label all samples and remember to label all solutions that are to be saved for later tests.

You will work with very small samples, so any contamination is likely to be significant. All glassware must be thoroughly cleaned and rinsed with distilled water before use.

Keeping clean stirring rods and dropping pipets in a small beaker of distilled water during your experiments helps to prevent accidental contamination.

Be especially careful not to contaminate one solution with another. Keep clean and dirty test tubes and other equipment well separated on your benchtop. Clean dirty equipment as frequently as possible.

It is very important to use distilled or deionized water for cleaning equipment, making solutions, and washing precipitates. The ions present in tap water almost certainly will interfere with your analysis. This is especially true when analyzing for Cation Group 1, which are precipitated as chlorides, because nearly all water supplies are chlorinated.

Never use tap water in qualitative analysis experiments.

VOLUME MEASUREMENTS

Precise volume measurements are not necessary in these experiments. Before starting the experiment, use a small graduated cylinder to measure 1, 2, 3, and 4 mL portions of water into a test tube, in order to observe the liquid heights of these volumes in your test tubes. Use your dropper to add water to the graduated cylinder, in order to determine the average number of drops per mL. Then, estimate volumes from the height of liquid in your test tube, or by counting drops.

MIXING SOLUTIONS

Stir and mix solutions with a stirring rod or by holding the top of the test tube in your fingers and carefully shaking the tube bottom sideways.

Never close the top of the test tube with your thumb or with a stopper, before shaking. Doing so will introduce contaminants that might interfere with your analysis.

HEATING SOLUTIONS

It is difficult to heat small test tubes over a flame without "bumping" or spattering the solutions out of the tube. For this reason, all solutions are to be heated in a water bath (see Figure 31.2).

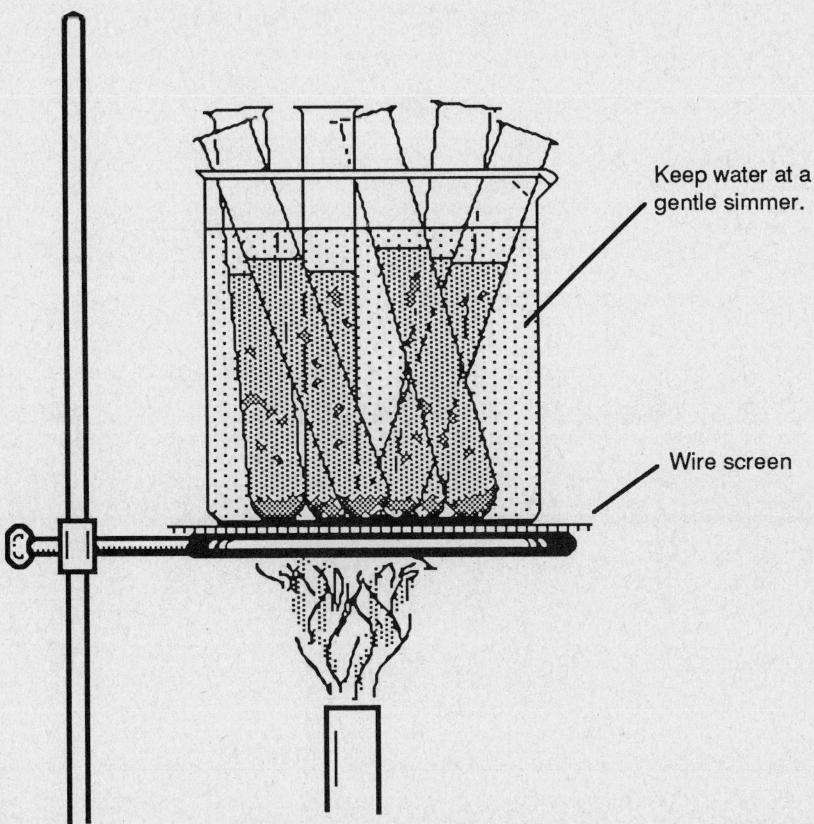

Figure 31.2.   Solutions in test tubes are heated in a water bath to avoid "bumping" and spattering.

HANDLING PRECIPITATES

Most experiments in qualitative analysis require you to form and separate precipitates. The basic techniques are described in Sections VII, VIII, and IX of *Chemical Laboratory Methods.* A few additional suggestions are especially helpful for qualitative analysis:

Formation of Precipitates

1. When trying to form a precipitate, add the precipitating reagent slowly with a dropper while stirring the solution continuously. Warming the solution in a hot water bath helps to grow large particles that settle out faster and are easier to filter.

2. After the precipitate has formed check for completeness of precipitation by adding another drop or two of the precipitating reagent. If more precipitate develops, add a few more drops of precipitating reagent and warm the solution again.

    Avoid adding too much precipitating reagent because a great excess can redissolve some precipitates by forming complex ions.

Centrifuging Precipitates

1. A centrifuge spins the sample tubes very rapidly to speed the settling out of precipitates from a solution. By spinning the sample at speeds greater than 1000 rpm, centrifugal forces around 400 times the force of gravity are generated and even the finest precipitates usually require no more than 2-3 minutes to settle completely. Coarse precipitates may take only 30 seconds. Your instructor will demonstrate the correct use of the centrifuge.

2. A centrifuge must always be balanced with equally full tubes placed opposite one another. You often will centrifuge more than one sample at the same time or might share the centrifuge with other students.

    Be sure your samples are labelled for easy identification.

3. After centrifuging is complete, the supernatant is removed with a capillary pipet.

    Always test the supernatant for completeness of precipitation.

Washing the Precipitate

1. After the supernatant is removed, the precipitate must be washed to remove the last traces of solution and any adsorbed ions that might interfere with later tests.

2. Several washings with small portions of the wash liquid (usually distilled water, but sometimes other solutions are used) are more effective than one washing with a large quantity of wash liquid.

3. Centrifuging often packs precipitates tightly in the bottom of the tube.

    Before washing, break up the packed precipitate with a stirring rod.

    Then, thoroughly stir the precipitate with the wash liquid, centrifuge again, and pipet off the wash liquid. Repeat the washing at least three times.

EXPERIMENT 32
Analysis of Group 1 Cations:
Ag^+, Hg_2^{2+}, Pb^{2+}

PRELABORATORY PREPARATION
1. Do the Prelaboratory Exercise and turn it in at the beginning of your laboratory period.
2. Read "Introduction to Semimicro Qualitative Analysis," page 383.

INTRODUCTION
Of all the cations that we will consider, only Ag^+, Hg_2^{2+}, and Pb^{2+} form insoluble chlorides. Therefore, they are separated first in our cation analysis scheme by precipitation in dilute HCl.

$$Ag^+ + Cl^- \longrightarrow AgCl(s) \qquad\qquad K_{sp} = 1.8 \times 10^{-10} \qquad (32.1)$$
$$Hg_2^{2+} + 2\,Cl^- \longrightarrow Hg_2Cl_2(s) \qquad\qquad K_{sp} = 1.1 \times 10^{-18} \qquad (32.2)$$
$$Pb^{2+} + 2\,Cl^- \longrightarrow PbCl_2(s) \qquad\qquad K_{sp} = 1.7 \times 10^{-5} \qquad (32.3)$$

Silver and mercury(I) chlorides have such low solubilities that they precipitate almost completely if only a slight excess of HCl is added to the sample solution. Lead(II) chloride is much more soluble, as indicated by its larger solubility product, and will require a higher concentration of chloride ion to precipitate. However, too great an excess of Cl^- must be avoided, because all the cations in Group 1 can form complexes that have high solubility, Reactions 32.4-32.6.

$$Ag^+ + 2\,Cl^- \longrightarrow AgCl_2^-(aq) \qquad\qquad (32.4)$$
$$Hg_2^{2+} + 4\,Cl^- \longrightarrow HgCl_4^{2-}(aq) + Hg(l) \qquad\qquad (32.5)$$
$$Pb^{2+} + 3\,Cl^- \longrightarrow PbCl_3^-(aq) \qquad\qquad (32.6)$$

Reactions 32.4-32.6 would prevent Group 1 cations from precipitating. Therefore, it is best to precipitate $PbCl_2$ from a cold solution in order to minimize its solubility. This will keep the amount of Cl^- that is needed to obtain a detectable precipitate as small as possible. Some Pb^{2+} will, nevertheless, remain in solution and carry over to cation Group 2, giving a precipitate of lead sulfide when the sample is analyzed for Group 2 cations. Hydrochloric acid is used to supply Cl^-, instead of a soluble chloride salt such as NaCl, to avoid introducing additional metal cations and because the solution must be acidic. The acidic solution prevents precipitation of the oxychlorides of any Bi(III) or Sb(III) that might be present:

$$Bi^{3+} + Cl^- + H_2O \longrightarrow BiOCl(s) + 2\,H^+ \qquad\qquad (32.7)$$

Bismuth and antimony will be precipitated as sulfides with Group 2.

Separation and Identification Within the Group
If the addition of HCl produces a precipitate, it indicates that cations of Group 1 are present. The next step is to identify which of the Group 1 cations have been precipitated. First, the precipitate is centrifuged so that the supernatant, which may contain cations of other groups, can be separated easily from the solid precipitate which contains the Group 1 cations. The supernatant is saved for later analysis of cations in other groups. Then, the precipitate is tested for the cations of Group 1.

Test for Pb^{2+}
The addition of hot water to the precipitate dissolves soluble $PbCl_2$, but not the less soluble chlorides of Ag(I) and Hg(I). If the hot water has dissolved any Pb^{2+}, adding chromate ion to the solution will form a yellow precipitate:

$$Pb^{2+} + CrO_4^{2-} \longrightarrow \underset{\text{yellow ppt.}}{PbCrO_4(s)} + 2\,Cl^- \qquad\qquad K_{sp} = 1.8 \times 10^{-14} \qquad (32.8)$$

Test for Ag^+
AgCl dissolves in 6 M NH_4OH, forming the soluble silver-ammonia complex:

$$AgCl(s) + 2\,NH_3 \longrightarrow Ag(NH_3)_2^+(aq) + Cl^- \qquad\qquad (32.9)$$

If the silver-ammonia complex is present, acidifying the solution with HNO_3 reverses Reaction 32.9 and causes the white precipitate of AgCl to form, confirming the presence of Ag^+.

$$Ag(NH_3)_2^+(aq) + Cl^- + 2\,H^+ \longrightarrow \underset{\text{white ppt.}}{AgCl(s)} + 2\,NH_4^+ \qquad\qquad (32.10)$$

Test for Hg_2^{2+}

Adding NH_4OH, in the test for Ag^+, causes Reaction 32.11 to occur also:

$$Hg_2Cl_2 + 2\,NH_3 \longrightarrow HgNH_2Cl(s) + Hg(l) + NH_4Cl \qquad (32.11)$$
$$\text{white ppt.} \qquad \text{black ppt.}$$

If the original solution contained Hg_2^{2+}, a gray-black precipitate is obtained, which is a mixture of the two precipitates from Reaction 32.11. This precipitate will remain after solubilizing $PbCl_2$ and $AgCl$, confirming the presence of Hg_2^{2+}.

PLAN OF EXPERIMENT

First a series of preliminary experiments are performed on known solutions, to observe the behavior of Group 1 cations. Then, an unknown solution, containing one or more cations from Group 1, is analyzed qualitatively.

SAFETY

1. Wear approved eye protection.

2. Always add acids to your solutions slowly and with caution, to avoid spattering. Hold the solutions well away from your face.

3. Avoid inhaling fumes from HNO_3 and NH_4OH, which are very irritating.

4. Lead and mercury solutions are toxic. Avoid contact with them and wash your hands with soap and water at the end of the laboratory period.

PROCEDURE
Reagents used for Group 1 analysis

3 M HCl	6 M $HC_2H_3O_2$	3 M $NH_4C_2H_3O_2$
3 M HNO_3	15 M NH_4OH (conc.)	blue litmus paper
18 M H_2SO_4 (conc.)	1 M K_2CrO_4	

A.   Preliminary Experiments with Separate Solutions, Each Containing only 1 Cation
(Your instructor may prefer that you perform preliminary experiments with a single solution containing all 3 cations. In this case, use the procedure for analysis of an unknown, part B., on a sample known to contain all the cations of Group 1)

1.   Obtain known test solutions, each containing one cation, in dropper bottles. Place 3 test tubes in marked positions in a test tube rack. Put 1 mL of each known solution into different test tubes. Add 2-3 drops of 3 M HCl to each test tube and shake. Write your observations on the data sheet.

2.   Centrifuge each sample and decant the supernatant liquid. Add about 2 mL of water to each precipitate, stir, and heat in a water bath. $PbCl_2$ should dissolve. Note your observations on the data sheet. Save the Pb^{2+} solution for later tests.

3.   Centrifuge and decant the water from the Ag^+ and Hg^{2+} samples. Add about 1 mL 15 M NH_4OH to each sample. The AgCl precipitate should dissolve and a gray-black precipitate should remain in the Hg_2^{2+} sample tube. The precipitate might be more easily seen if the solution is filtered. Note your observations on the data sheet.

4.   To the solution in the Ag^+ sample tube, add 16 M HNO_3 dropwise while stirring, until the solution just turns acidic. Test by touching the stirring rod to blue litmus paper after each drop. Note your observations on the data sheet.

5.   Divide the solution in the Pb^{2+} sample tube into 2 equal portions. Add 2-3 drops of 18 M H_2SO_4 to one portion and about 1 mL of 1 M K_2CrO_4 solution to the other. Different kinds of precipitates should form in each portion. Record your observations on the data sheet.

Figure 32.1.   Flow Chart for Analysis of Group 1 Cations.

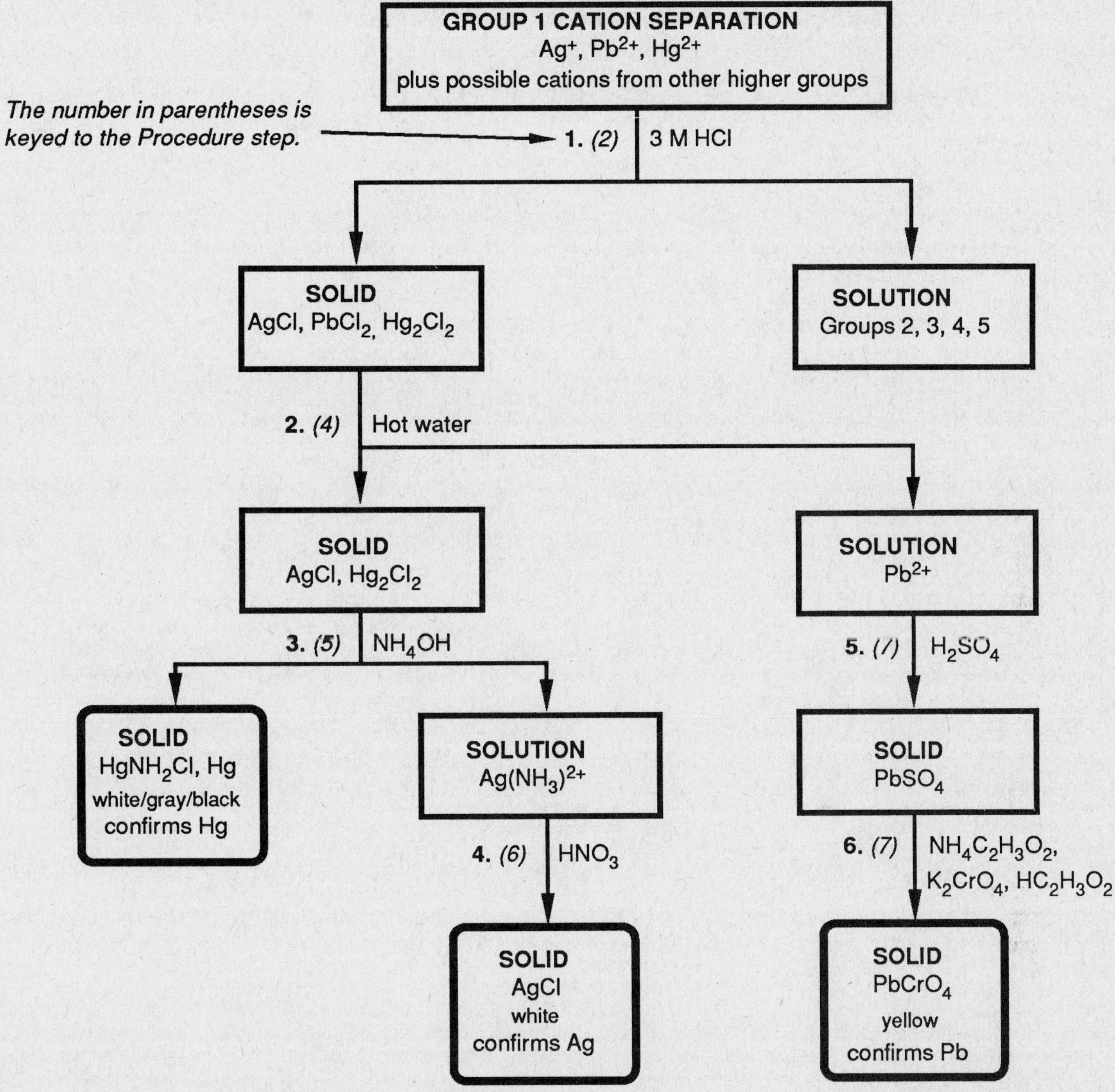

B. Unknown Analysis

1.   Obtain an unknown sample from your instructor. It will contain one or more cations from Group 1. Record the unknown number on your data sheet above the blank sample analysis flow chart.

2.   Put 5 drops of unknown sample into a 75 mm test tube and dilute it with 15 drops of distilled water. Add 3 drops of 3 M HCl and mix well. Centrifuge the sample. Add 1 drop 3 M HCl to the supernatant to test for complete precipitation. If supernatant turns cloudy, add another drop of HCl and centrifuge again. Test the supernatant as before. When the supernatant tests clear, it may be decanted and saved for Group 2, 3, 4, and 5 analysis. (If your instructor tells you that your unknown contains only Group 1 cations, the decanted liquid may be discarded.)

3.   After decanting, wash the remaining precipitate with 2 drops distilled water and 1 drop 3 M HCl. Stir and centrifuge. Decant and discard the wash liquid.

4.   Add 1 mL distilled water to the washed precipitate. Stir and heat the test tube in a water bath while continuing to stir. When solution is hot, remove it from the bath and carefully centrifuge it while still hot. Decant and save the supernatant liquid: it contains any Pb^{2+} that might be present. save the remaining precipitate: it contains any Ag^+ and Hg_2^{2+} that might be present.

5.   To the precipitate, add 5 drops distilled water and 5 drops 15 M NH_4OH. Stir thoroughly and centrifuge. Decant the supernatant and save it. If a gray-black precipitate remains, there is Hg_2^{2+} in your unknown.

6.   To the supernatant from step 5, add 3 M HNO_3 dropwise with stirring until the solution just turns acidic. Test by touching the stirring rod to blue litmus paper after each drop. The appearance of a white precipitate confirms the presence of Ag^+.

7.   To the supernatant from step 4, carefully add 4 drops of 18 M H_2SO_4. The solution will become warm. Stir until solution is cool. If Pb^{2+} is present, a white fluffy precipitate should *slowly* form. Centrifuge and discard the supernatant. Add 0.5 mL 3 M $NH_4C_2H_3O_2$, stir and heat in the water bath. The precipitate should dissolve. this test is made conclusive by adding 1 drop of 1 M K_2CrO_4 to the solution. The appearance of a yellow precipitate confirms the presence of Pb^{2+}.

Name ____   Date ____________

EXPERIMENT 32
PRELABORATORY EXERCISE

1.   Why must the chloride solution from which Group 1 cations are precipitated be kept cold?

2.   Why must the chloride solution from which Group 1 cations are precipitated be acidic?

3.   In the precipitation of Group 1 cations, why is it important that the chloride ion concentration not be too high?

4.   Why is it especially important to never use tap water as a solvent in the analysis of Group 1 cations?

5.   How are $AgCl$ and Hg_2Cl_2 precipitates separated from one another?

6.   Why should you heat qualitative analysis samples in a water bath rather than over a burner?

7.   Describe the proper techniques for mixing solutions.

Name ____   Date _____________

EXPERIMENT 32
DATA

A. Preliminary Experiments

Sample No.	Cations present	Reagent added	Results and Observations	Ppt. contains	Sol'n contains
1.					
2.					
3.					

B. Unknown Analysis

Unknown Sample No. __________, analyzed for Group 1 cations.

Cations identified in sample: ____

FLOW CHART FOR ANALYSIS OF GROUP 1 UNKNOWN SAMPLE

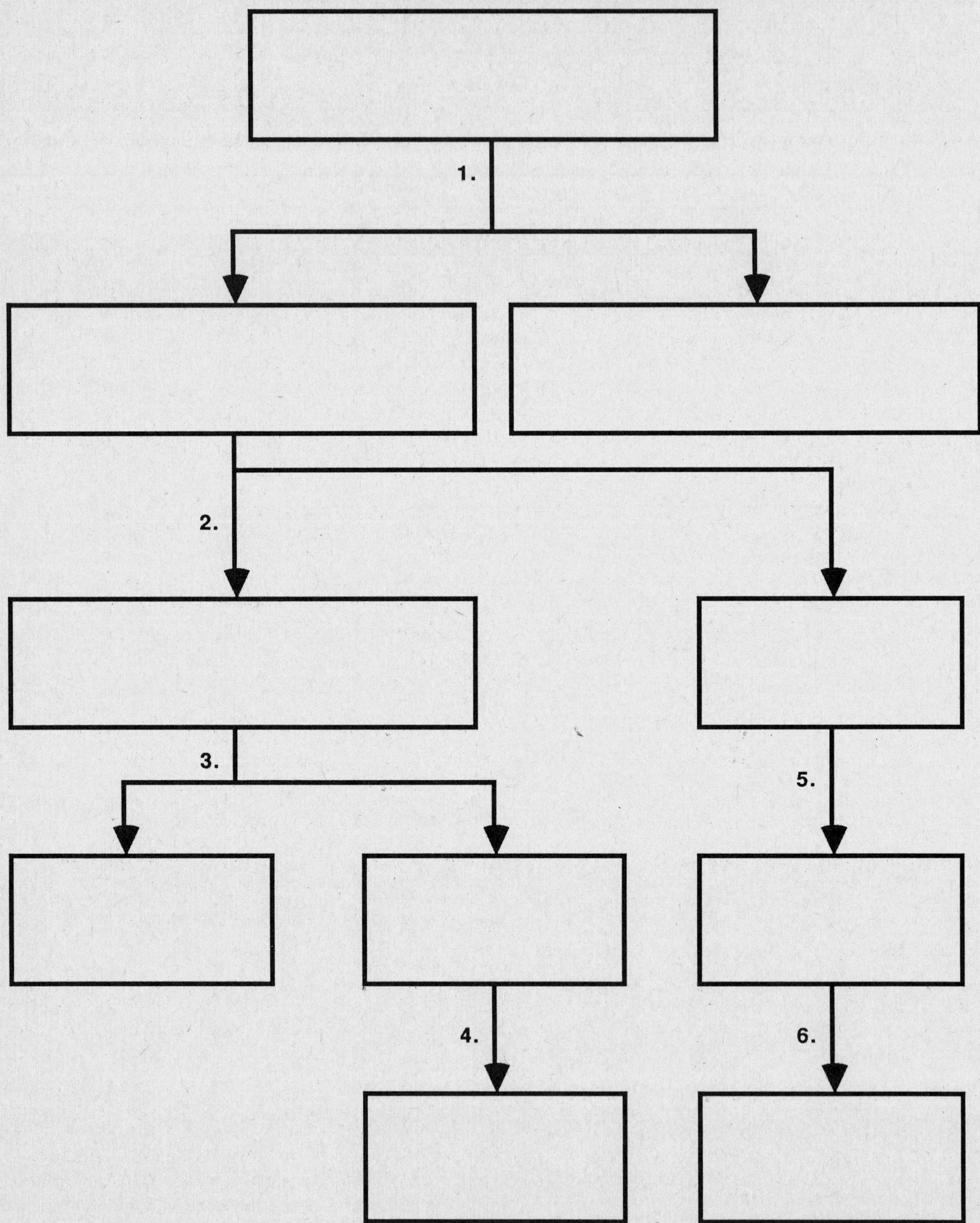

EXPERIMENT 33
Analysis of Group 2 Cations:
Cu^{2+}, Cd^{2+}, Hg^{2+}, Pb^{2+}, Bi^{3+}, Sn^{2+}, Sn^{4+}*

PRELABORATORY PREPARATION
1. Do the Prelaboratory Exercise and turn it in at the beginning of your laboratory period.
2. Read "Introduction to Semimicro Qualitative Analysis," page 383.

INTRODUCTION
Group 2 is composed of those cations that form soluble chlorides but extremely insoluble sulfides. Group 2 is precipitated as sulfides in an acidic H_2S solution, after the insoluble chlorides of Group 1 have been precipitated. Group 3 cations also form sulfides of low solubility, but they are more soluble than the sulfides of Group 2, and, with care, can be prevented from precipitating with Group 2.

The sulfides of Group 2 cations have smaller solubility products than the sulfides of Group 3. This allows Group 2 to be separated because they will precipitate at a lower sulfide ion concentrations than is needed for Group 3.

Solubility products for Group 2 sulfides are listed in Table 33.1.

Table 33.1. Solubility products at 25°C for sulfides of Group 2 cations

Equilibrium Reaction	K_{sp}
$PbS \rightleftharpoons Pb^{2+}(aq) + S^{2-}(aq)$	8.4×10^{-28}
$CdS \rightleftharpoons Cd^{2+}(aq) + S^{2-}(aq)$	3.6×10^{-29}
$SnS \rightleftharpoons Sn^{2+}(aq) + S^{2-}(aq)$	8×10^{-29}
$CuS \rightleftharpoons Cu^{2+}(aq) + S^{2-}(aq)$	8.7×10^{-36}
$Hg_2S \rightleftharpoons 2\,Hg^{+}(aq) + S^{2-}(aq)$	1×10^{-45}
$HgS \rightleftharpoons Hg^{2+}(aq) + S^{2-}(aq)$	3×10^{-53}
$Bi_2S_3 \rightleftharpoons 2\,Bi^{3+}(aq) + 3\,S^{2-}(aq)$	1.6×10^{-72}

For comparison, the most insoluble cation of Group 3 is ZnS, with $K_{sp} = 1.6 \times 10^{-23}$.

Success in separating Group 2 cations depends upon maintaining the sulfide ion concentration low enough that cations from Group 3 will not precipitate, but high enough for essentially complete precipitation of Group 2. Sulfide ion concentrations are controlled by the equilibria of Reactions 33.1a and 33.1b:

$$H_2S(aq) \rightleftharpoons H^{+}(aq) + HS^{-}(aq) \qquad K_{a1} = 1.0 \times 10^{-7} \qquad (33.1a)$$
$$HS^{-}(aq) \rightleftharpoons H^{+}(aq) + S^{2-}(aq) \qquad K_{a1} = 1.3 \times 10^{-13} \qquad (33.1b)$$

If the solution is kept acidic, the high values of $[H^{+}(aq)]$ will hold the equilibria of Reactions 33.1a, b to the left side and $[S^{2-}(aq)]$ will remain very low.

It is important to adjust the pH, using HCl, of the Group 2 solution to between 0.3 and 0.5, in order to establish a value for $[S^{2-}(aq)]$ that is high enough to precipitate Group 2, but too low to precipitate Group 3.

---

*Arsenic and antimony also fall into Group 2, but will not be used in our experiments because their high toxicity makes disposal of laboratory wastes a serious problem.

PLAN OF EXPERIMENT

Group 2 is the largest of all the cation groups and its analysis is longer and more complicated. After the initial precipitation, it is broken down into two subgroups in order to simplify separating the individual metals.

Initial Precipitation

Group 1 cations, if present, must first be separated.[1] The remaining solution then is made about 0.5 M in HCl. Hydrogen sulfide, H_2S, is formed in the solution by adding thioacetamide[2] and warming the solution in a boiling water bath. All the Group 2 cations should precipitate during the heating period. A typical sulfide precipitation reaction is:

$$Hg^{2+}(aq) + H_2S \longrightarrow HgS(s) + 2\,H^+(aq)$$

Separation of Subgroups 2a and 2b.

The solution is made basic by adding KOH. Subgroup 2a is composed of those sulfides that are insoluble in basic solutions containing the anion S^-. Subgroup 2b is composed of those sulfides that form sulfide complexes in basic solution and, therefore, are soluble in such solutions.

The sulfides of *Subgroup 2a* are CuS, Bi_2S_3, CdS, and PbS, which do not dissolve in basic solutions containing S^-. It may also contain some HgS that did not dissolve as an S^- complex.

The sulfides of *Subgroup 2b* form the soluble complexes HgS_2^{2-}, AsS_2^{2-}, AsO_2^-, SbS_2^-, $Sb(OH)_4^-$, SnS_3^-, and $Sn(OH)_6^{2-}$ in basic solutions containing S^-. (Arsenic and antimony are not used in this experiment.)

After the subgroups are separated, the individual cations are separated and identified with a series of specific tests described in the procedure section.

<table><tr><td>

SAFETY

1. Wear approved eye protection.

2. Always add strong acids and bases to solutions, slowly and with caution, to avoid spattering. Hold the solutions well away from your face.

3. Although very little H_2S gas should escape from the solutions, it is very toxic and foul smelling. Any solution containing more than 3 drops of thioacetamide should be warmed only in a hood. Do not sniff at your solutions and avoid inhalation of any vapors.

4. Many metal cation solutions are toxic. Avoid contact with them and wash your hands with soap and water at the end of the laboratory period.

</td></tr></table>

PROCEDURE
Procedure Notes:

1.   It is important to adjust the pH carefully to between 0.3 and 0.5, as explained in the Introduction. Procedure steps 1-4 perform this process.

2.   You frequently are told to save a solution or precipitate for later use. It is essential to label these samples carefully.

It will help you identify your samples if your label always includes the number of the Procedure Step that generated the sample, e.g., "Step 28 supernatant."

---

[1] Ag^+ from Group 1 would precipitate with Group 2 if not removed in the Group 1 analysis.

[2] Thioacetamide, CH_3CSNH_2, hydrolyzes when heated in acid solution to form H_2S:

$$CH_3CSNH_2 + H_2O \rightleftharpoons CH_3CONH_2 + H_2S$$

Thioacetamide is used instead of H_2S gas because it is much safer to handle and stores well. Since it forms H_2S uniformly and slowly throughout the solution when warmed, the resulting precipitate forms more slowly and evenly than if H_2S gas were bubbled through the solution. This yields larger and purer precipitate particles that are easier to centrifuge.

Figure 33.1.   ANALYSIS FLOW CHART FOR CATION GROUP 2

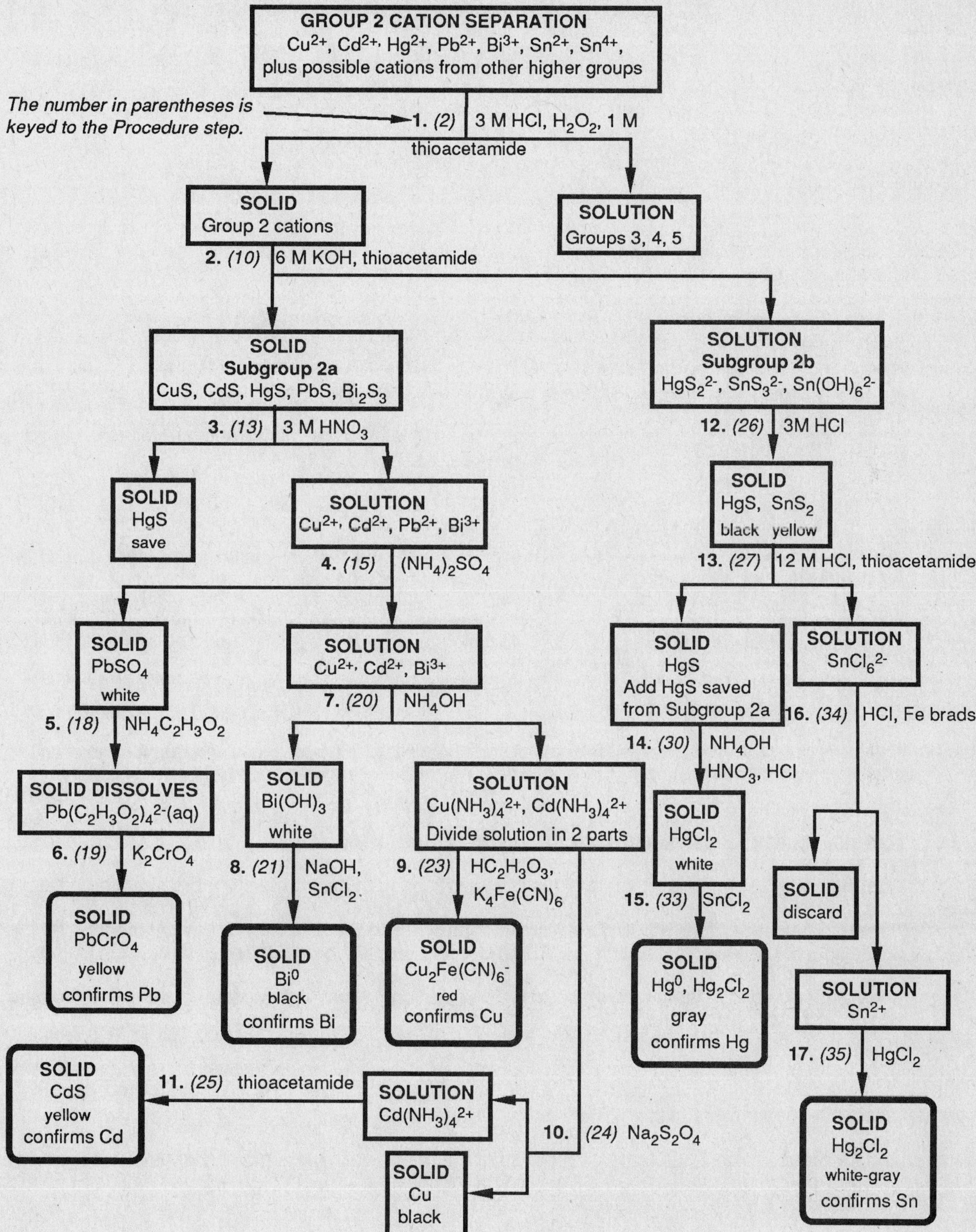

Reagents Used for Group 2 Analysis

3 M HCl	6 M $HC_2H_3O_2$	3 M $NH_4C_2H_3O_2$	$(NH_4)_2SO_4$
6 M HCl	6 M KOH	0.25 M $SnCl_2$	$Na_2S_2O_4$
12 M HCl	6 M NH_4OH	0.2 M $HgCl_2$	3% H_2O_2
3 M HNO_3	15 M NH_4OH (conc.)	0.5 M $K_4Fe(CN)_6$	rust free iron nails (brads)
16 M HNO_3	4 M NH_4Cl	1 M thioacetamide	

INITIAL SEPARATION OF GROUP 2:

(If you are not continuing with a sample already analyzed for Group 1 cations, you will be given an unknown containing only Group 2 cations.)

1.   As your Group 2 sample, use either the supernatant left from your Group 1 analysis or a known or unknown sample provided by your instructor. Place 8 drops of your Group 2 sample in a 10 mL beaker. If your sample is an unknown, record its identification number.

2.   Acidify the sample by adding 2 drops of 3 M HCl. Then add 3 drops of H_2O_2, which oxidizes all tin in the sample to the +4 oxidation state.
> **In acidic sulfide solution, Sn^{+4} precipitates as SnS_2, which will dissolve in basic sulfide solution, a necessary process for later separation of Subgroups 2a and 2b.**

Evaporate the liquid until about 2 drops remain.
> <u>Do not let the sample go dry,</u> **because some Group 2 chlorides are volatile and you may lose them. Be prepared to add a drop or two of HCl if drying seems imminent.**

3.   Cool the sample and then add 6 drops of 6 M HCl. Hold the beaker with tongs and pass it in and out of the burner flame to evaporate the sample to a paste.
> <u>Do not let the sample go dry.</u> **It is best to stop heating before the solution turns pasty and allow the warm beaker to complete the evaporation. Add more HCl if necessary.**

4.   To the paste from Step 3, add 15 M NH_4OH dropwise until the sample is just alkaline to litmus. Then add 3 M HCl until the solution *just* turns acidic. The solution volume at this point should be about 0.75 mL. If not, add a drop or two of deionized water. Now add *exactly* 3 drops of 3 M HCl.
> **The pH now should be correct for precipitation of Group 2 as sulfides.**

Transfer the solution to a 75 mm test tube.

5.   Add 10 drops of 1 M thioacetamide solution and stir well. **In a hood,** place the test tube in a boiling water bath for at least 10 minutes. After boiling, add enough deionized water to double the liquid volume and add 2 more drops of thioacetamide, with stirring. Then heat the sample in the water bath for another 5 minutes.
> **This step accomplishes the initial precipitation of Group 2 sulfides.**

6.   Centrifuge the sample and decant the supernatant into a clean test tube, labelled "Step 6 supernatant."
> **The supernatant contains any cations from Groups 3, 4, and 5. It also may contain some unprecipitated cations from Group 2. You will test for complete precipitation of Group 2 in Step 9.**

7.   **The solid may contain large amounts of adsorbed and entrapped higher group cations that must be washed off of the precipitate.**
To make a wash solution, add 5 drops of 4 M NH_4Cl and 1 drop of thioacetamide solution to 1 mL of deionized water and heat just to boiling. Thioacetamide forms S^{2-} in the wash water to prevent any of the Group 2 sulfides from redissolving during washing. NH_4Cl helps prevent the precipitate from forming a colloidal suspension during washing.

8.   Add 10 drops of wash solution to the precipitate and stir well with a glass rod. Let the solid settle and decant the wash water into the Step 6 supernatant, to collect additional higher group cations together. Repeat the washing but discard the second wash water.

9.   To test for complete precipitation of Group 2, put 1 mL from the combined wash water and Step 6 supernatant into a clean test tube. Add 1 mL of deionized water and 2 drops of thioacetamide. Heat in a boiling water bath. A yellow, orange, or brown precipitate indicates that precipitation was incomplete.

> **No precipitate at all, or a precipitate that is white, black, or gray (from S or Group 3 cations), indicates that the Group 2 precipitation was complete.**

If incomplete Group 2 precipitation is indicated, dilute the total wash and supernatant solution sufficiently to double its volume, add 5 drops of thioacetamide and, in a hood, heat in a boiling water bath for 5 minutes. Centrifuge any precipitate obtained, decant the supernatant into a clean test tube.

> **Your instructor will tell you whether or not to save the liquid for analysis of Groups 3, 4, and 5. If you save the liquid, heat it in the boiling water bath for 10 more minutes to complete decomposition of thioacetamide and boil off all H_2S, in order to prevent slow oxidation to SO_4^{2-}, which can interfere with later analyses. Then, if your instructor directs, stopper and label the tube as the sample for analysis of Groups 3, 4, and 5.**

Combine the solid with the Step 6 precipitate.

SEPARATION OF SUBGROUPS 2A AND 2B:
10.   To the solid from Step 9, add 15 drops of 6 M KOH, 2 mL of deionized water, and 1 drop of thioacetamide solution. Stir well and heat in the boiling water bath for 3 minutes.

> **Concentrated KOH solutions are prone to bumping and spattering during heating. Stir the solution slowly while heating to prevent this. Keep your face away from the tube mouth during this process. *Handle the hot alkaline solution with great care.***

11.   Centrifuge the tube immediately after removing it from the bath. Use a micropipet to withdraw the supernatant liquid into a clean test tube.

12.   Wash the solid twice, stirring with 5 drops of deionized water each time. Add the wash water to the Step 11 supernatant .

> **The solid contains the sulfides of Subgroup 2a: CuS, Bi_2S_3, CdS, and PbS. The supernatant liquid contains Subgroup 2b: HgS_2^{2-}, SnS_3^{2-}, and $Sn(OH)_6^{2-}$.**

Stopper and label the tube containing the supernatant .

DETECTION OF THE METALS IN SUBGROUP 2A:
13.   To the Step 12 precipitate, add 1.5 mL of 3 M HNO_3 and heat it in the boiling water bath for 2 minutes.

14.   Remove the sample from heat, centrifuge, and use a micropipet to withdraw supernatant liquid into a 10 mL beaker for analysis of Cu^{2+}, Bi^{3+}, Cd^{2+}, and Pb^{2+}.

> **Any residue in the tube is mercury that did not form a soluble sulfide complex. In their various forms as oxide or sulfide, the mercury solids may be red, black, white, or shades in between.**

Wash the precipitate with a solution of 3 drops of water and 2 drops of 3 M HNO_3. Add the wash water to the supernatant liquid in the 10 mL beaker. Label and save the precipitate to add later to HgS obtained from analysis of Subgroup 2b.

Detection of Pb:
15.   To the Step 14 supernatant, add about 0.4 g $(NH_4)_2SO_4$ for each 1.5 mL of solution. Stir to dissolve the salt and let it stand for a few minutes, watching for the formation of a precipitate.

> **Of the cations in the solution, only Pb^{2+} forms an insoluble sulfate. However, it easily develops a supersaturated solution and it may be necessary to force precipitation.**

If a precipitate does not form, chill the tube in an ice bath. The solubility of $PbSO_4$ decreases markedly at lower temperatures. If there still is no precipitate, try adding additional sulfate in the form of Na_2SO_4, which dissolves more easily than $(NH_4)_2SO_4$

> **The amount of precipitate may be very scant. Even if you cannot see any, continue with the confirmatory test for Pb, Steps 16-18.**

16.   Centrifuge the tube, visible precipitate or not, and micropipet the supernatant into a clean test tube for analysis of Cu^{2+}, Bi^{3+}, and Cd^{2+}.

17.   Wash the precipitate, visible or not, twice with 5 drop portions of 1 M $(NH_4)_2SO_4$, adding the first wash water to Step 16 supernatant. Discard the second wash water.

18.   Add 0.5 mL $NH_4C_2H_3O_2$ solution to the precipitate, visible or not, and place the tube in a boiling water bath for 2 minutes, stirring slowly.
> **The solution should be clear. If it is not, centrifuge it and then micropipet the clear supernatant into a clean test tube.**

19.   To the clear solution, add 1 drop of 6 M acetic acid ($HC_2H_3O_2$) and 2 drops of K_2CrO_4 solution.
> **A yellow precipitate of $PbCrO_4$ confirms the presence of Pb.**

Detection of Bi:
20.   To the Step 16 supernatant, add 15 M NH_4OH, dropwise, until the solution is just alkaline to litmus. Then add 2 drops more. A deep blue color indicates Cu^{2+}, but does not prove it.[3] A gelatinous precipitate, not always easily observed, indicates $Bi(OH)_3$. Centrifuge the tube and save the supernatant for analysis of Cu^{2+} and Cd^{2+}.

21.   Wash the Step 20 precipitate two times with 5 drop portions of deionized water, discarding the wash water. Add 6 drops of 6 M NaOH and 3 drops of $SnCl_2$ solution and stir. If a milky-white precipitate forms, add more NaOH and stir.
> **An immediate black precipitate of metallic Bi proves the presence of Bi.**
A slower-forming brown precipitate is probably metallic Sn, not completely separated from Subgroup 2b.

Detection of Cu:
22.   The Step 20 supernatant may contain $Cu(NH_3)_4^{2+}$ and $Cd(NH_3)_4^{2+}$.
> **Cu can be identified in the presence of Cd, but Cd cannot be identified in the presence of Cu. Therefore it is necessary to separate the solution into 2 parts so that Cu can be removed from one of them.**
Divide the supernatant into 2 portions of 1/3 and 2/3 the original volume. Use the 1/3 portion for the Cu test and save the 2/3 portion for the Cd test.

23.   Use 6 M $HC_2H_3O_2$ dropwise to make the 1/3 portion just acidic to litmus. Add 2 drops of $K_4Fe(CN)_6$ solution.
> **A brick-red precipitate of $CuFe(CN)_6$, or reddish cloudiness, proves the presence of Cu.**

Detection of Cd:
24.   To the 2/3 portion of the solution from Step 22, use a small spatula to add a "pinch" (about the volume of a BB pellet) of $Na_2S_2O_4$, and heat the solution in a hot water bath for for 3 minutes.
> **Metallic copper precipitates quickly as finely divided, black particles.**
Centrifuge the sample and decant the supernatant, which should be clear and colorless, into a clean test tube. Discard the precipitate.

25.   To the supernatant, add 5 drops of thioacetamide solution and heat in the water bath for 2 minutes.
> **A yellow precipitate of CdS confirms the presence of Cd.**

DETECTION OF THE METALS IN SUBGROUP 2B:
26.   Transfer the Step 12 supernatant to a 10 mL beaker. Include in the transfer any precipitate that might have formed while the solution was standing. Add 3 M HCl, dropwise with stirring, until the solution is just acidic to litmus.

27.   Add 2 drops of thioacetimide solution and heat in the water bath for 3 minutes.
> **A yellow or brown precipitate indicates sulfides of Hg and Sn. A fine, white precipitate of S only, indicates that Subgroup 2b is absent.**
Centrifuge the tube and discard the supernatant. Add 10 drops of 12 M HCl to the precipitate, stir well, and heat for 2 minutes in the water bath while stirring.

28.   Add 2 drops of thioacetamide and heat for 2 minutes more, to precipitate any Hg that might have dissolved. Remove the tube from the bath and immediately centrifuge and separate the supernatant into a clean test tube.

---

[3]If any NiS from Group 3 was accidently precipitated with Group 2, it will give a blue color at this point. Proof of Cu must await a more definitive test.

29.   Wash the precipitate with 3 drops of 12 M HCl and 2 drops of deionized water. Add the wash water, which might contain Sn^{2+}, to the Step 28 supernatant .

Detection of Hg:
30.   Wash the Step 29 precipitate with 5 drops of water, discarding the wash water. Then, add 10 drops of 15 M NH_4OH and 2 drops of water and stir well. Centrifuge the tube and discard the supernatant. Combine the precipitate with that saved from Step 14. Wash the combined precipitates with 5 drops of water, discarding the wash water.

31.   Add 5 drops of 12 M HCl and 1 drop of 16 M HNO_3. Heat in the water bath until a visible reaction begins, then transfer the sample to a 5 mL beaker, rinsing the tube with a drop or two of water to collect all the sample.

32.   **In the hood,** boil the sample gently to concentrate it down to a few drops.
        Do not boil to dryness.
This evaporates most of the excess Cl^-, which forms $HgCl_4^{2-}$, slowing the precipitation reactions of mercury. Rinse the concentrate into a test tube with 2 drops of water. Centrifuge and separate the supernatant into a clean 5 mL beaker.

33.   To the supernatant, add 2 drops of 0.25 M $SnCl_2$ solution.
        The presence of Hg is confirmed by the formation of an opalescent white precipitate of Hg_2Cl_2, which may turn gray or black due to metallic mercury.

Detection of Sn:
34.   Gently boil the Step 32 supernatant to drive off all the excess H_2S. Test the steam coming off the solution with lead acetate paper, which blackens as long as H_2S is present in the vapor. When H_2S is gone, add 5 drops of water, 1 drop of 12 M HCl, and 2 rust-free clean iron brads. Heat for 5 minutes in the water bath.

35.   Centrifuge the sample and micropipet the supernatant to a clean test tube. To the supernatant, add 2 drops of 0.2 M $HgCl_2$ solution.
        A white, gray, or black precipitate of Hg_2Cl_2 and metallic Hg confirms the presence of Sn.
Note that the precipitate does not contain tin. The presence of Sn^{2+} in the solution causes the reduction of Hg^{2+} to $Hg(0)$ and $Hg(I)$.

Name ____ Date ____________

EXPERIMENT 33
PRELABORATORY EXERCISE

1. What pH range is required for separation of the Group 2 cations? Why is it important to maintain this pH range fairly closely?

2. Why is thioacetamide used to generate H_2S instead of using H_2S gas?

3. Why is it important, in the initial separation of Group 2, never to heat the sample to dryness?

4. In the detection of Pb, why is it sometimes necessary to chill the sample solution?

5. Describe the precautions necessary when generating the gas H_2S.

Name ____   Date ____________

EXPERIMENT 33
DATA

Unknown Analysis

Unknown sample No. ______________, analyzed for Group 2 cations.

Cations identified in sample: ____

Draw a flow chart below, indicating your procedure and results.

EXPERIMENT 34
Analysis of Group 3 Cations:
Fe^{2+}, Fe^{3+}, Co^{2+}, Ni^{2+}, Al^{3+}, Cr^{3+}, Mn^{2+}, Zn^{2+}

PRELABORATORY PREPARATION
1. Do the Prelaboratory Exercise and turn it in at the beginning of your laboratory period.
2. Read "Introduction to Semimicro Qualitative Analysis," page 383.

INTRODUCTION
The cations in Group 3 will precipitate as sulfides or hydroxides from a moderately alkaline solution (pH = 9). The divalent cations Fe^{2+}, Co^{2+}, Ni^{2+}, Mn^{2+}, and Zn^{2+}, precipitate as sulfides but require a higher S^{2-} concentration than was used for Group 2. The higher $[S^{2-}]$ is attained with H_2S (from thioacetamide) by raising the pH higher than the Group 2 conditions. The higher pH shifts the equilibrium of:

$$H_2S \rightleftharpoons 2H^+ + S^{2-}$$

to the right. At the higher pH there is of course a higher concentration of OH^-, causing the trivalent cations Fe^{3+}, Al^{3+}, and Cr^{3+} to precipitate as insoluble hydroxides. The solution then is made alkaline with $NH_3(aq)$. Adding additional NH_4Cl shifts the equilibrium of:

$$NH_3 + H_2O \rightleftharpoons NH_4^+ + OH^-$$

to the left and, at the same time, buffers the solution at about pH 9. Buffering the solution at pH 9 is important, because if $[OH^-]$ becomes too large, $Mg(OH)_2$ from Group 4 will precipitate if it is present.

PLAN OF EXPERIMENT
Like Group 2, Group 3 is large and it is useful to break it down into two subgroups, 3a and 3b, in order to simplify separating the individual metals. However, the analyses are straightforward and generally easier than for Group 2. Especially helpful is the fact that many of the Group 3 precipitates and solutions have characteristic colors that aid identification.

Initial Precipitation
Group 1 and Group 2 cations, if present, must first be separated.[1] Then Group 3 can be separated from Groups 4 and 5 by precipitation as insoluble sulfides or hydroxides.

SAFETY
1. Wear approved eye protection.
2. Always add strong acids and bases to solutions, slowly and with caution, to avoid spattering. Hold the solutions well away from your face.
3. Although very little H_2S gas should escape from the solutions, it is very toxic and foul smelling. Any solution containing more than 3 drops of thioacetamide should be warmed only in a hood. Do not sniff at your solutions and avoid inhalation of any vapors.
4. Many metal cation solutions are toxic. Avoid contact with them and wash your hands with soap and water at the end of the laboratory period.

---

[1] Ag^+ would precipitate with Group 2 if it had not been removed in the Group 1 analysis.

Figure 34.1.   Flow chart for analysis of Group 3 cations.

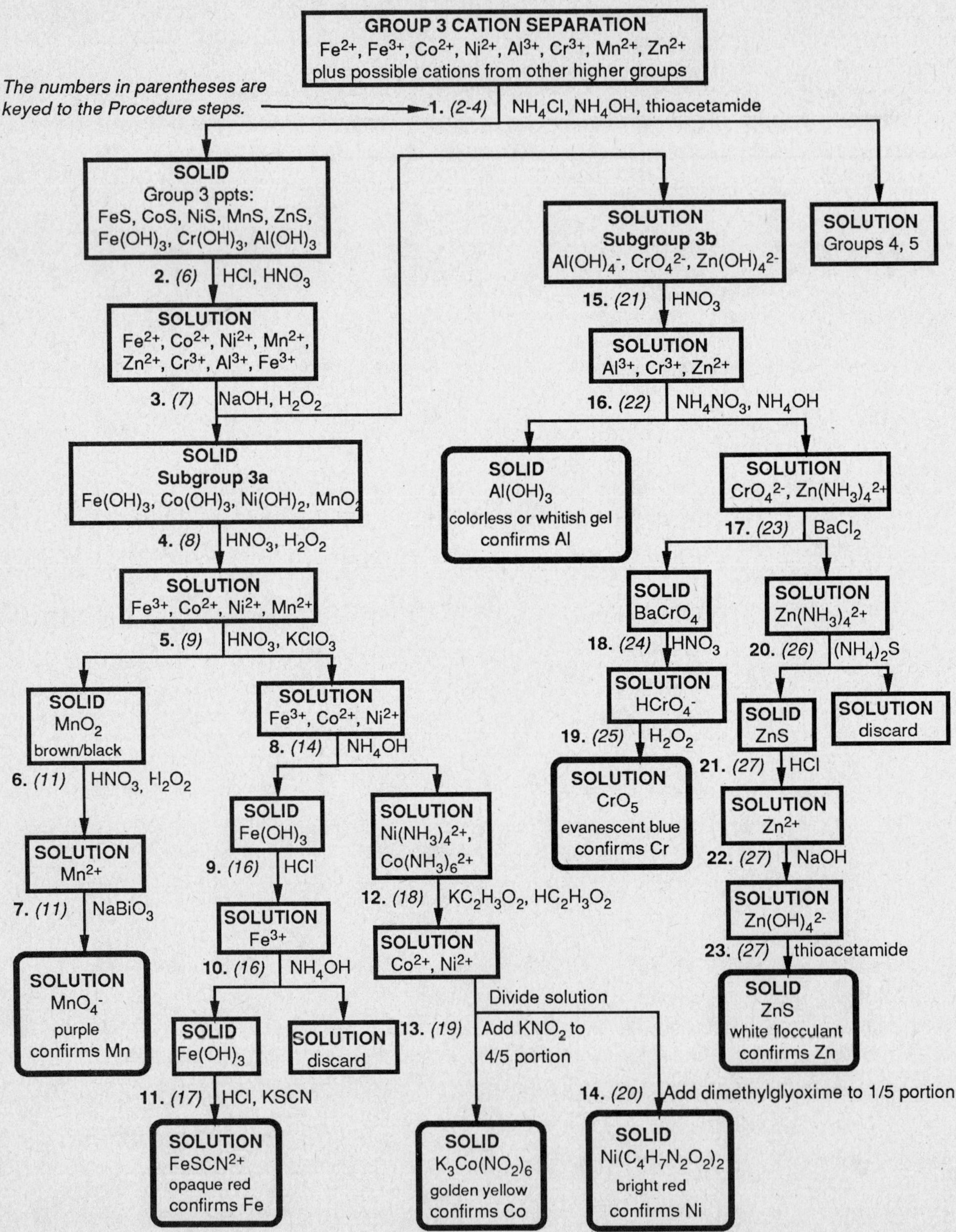

PROCEDURE
Procedure Notes:

1. It is important to adjust the pH carefully to around pH 9, as explained in the Introduction. Procedure steps 2 and 3 perform this process.

2. You are frequently told to save a solution or precipitate for later use. Label these samples carefully. **It will help you identify your samples if your label always includes the number of the Procedure Step that generated the sample, e.g., "Step 28 supernatant."**

Reagents Used for Group 3 Analysis

6 M NaOH	4 M NH_4Cl	1 M $BaCl_2$	
3 M HCl	6 M NH_4OH	3% H_2O_2	dimethylglyoxime: $(CH_3)_2C_2(NOH)_2$
12 M HCl	15 M NH_4OH (conc.)	$NaBiO_3(s)$	aluminon reagent
3 M HNO_3	2 M $KC_2H_3O_2$	$NH_4NO_3(s)$	litmus paper
16 M HNO_3	6 M KNO_2	$KClO_3(s)$	
6 M $HC_2H_3O_2$	1 M thioacetamide	0.1 M KSCN	

$(NH_4)_2S$ solution {Add 1 vol. reagent grade $(NH_4)_2S$ to 2 vols. water}

Table 34.1. Colors of Group 3 cations under various conditions.

Metal	Ion in neutral or acid aqueous solution		Ion in ammoniacal or basic aqueous solution		Solid precipitate	
Al	Al^{3+}	colorless	$Al(OH)_4^-$	colorless	$Al(OH)_3$	colorless-whitish gel
Cr	Cr^{3+}	green	$Cr(NH_3)_6^{2+}$	yellow	$Cr(OH)_3$	gray-green
	$Cr_2O_7^{2-}$	orange	$Cr(OH)_4^-$	green	$BaCrO_4$	yellow
	$CrO_5(aq)$	vanishing blue	CrO_4^{2-}	yellow		
Mn	Mn^{2+}	pink			MnS	salmon
	MnO_4^-	purple			MnO_2	black
					$Mn(OH)_2$	tan
Fe	Fe^{3+}	yellow			FeS	black
	$FeSCN^{2+}$	opaque red			$Fe(OH)_2$	green
					$Fe(OH)_3$	red-brown
Co	Co^{2+}	pink	$Co(NH_3)_6^{2+}$	pink	CoS	black
			$Co(NH_3)_6^{3+}$	brown	$Co(OH)_2$	blue
					$Co(OH)_3$	black
					$K_3Co(NO_2)_6$	golden yellow
Ni	Ni^{2+}	green	$Ni(NH_3)_6^{2+}$	blue	NiS	black
	$NiCl_4^{2-}$	blue			$Ni(C_4H_7N_2O_2)_2$	bright red
					$Ni(OH)_2$	green
Zn	Zn^{2+}	colorless	$Zn(NH_3)_4^{2+}$	colorless	ZnS	white
			$Zn(OH)_4^{2-}$	colorless		

INITIAL SEPARATION OF GROUP 3:
(If your sample was previously analyzed for Group 2 anions, begin with Step 1a. If your sample never contained any anions except from Group 3, begin with Step 1b.)

1a. Only For a Sample Left from a Group 2 Analysis
If you are using supernatant left from the analysis of Group 2 cations as your Group 3 sample, you must first evaporate away the excess HCl. If any precipitate has formed during storage, centrifuge the sample, discard the precipitate and proceed with the remaining supernatant.
Boil the supernatant sample down to about 5 drops in a 10 mL beaker. To avoid spattering, do not heat the beaker directly. Place the beaker on a wire gauze over a burner and direct the flame to a spot on the gauze about 2-3 cm to the side of the beaker. Do not heat to dryness.
When finished, transfer the sample to a 10 × 75 mm test tube.

1b. For a Sample Containing Group 3 Only
If you are using a known or unknown obtained from your instructor that contains only Group 3 cations, place about 5 drops in a 10 × 75 mm test tube.

(All samples continue with Step 2.)

2.   Dilute the sample in the test tube with 5 mL deionized water and 0.5 mL of 4 M NH_4Cl. Heat to boiling in the water bath and then remove the sample from the bath.

3.   To the hot solution, add 15 M NH_4OH dropwise until just basic to litmus (usually, only a few drops are needed). Then add 2 more drops of NH_4OH.

> **Record on your flow chart whether a precipitate has formed and the colors of both the precipitate and solution. Centrifuging the sample can help distinguish the colors.**

If the amount of precipitate is so great that the solution becomes pasty, add a little water to keep a liquid layer. NH_4OH and NH_4Cl buffer the solution at around pH 9, high enough to precipitate $Fe(OH)_3$, $Al(OH)_3$, and $Cr(OH)_3$, but too low to precipitate $Mg(OH)_2$ from Group 4. Also, NH_4Cl makes the solution more ionic, helping to coagulate sulfide precipitates, which tend to form colloids. The sulfide precipitates are made in the next step.

> **No precipitate in Step 3 means that Fe, Al, and Cr are not in the sample.**

4.   Reheat the sample, including any precipitates, to boiling in the bath with constant stirring. Continue stirring as you slowly add 10 drops of thioacetamide solution.

> **Heat for 5 min, because sulfide precipitates may form slowly. Lower the temperature a bit, if necessary, to prevent frothing.**

Centrifuge and add one more drop of thioacetamide to see if precipitation was complete. If any additional solid forms, add thioacetamide a drop at a time to the solution in the hot water bath, centrifuging after each drop, until precipitation is complete. The supernatant should be clear and nearly colorless.

    When precipitation is complete, wash any solids from the sides of the test tube with a few drops of water. Centrifuge and separate the precipitate and supernatant. Note the colors of the precipitate and supernatant on your Data sheet flow chart.

> **Only For Samples That Also Contain Cations From Groups 4 and 5:**
>   If you are instructed to save the supernatant for analysis of Groups 4 and 5, S^{2-} must be removed to prevent formation of SO_4^{2-}, which forms insoluble sulfates with Ba^{2+} and Sr^{2+} from Group 4. To remove S^{2-} and save the supernatant, add acetic acid, $HC_2H_3O_2$, until the solution is acidic. Centrifuge and discard any precipitate. Then, boil the supernatant down to about 1/2 its original volume in the water bath. Stopper and label the tube for later analysis of Groups 4 and 5.

5.   Make a solution for washing the Step 4 precipitate by mixing in a clean test tube, 3 drops of 15 M NH_4OH, 3 drops of 4 M NH_4Cl, 1 drop of thioacetamide, and 1 mL of water. Pour the wash solution over the precipitate and stir well. Warm the sample for a few minutes in the water bath and then centrifuge. Discard the supernatant.

> **The precipitate contains $Fe(OH)_3$ (red-brown), $Al(OH)_3$ (white), $Cr(OH)_3$ (green), FeS (black), CoS (black), NiS (black), MnS (pink), and ZnS (white), if the cations were present.**

6.   To the Step 5 precipitate, add 5 drops of concentrated HCl and stir. Any black precipitate remaining is NiS and CoS. If all the precipitate dissolves, Ni and Co are absent.

> **If any dark precipitate of NiS and CoS remains, add 5 drops of concentrated HNO_3 and heat in a water bath until all solids have dissolved.**

Separation of Subgroups 3a and 3b

7.   To the solution of Step 6, add 6 M NaOH dropwise until just basic. Then add 4 drops more. *Slowly*, add 8 drops of 3% H_2O_2. Heat the solution near boiling in a water bath for at least 2 minutes or until all oxygen gas evolution has stopped. If necessary, add deionized water during heating to keep the liquid level constant. Then centrifuge and separate. Subgroup 3a cations are contained in the precipitate and Subgroup 3b cations are in the supernatant solution. Note on your flow chart the colors of the solution and precipitate.

Precipitate Subgroup 3a		Solution Subgroup 3b	
$Fe(OH)_3$	red-brown	$Al(OH)_4^-$	colorless
$Co(OH)_3$	black	$Zn(OH)_4^{2-}$	colorless
$Ni(OH)_3$	green	CrO_4^{2-}	yellow
MnO_2	black		

Label, stopper, and save the supernatant solution for later analysis.

SUBGROUP 3a (Mn, Fe, Co, and Ni)
Separation and Identification of Mn

8. Dissolve all the Step 7 precipitate (Subgroup 3a) in 5-10 drops of 16 M HNO_3 and 2 drops of H_2O_2. Heat the clear solution in a water bath until it evaporates down to about 5 drops or until solid begins to form. Add 10 drops more of 16 M HNO_3 and and evaporate once more down to about 5 drops. This process removes the H_2O_2 that was used to help dissolve MnO_2 and $Co(OH)_3$. If any H_2O_2 remains in the sample, the purple solution color that confirms Mn may not develop.

> **Do not evaporate to dryness. Have some concentrated HNO_3 handy, so you can *carefully* add a drop or two to the sample in case the hot test tube continues to evaporate liquid after you have stopped heating.**

9. Add 16 M HNO_3 to bring the volume to about 1 mL. Heat the test tube to just below boiling and add 0.15 g of $KClO_3$ in several small portions. Continue to heat and stir for another minute after the $KClO_3$ has been added, adding HNO_3 if needed to keep the liquid level constant. A dark brown or black precipitate indicates MnO_2. Centrifuge and separate the solid. Label and save the supernatant for determination of Fe, Co, and Ni.

10. Wash the Step 9 precipitate in the test tube 3 or 4 times with 5-drop portions of hot deionized water.

> **Add the first washing to the Step 9 solution and discard the others.**

11. Dissolve the precipitate in about 5 drops of 3 M HNO_3 and 3 drops of H_2O_2, using more of these reagents if needed. Heat the tube in the boiling water bath for 5 minutes to eliminate the H_2O_2 and then cool to room temperature. Add about 0.2 g of solid sodium bismuthate, $NaBiO_3$, and stir well. Let the mixture stand for about 1 minute and then centrifuge it.

> **A purple solution phase indicates MnO_4^- and confirms the presence of Mn.**

Separation and Identification of Iron

12. Examine the Step 9 solution. If any white or colorless crystals have formed, they are KCl and must be removed. Centrifuge and separate the sample, wash the crystals with 5 drops of water, add the wash water to the supernatant solution, and discard the solid.

13. Evaporate the Step 12 solution down to about 0.5 mL, or until solid starts to form. Then cool the sample to room temperature.

14. Place 1 mL 15 M NH_4OH in *another* test tube. Then carefully add to this test tube, dropwise with stirring, the Step 13 sample solution. Rinse the Step 13 test tube with 2-3 drops of water and add the rinse water to the solution. Stir thoroughly.

> **It is necessary to add the acidic sample to the NH_4OH solution, instead of vice versa, so that the sample becomes strongly basic instantly. This prevents premature precipitation of $Ni(OH)_2$ and $Co(OH)_2$ that might be difficult to redissolve completely.**

15. Test the solution with litmus. If it is not strongly basic, add 16 M NH_4OH until the sample is basic, then add 3 drops more. Stir, centrifuge and separate. Label and save the supernatant for analysis of Ni and Co. A reddish–brown gelatinous precipitate may be $Fe(OH)_3$. Wash the precipitate with a few drops of water and add the wash water to the supernatant solution.

16. Dissolve the precipitate in 15 drops of 3 M HCl, heating the sample in the water bath until the precipitate is dissolved. Add 6 M NH_4OH with stirring, until the solution is basic to litmus, and then add 10 drops more. Any precipitate is probably $Fe(OH)_3$. Centrifuge, separate, and wash the solid once with water, discarding the wash water.

17. Dissolve the Step 16 precipitate in 10 drops of 3 M HCl, heating the sample in the water bath if necessary. Cool to room temperature and add 2-3 drops of 0.1 M KSCN.

> **Formation of an opaque red solution indicates $FeSCN^{2+}$ and confirms the presence of Fe.**

Separation and Detection of Cobalt

18.   Evaporate the Step 15 solution down to about 1 drop to remove most of the NH_3, adding water, if necessary, to prevent loss of all liquid. Add 4 drops of 2 M $KC_2H_3O_2$ and enough 6 M $HC_2H_3O_2$ to make the solution just acidic to litmus. Then add 2 more drops of $HC_2H_3O_2$. Pour 1/5 of the solution into another test tube, label it, and set it aside for analysis of Ni.

19.   Double the volume of the remaining 4/5 portion by adding 6 M KNO_2. Warm the sample in the water bath and then let it cool for at least 15 minutes.

>   **A golden yellow precipitate, which may form slowly, is $K_3Co(NO_2)_6$ and confirms the presence of Co.**

Separation and Detection of Nickel

20.   Add 1 drop of 15 M NH_4OH to the 1/5 portion of solution from Step 18 and dilute the sample to 1 mL with water. Add 5 drops of dimethylglyoxime reagent $(CH_3CNOH)_2$. Stir and let the sample stand for about 10 min. It may be necessary to centrifuge the sample in order to determine the color of any precipitate that formed.

>   **A bright red precipitate is $Ni(C_4H_7N_2O_2)_2$ and confirms the presence of Ni. A red color in the solution indicates the presence of Co or Fe, but not Ni.**

SUBGROUP 3b (Al, Cr, and Zn)

Separation and Identification of Al

21.   To the solution saved from Step 7 (Subgroup 3b), add dropwise 16 M HNO_3 until the solution is just acidic to litmus. This changes chromate ion, CrO_4^{2-} (yellow), to dichromate ion, $Cr_2O_7^{2-}$ (orange), and serves as a preliminary test for Cr. Note any change in the color of the solution on your Data sheet flow chart.

22.   Evaporate the solution in the water bath down to about 3 mL and let cool. Then add 0.1 g of powdered NH_4NO_3, heating and stirring until it is all dissolved. Now add dropwise 15 M NH_4OH until the solution is just basic to litmus and then add 5 drops more.

>   **It is important to add sufficient NH_4NO_3 and NH_4OH. Otherwise Zn may form the insoluble $Zn(OH)_2$ precipitate instead of soluble $Zn(NH_3)_4^+$, and precipitate along with Al.**

Warm the solution again and stir thoroughly. At this point, if you hold your test tube up to the light, you may detect a colorless gel that is difficult to see. This is $Al(OH)_3$. It is more easily recognized if you centrifuge the sample and decant the supernatant.

>   **Save the supernatant for analysis of Cr and Zn.**

Wash the precipitate twice with hot water containing 2 drops of 4 M NH_4Cl.

>   **A colorless or whitish gel in the bottom of the test tube is $Al(OH)_3$ and confirms the presence of Al.**

If the presence of Al is uncertain, perform the following more sensitive test:

>   **a.** To the centrifuged and decanted test tube that might contain an $Al(OH)_3$ precipitate, add 10 drops of water and stir well.
>   **b.** Add 6 M $HC_2H_3O_2$ until the solution is just acidic.
>   **c.** Add 4 drops of aluminon reagent and heat to boiling in a water bath.
>   **d.** Add a few drops of 6 M NH_4OH until the hot solution is just basic.
>   **e.** Stir thoroughly and centrifuge.

>   **A cherry-red precipitate indicates that $Al(OH)_3$ was adsorbed to the aluminon reagent and confirms the presence of Al.**

Separation and Identification of Cr

23.   To the supernatant solution of Step 22, which may be yellow due to the presence of CrO_4^{2-}, add dropwise 1 M $BaCl_2$ solution until precipitation is complete. This should not require more than about 5-6 drops. Add at least one drop of $BaCl_2$ solution even if the solution shows no yellow color. Centrifuge and separate. Label and save the supernatant for later analysis of Zn. The precipitate may contain yellow $BaCrO_4$ with a little white $BaSO_4$ and $BaCO_3$. (SO_4^{2-} may be present from oxidation of S^{2-} that was not removed completely, and CO_3^{2-} can arise from CO_2 dissolved from the atmosphere.)

24.   Add 10 drops of hot water to the Step 23 precipitate, centrifuge and discard the supernatant. Add 3 drops of 3 M HNO_3 and stir thoroughly to dissolve the precipitate . Add 10 drops of cold water and stir again.

25.   Now, it is necessary to add H_2O_2 to form the temporary blue color from CrO_5, which is very unstable so that its color fades fairly quickly. The color is most easily seen if the H_2O_2 is added suddenly. Use the following procedure to add the correct amount of H_2O_2 all at once.

    a. Raise the dropper above the H_2O_2 dropper bottle and empty it. Release pressure on the bulb and insert it into the H_2O_2 solution.

    b. Carefully squeeze the bulb until just 3 bubbles of air come out.

    c. Release the bulb and allow H_2O_2 to replace the volume vacated by the 3 air bubbles. This places approximately 3 drops of H_2O_2 in the dropper.

    d. Now squirt the H_2O_2 all at once into the solution of Step 24.

A blue color that disappears rapidly indicates unstable CrO_5 and confirms the presence of Cr.

Separation and Identification of Zn

26.   To the solution from Step 23, add 1 drop of $(NH_4)_2S$ solution. If a white or gray precipitate forms, it probably is ZnS with possibly some free sulfur. Centrifuge and separate, discarding the supernatant. Wash the precipitate with 3 small portions of water, discarding the wash water.

27.   Dissolve the precipitate in 1 mL of 3 M HCl, warming the solution a little. This dissolves ZnS but not free sulfur. Evaporate the solution down to about 0.5 mL to remove H_2S and HCl. Cool the solution and carefully add 6 M NaOH dropwise with constant stirring until the solution is basic to litmus. Then add 5 more drops. If the solution has any sediment or cloudiness, centrifuge and discard the solid. To the clear solution add 2 drops of thioacetamide solution and heat in a boiling water bath for 5 min.

A white precipitate is ZnS and confirms the presence of Zn.

Name ____   Date ____________

EXPERIMENT 34
PRELABORATORY EXERCISE

1.   Several cations in both Groups 2 and 3 precipitate as sulfides. How are the Group 2 cations made to precipitate separately from the Group 3 cations?

2.   What non-sulfide precipitates are used in the Group 3 separations?

3.   Suppose your sample did not contain any Fe, Al, or Cr. What is the earliest step in the analysis that you would know this?

4.   Suppose your sample did not contain any Ni or Co. What is the earliest step in the analysis that you would know this?

5.   Why is it necessary, when doing the confirming test for Cr, to add the H_2O_2 reagent suddenly and to watch the sample very closely.?

Name ____   Date ____________

EXPERIMENT 34
DATA

Unknown Analysis

Unknown sample No. ____________, analyzed for Group 2 cations.

Cations identified in sample: ____

Draw a flow chart below, indicating your procedure and results.

Name ____   Date ____________

EXPERIMENT _____

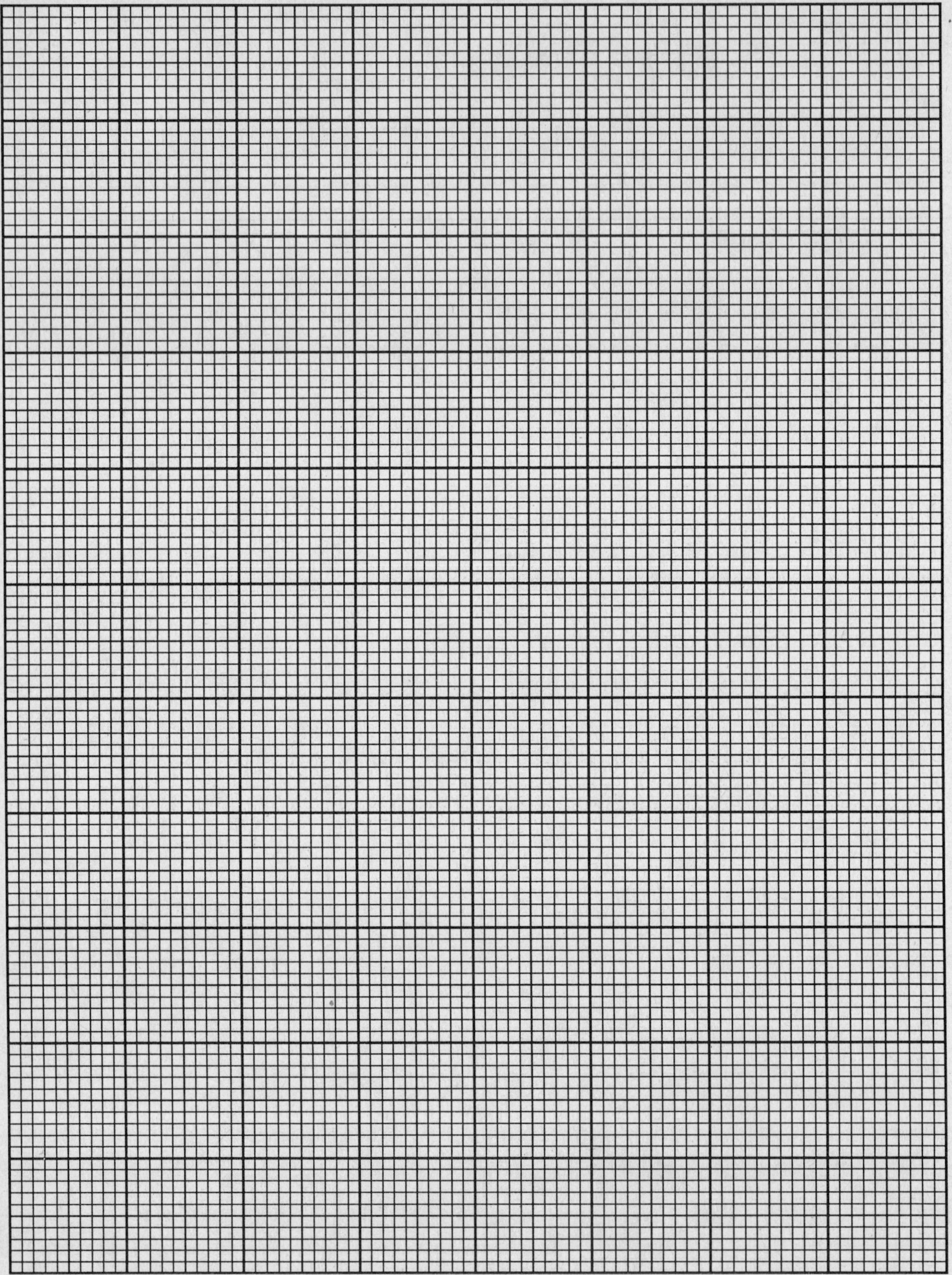

Name _______________________________________ Date ____________

EXPERIMENT _____

Name ____   Date ____________

EXPERIMENT _____

  Supplemental Graph Paper

Name ______________________________________   Date ____________

EXPERIMENT _______

 Supplemental Graph Paper

Name ___________________________________   Date ___________

EXPERIMENT _____

Name ____   Date ____________

EXPERIMENT _____

Name _______________________________________   Date ___________

EXPERIMENT _____

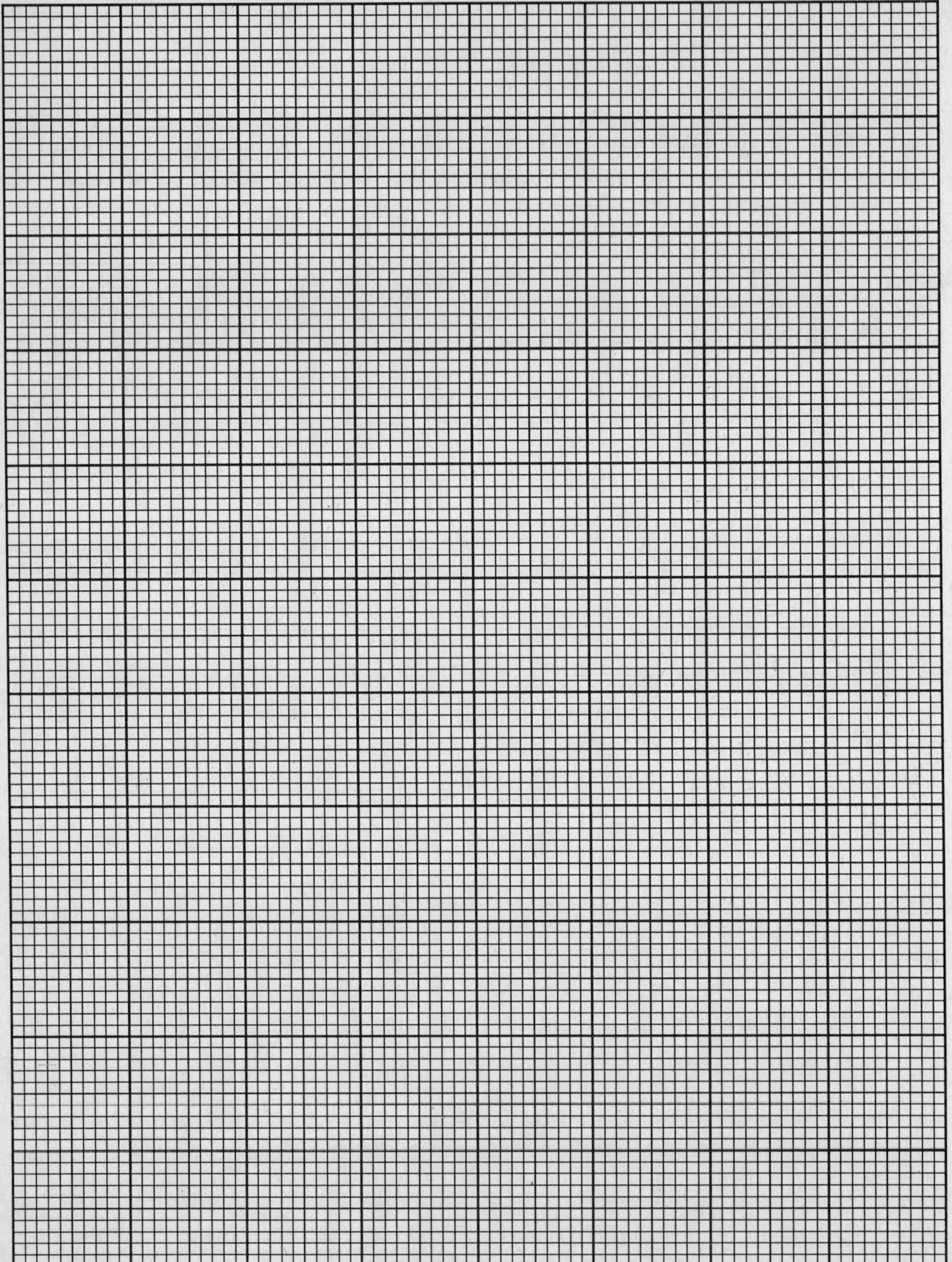

Name ____   Date ____________

EXPERIMENT _____

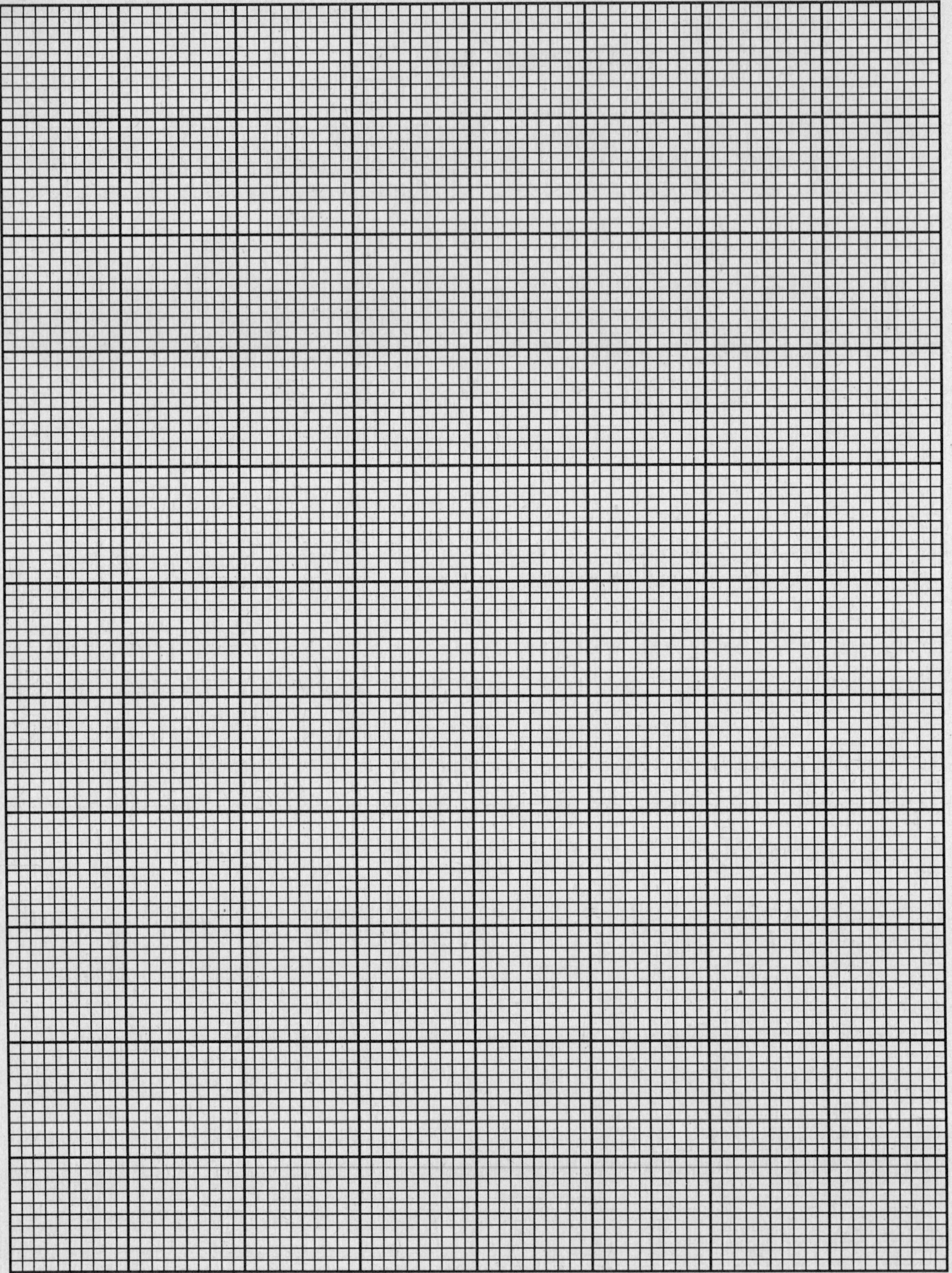

Name ____   Date ____________

EXPERIMENT _____

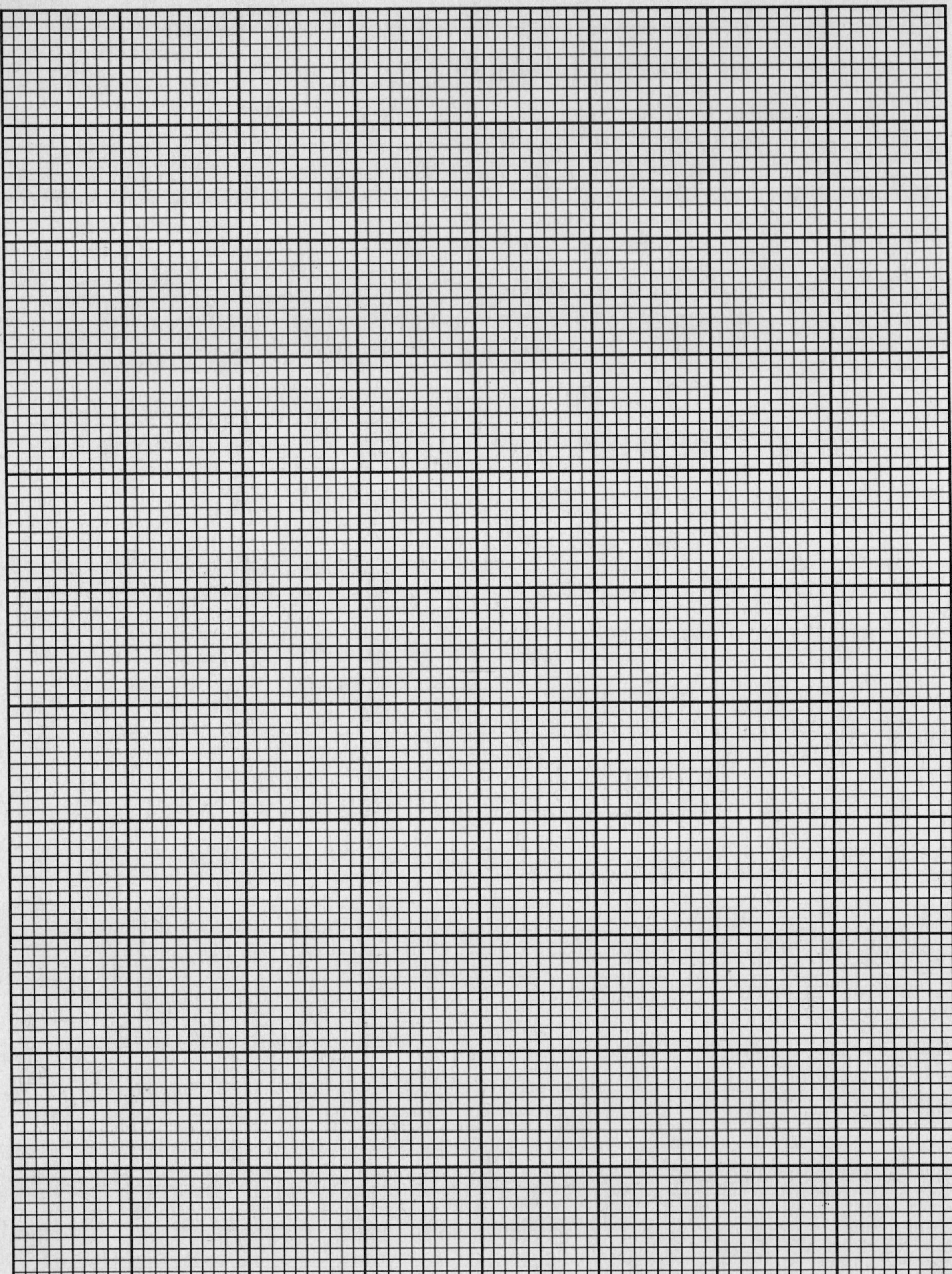